UNCONVENTIONAL METHODS IN EXPLORATION FOR PETROLEUM AND NATURAL GAS

II

UNCONVENTIONAL METHODS IN EXPLORATION FOR PETROLEUM AND NATURAL GAS

II

A SYMPOSIUM UNDER THE AUSPICES OF THE
INSTITUTE FOR THE STUDY OF
EARTH AND MAN
SOUTHERN METHODIST UNIVERSITY

Edited by
BENJAMIN M. GOTTLIEB

SMU PRESS • DALLAS

Library of Congress Cataloging in Publication Data

Main entry under title:

Unconventional methods in exploration for petroleum and natural gas II.

Proceedings of the Second Symposium on Unconventional Methods in Exploration for Petroleum and Natural Gas held in Dallas, Tex. Sept. 13-14, 1979.

1. Prospecting—Congresses. 2. Petroleum—Congresses. 3. Gas, Natural—Congresses. I. Gottlieb, Benjamin M., 1912- II. Symposium on Unconventional Methods in Exploration for Petroleum and Natural Gas (2nd : 1979 : Dallas, Tex.) III. Southern Methodist University. Institute for the Study of Earth and Man.

TN271.P4U53 622′.1828 81-8805
ISBN 0-87074-179-9 AACR2

INSTITUTE FOR THE STUDY OF EARTH AND MAN

"Nature and history, the earth and man, stand in the closest relations to each other, and form only one grand harmony."

—Arnold Guyot, 1849

The Institute, under whose auspices this symposium was arranged, is organized as an affiliate of Southern Methodist University "to encourage, foster, promote and advance research and development and education in the field of the anthropological and geological sciences and other related fields of scholarship."

Incorporated in Texas on May 11, 1966

The officers and trustees of the Institute as of October, 1981, are as follows:

CONTENTS

PREFACE

The Institute's second symposium on unconventional methods in exploration for petroleum and natural gas was held at the Dallas Hilton Inn on September 13-14, 1979. More than two hundred persons attended—about a third more than the number present at the first symposium on this same subject staged in the fall of 1968. Included among the registrants were independent and consulting geologists, geoscientists affiliated with universities and governmental agencies, and representatives of some ninety different oil and oil-related companies. A bare majority of the participants came from various parts of Texas, but eighteen other states and five foreign countries were represented as well.

Planning of the symposium was the work of a committee chaired by Benjamin M. Gottlieb and including as members William B. Heroy, Jr., M. J. Davidson, R. C. Weart, C. F. Dodge, and Claude C. Albritton, Jr.; James E. Brooks and Marion Y. C. Hustis were also active in the deliberations of the committee.

Expenses of presenting the program and publishing the proceedings were underwritten by twenty-one individuals or corporations. These are American Petrofina, Inc., AMOCO Oil Company, ARCO Oil and Gas Company, Conoco, Inc., Core Laboratories, Inc., Daniel Oil Company, Dresser Industries, Inc., Enserch Corporation, Exxon Company, General Crude Oil Company, H. J. Gruy and Associates, Jake L. Hamon, Hunt Oil Company, Marathon Oil Company, D. R. McCord and Associates, Petroconsultants, S. A., Schlumberger Technology Corporation, Sunmark Exploration Company, Tenneco Oil Company, Texas Instruments, Inc., and Texas Pacific Oil Company.

After the symposium, the speakers were given opportunities to clarify or amplify the written versions of their presentations in the light of the questions and discussions that followed their talks. Also, many of the illustrations, originally drafted for viewing as slides, had to be redrafted for publication. Consequently, the printing of these proceedings has been delayed longer than was intended.

For sake of economy, some of the illustrations drafted in color are here reproduced in black and white. Others, which would have been unintelligible without reproduction in color, are gathered in a separate atlas, together with some that would not stand reduction to page size.

The members of the organizing committee extend their thanks to Margaret Hartley and Charlotte Whaley for guiding the manuscript through the press, and to Martin Reese for supervising the printing.

Claude C. Albritton, Jr.

Dallas, Texas
October 26, 1981

UNCONVENTIONAL METHODS IN EXPLORATION FOR PETROLEUM AND NATURAL GAS

II

SYMPOSIUM II ON UNCONVENTIONAL METHODS IN EXPLORATION FOR PETROLEUM AND NATURAL GAS

Opening Remarks by William B. Heroy, Jr.

INSTITUTE FOR THE STUDY OF EARTH AND MAN

Members of the Symposium, on behalf of Southern Methodist University and the Institute for the Study of Earth and Man, it is my privilege to welcome you.

The recent energy crisis has served to reinforce our belief that it is indeed time for another symposium on "unconventional" methods. As we might expect, many methods and ideas held to be "unconventional" in the past have moved up on the scale of respectability and have taken their rightful place.

Your presence here is a clear indication that, although we have learned much about the origin, migration, and occurrence of petroleum and natural gas, we have somehow come up with more questions than answers. There is still a great need for men with imagination, perseverance, and the courage to pursue their ideas despite doubt, risk, discouragement, and failure. It will be a sad day for our nation and mankind when there is no longer an interest in the "unconventional."

The history of the petroleum industry is more than proof that truth is stranger than fiction. It is full of heroes and villains, luck and adventure, tragedy and success. It is a colorful example of the characteristics that made our country great. No other industry has had more impact on the world or contributed more to its growth and progress. Yet, no business is more misunderstood.

Between 1970 and 1978, the U.S. consumption of oil increased from 14,790,000 barrels daily to 18,822,000 barrels per day, an increase of 3,470,000 barrels. During this same period our domestic production decreased over 1,000,000 barrels per day despite the addition of 1,200,000 barrels per day from the North Slope. Thus, our consumption increased over 23% in eight years, and our production decreased over 9%, and most Americans refused to believe that there was a problem at all.

During this same period, 1970-78, U.S. coal production increased from 603,000,000 tons per year to 653,000,000 tons per year, an increase of only 8%. However, these numbers must be qualified in terms of the fact that total production for the years 1975, 1976, and 1977 was somewhat greater than that for 1978. The drop was the result of labor problems in mining and transportation, governmental indecision, and delays related to environmental factors.

Coincidentally, most of the major efforts to increase coal production have come from oil companies. Yet here again, instead of appreciating that someone is responding to the nation's critical energy needs, the government and many of our so-called political leaders have made repeated threats against horizontal diversification, as if there were something wrong about being in the energy business instead of just in the coal, oil, gas, or uranium business.

Another aspect of the energy situation which puzzles me even more is all the talk and governmental concern about oil company profits. Anyone expressing concern about the magnitude of oil company profits demonstrates one thing very clearly. They know absolutely nothing about investments and the stock market. Every year *Forbes* magazine has a special issue devoted to the profitability, growth, and stock market performance of the largest U.S. corporations. From these figures, I have tabulated the profitability rankings of the twelve major U.S. oil companies for the 1970-1978 period. This is a ranking based on the percentage of profit after taxes in relation to total investment or net worth. Not a single one of these twelve companies averaged in the top two hundred most profitable U.S. companies. In relation to total assets and earnings, after an adjustment for inflation, most of them had negative growth.

I would be interested in seeing some figures on the total tax burden of the petroleum industry. It is quite obvious that a great majority of the governments of the world lean heavily on the taxes siphoned off from the flow of oil from the well to the consumer's gas tank.

I find these things very disturbing because so much depends on our success, while so many are failing to understand how we got where we are and what it will take to stay there, much less move even higher.

Your presence here is encouraging. No idea should be discarded lightly. We wish you "good hunting."

Philosophy of Symposium II
THE NEED FOR INNOVATION IN EXPLORING FOR OIL AND GAS IN THE UNITED STATES

Benjamin M. Gottlieb

SYNERGY OIL & GAS, INC.

Mr. Chairman, Dr. Heroy, fellow geoscientists, and energy hunters from all phases of industry. I know you have all assembled here in search of the answer to our nation's energy crisis. I have caucused with our distinguished speakers and panelists, and they have assured me that you won't be disappointed.

It is indeed gratifying to see the caliber of the attendants gathered here, from the petroleum and other allied and competitive businesses, colleges and universities, and various government agencies.

Approximately eleven years ago, a small group of creative and courageous geoscientists assembled here in order to expedite the "innovation cycle" of improving our conventional exploration techniques.

Symposium I was established as an interdisciplinary forum with the following objectives: (1) to assemble the most creative and authoritative thinking of speakers, panelists, and attendants of all the earth science disciplines; (2) to coordinate and document all available research data, both basic and applied; and (3) to strengthen communications between the various scientific disciplines (geologists, geophysicists, geochemists, engineers) and develop a team effort to evaluate and innovate unconventional exploration methods.

As we planned Symposium II (conversing with and screening speakers and reviewing the literature), it was evident that marked progress has been made in objectives 1 and 2—we have established a "communications medium" for publishing new ideas and case histories in unconventional exploration methods.

In regard to objective 3, it would appear that the petroleum industry's concentration of its innovative energies on such matters as taxation, regulation, legislation, and diversification has served to retard technological advances. Countless billions of dollars have been invested by both industry and government in studies of "what makes our nation's *innovation cycle* tick." They all conclude that necessity is the mother of invention while taxation, finance, and competition probably control the birth rates.

Since Symposium I, it seems that conventional explorationists have tended to take unconventional tools out of the magic "black box" and put them in a sort of "pill box" as tranquilizers to ease the pain of drilling a dry hole on geological prospects.

The current state of the art of our oil and gas finding capability is depicted in Figure 1, which is self-explanatory; our oil-finding muscles have become flabby. In each year with the exception of 1968, when Prudhoe was discovered, we have found less oil than we have produced.

API statistics reveal that in 1978 we drilled 10,677 exploratory wells, of which 7,949 were dry holes; this involved drilling a footage of 63,234,745 ft, of which 45,625,092 ft led us to dry holes.

Dry hole drilling is a result either of failure to apply conventional scientific principles or of the inadequacy of our present methods. If present conventional methods are not capable of reducing this enormous unproductive effort, then our methods of finding should be examined in order to determine their weaknesses and deficiencies.

The large and increasing costs of geophysical surveying have renewed interest in the detection of subsurface accumulation of hydrocarbons by surface or near-surface indicators. Many authoritative explorationists feel that seismic surveying has reached a point of diminishing returns.

Figure 2, also taken from the 1979 API-

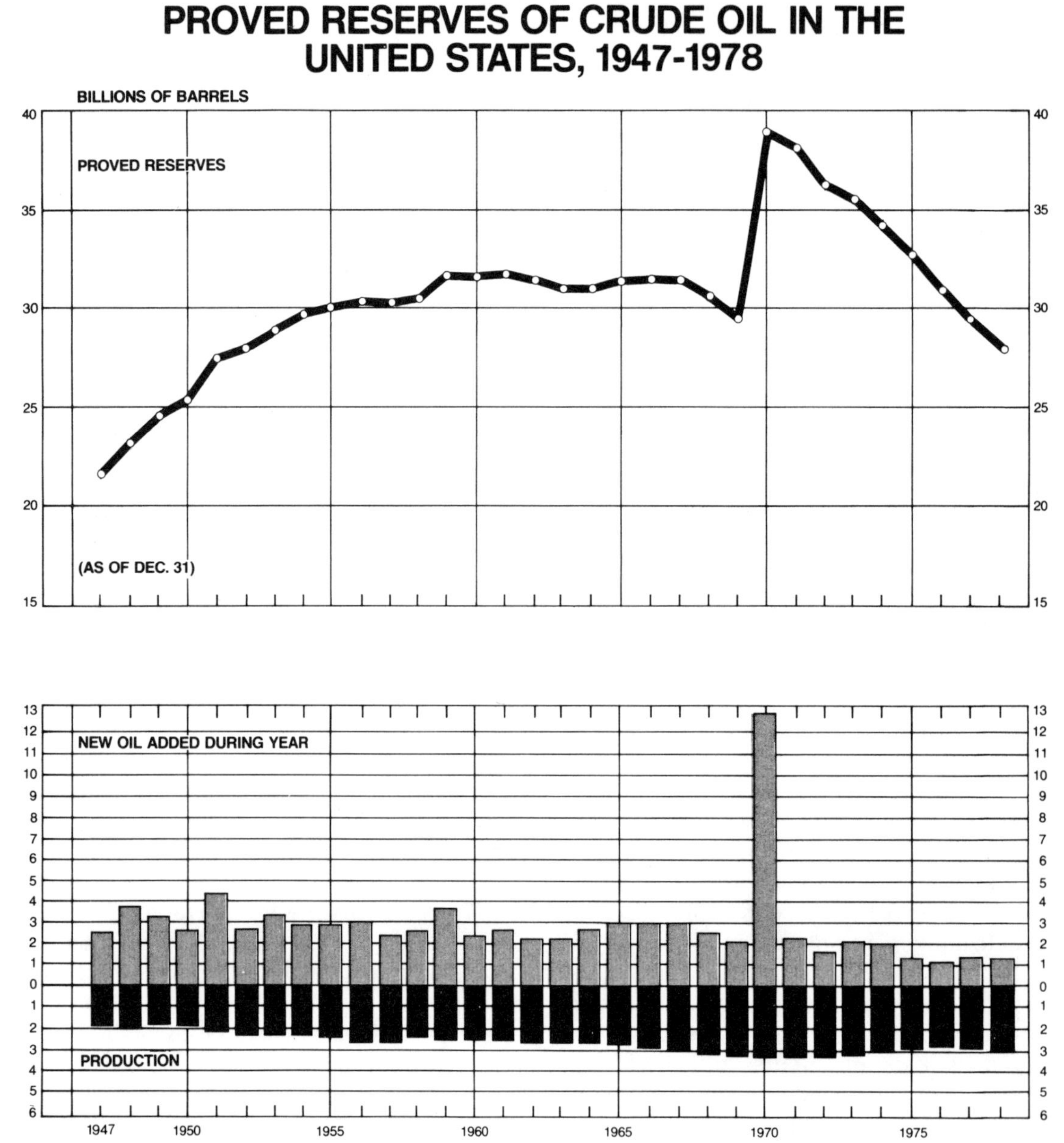

Fig. 1. From API Reserves Report, v. 33, June, 1979.

Reserves Report, breaks down the total of oil ultimately recoverable in the United States into "structural" and "stratigraphic" categories.

Approximately one-third, and in many areas one-half of the oil found to date has appeared in stratigraphic traps for which we have not, as yet, developed acceptable and recognizable methods of location.

Our conventional exploration efforts have been primarily concentrated on the identification of structures, most of which turn out to be dry. Why haven't we developed our capabilities to find hydrocarbons instead of structures?

Our philosophy is that the technology is now available to obtain better odds for success. This technology is dispersed among several of the scientific disciplines, and the real problem is to

ESTIMATED ORIGINAL OIL-IN-PLACE AND ULTIMATE RECOVERY OF CRUDE OIL
IN THE UNITED STATES BY TYPE OF ENTRAPMENT
DECEMBER 31, 1978
(Thousands of Barrels of 42 U.S. Gallons)

	Estimated Original Oil-In-Place		Estimated Ultimate Recovery	
State (1)	Structural (2)	Stratigraphic (3)	Structural (4)	Stratigraphic (5)
Alabama	748,131	27,330	227,846	6,008
Alaska	27,711,000	1,540	10,607,547	154
Arkansas	3,947,005	334,240	1,410,695	129,916
California /a	70,281,045	13,508,047	18,062,989	3,506,557
Coastal Region	13,707,605	2,709,331	3,286,549	471,267
Los Angeles Basin	29,668,045	1,889,195	7,556,507	475,558
San Joaquin Basin	26,905,395	8,909,521	7,219,933	2,559,732
Colorado	2,577,006	1,717,238	1,000,464	428,054
Florida	938,854	115,211	418,285	36,682
Illinois	5,340,144	3,747,446	2,110,857	1,127,232
Indiana	153,990	1,484,065	34,529	456,676
Kansas	6,804,901	9,366,840	2,130,145	3,009,934
Kentucky	320,489	1,795,396	108,426	569,714
Louisiana /a	36,057,361	5,100,126	17,716,880	1,032,809
North	3,226,069	5,076,767	1,170,698	1,019,739
South	32,831,292	23,359	16,546,182	13,070
Michigan	1,392,054	1.512,289	504,980	454,192
Mississippi	4,242,756	689,204	1,585,652	238,790
Montana	2,966,339	1,700,540	667,911	460,505
Nebraska	95,797	1,312,250	26,824	373,338
New Mexico	7,833,907	7,025,229	1,967,267	1,913,669
Northwest	102,573	835,015	25,335	163,213
Southeast	7,731,334	6,190,214	1,941,932	1,750,456
New York	–	1,117,739	–	238,088
North Dakota	1,827,126	1,016,784	466,976	219,206
Ohio	587,570	6,604,673	88,206	887,336
Oklahoma	16,879,165	21,565,300	5,887,372	6,784,128
Pennsylvania	–	6,668,990	–	1,343,095
South Dakota	25,400	21,645	5,480	3,539
Tennessee	–	35,424	–	8,088
Texas /a	76,364,757	76,731,685	27,465,219	23,229,701
District 1	2,295,069	1,188,413	683,209	212,861
District 2	5,606,476	56,884	2,538,353	15,883
District 3	15,662,572	305,118	7,472,078	84,584
District 4	6,366,891	2,489,939	2,146,006	910,800
District 5	1,469,142	391,585	800,584	177,037
District 6	4,257,834	8,010,267	2,109,992	5,987,329
District 7-B	1,116,137	6,766,038	305,652	1,548,672
District 7-C	1,044,909	10,781,094	252,672	1,348,390
District 8	27,917,356	14,551,898	8,546,004	2,955,422
District 8-A	6,615,174	16,597,517	1,658,327	6,232,606
District 9	3,822,195	8,293,686	918,942	2,126,321
District 10	191,002	7,299,246	33,400	1,629,796
Utah	508,821	3,329,519	126,435	679,713
West Virginia	–	2,640,476	–	549,739
Wyoming	11,132,302	5,141,032	3,981,892	1,245,933
Miscellaneous /b	25,710	167,724	4,908	29,393
Total U.S.	278,761,630	174,477,982	96,607,785	48,962,189
Gulf of Mexico /c	15,030,527	–	7,048,552	

/a Includes offshore reserves.
/b Includes Arizona, Missouri, Nevada, Virginia, and Washington.
/c Included with Texas and Louisiana.

Fig. 2. From API Reserves Report, v. 33, June, 1979.

develop a coordinated team effort to find oil, regardless of the geological classification of its location.

It is difficult for me to understand why a nation with the technological "muscle" to place men on the moon and televise their adventures to our living rooms has not "geared up" and accepted the direct finding techniques which have been developed and will be presented by the distinguished speakers who are here today.

It is our philosophy that the time is ripe for a technological breakthrough in the science and art of finding oil. This can best be accomplished by coordination of communications among the various scientific disciplines.

We sincerely hope Symposium II will improve your ability to drill more producers and fewer dry holes.

SOME USES OF TEMPERATURE DATA IN PETROLEUM EXPLORATION

W. H. Roberts III

GULF RESEARCH & DEVELOPMENT CO.

ABSTRACT

It has been known for at least 100 years that many oil and gas deposits are associated with positive temperature anomalies. Because of uncertainty as to the reasons for the association there has been little confidence in the use of temperature for petroleum exploration. As reliable temperature data have become more available, the reasons for the association are becoming clearer. This paper will explain how temperature is related to petroleum entrapment and how it can be used as a petroleum probability guide. First to be considered are some basic aspects of heat exchange, with special attention to sedimentary porous media. This will lead to a description of the essential hydrothermal functions of oil and gas traps. Finally are shown some well-documented examples of thermally delineated oil and gas fields.

We can think of our planet as a heat engine. The interior of the earth is an inexhaustible heat source. In a variety of ways the heat is continuously transmitted outward through the earth's crust. Things happen because of the heat. Also, the heat can tell us of things that are happening. We observe heat-related changes in solid, liquid, and gaseous earth materials. Where the earth's heat passes upward through sediment-covered areas of the earth's crust, the effects of temperature and temperature variation should be of special interest to explorationists seeking petroleum and mineral deposits. The temperature signal is shaped and modulated by the configuration of the rocks and the avenues of hydrothermal transmissibility.

In sedimentary media there is a functional relation between the movement of water and the movement of heat. There is also a functional relation between the movement of water and the movement of hydrocarbons. Consequently the heat and the hydrocarbons tend to move the water in the same direction by the same means. Oil and gas accumulations are generally found in areas of ascending water. As the upward flow of water is focused by the configuration of the rocks, both heat and hydrocarbons tend to accumulate in upward-reaching reservoir spaces.

The heat exchange capacity of water moving through sediments has been demonstrated in the laboratory and, of course, is quite obvious in the case of thermal springs and other examples of hydrothermal activity. It is important to note that the hydrothermal temperature signal is a live (dynamic) signal. Thus a water-dependent temperature anomaly will decay rapidly if water movement should cease. It follows that the strength of a temperature anomaly may be a measure of anomalous water movement. In the three-dimensional aqueous continuum of a sedimentary basin there are certain areas as well as certain depths in which favorable water movements (as revealed by temperature) are most conducive to hydrocarbon accumulation. The study of such associations, of course, should also give attention to pressures and water chemistry.

INTRODUCTION

The rationale for using temperature in oil finding is quite simple. In sedimentary basins, temperature difference and oil occurrence are both closely related to the existence and the movement of water. The purpose of this paper is to relate and illustrate these realities. First, we need to acknowledge that water movements *are* largely responsible for most of the temperature variations in sedimentary basins. On this point the evidence is pretty clear. The other point is more complicated. It will be argued, however, that water *is* the principal carrier of petroleum raw materials from the primary organic source to the traps where oil and gas deposits are formed. If, indeed, we are assured that the waters in basins *do* move, and that they *do* carry both heat and the raw materials of petroleum, then we will understand how temperature can be used to track the water, and thus show us where the hydrocarbons (HC's) are going. What we know so far indicates that the directions of movement don't change very often or very much through geologic time.

In setting up the rationale, let us be reminded that most of the sediments we deal with were deposited in or by water and that, for as long as they have remained below a water table, they have been thoroughly water-soaked. The water pervades the system. Almost every pore space contains some water. Because of its continuity, water can move through almost any kind of sedimentary material at some geologically appreciable rate. It is essentially a hydrophyllic system.

Most oil-seekers have been comfortable with the idea that the HC mixtures we know as oil and gas were pre-formed in the so-called "source rocks," and that because oil and water do not mix, the oil had to move from there to the traps as a distinct separate phase. It should be pointed out, however, that none of us have ever seen oil moving in that way and that the mechanical resistance to such movement in a water-soaked medium is formidable, if not prohibitive.

There is much evidence that oil and gas deposits *do* in fact coincide with temperature anomalies, and are accompanied by other signs of directed water movement. It is, thus, quite possible that water movement does have something to do with the movement and the entrapment of the raw materials of petroleum. Until we have more facts, we should probably stay uncommitted on what those raw materials really are.

There seem to be three basic modes in which the raw materials of petroleum *could* be moved from a source to a trap: (a) a molecular solution or colloidal dispersion of hydrocarbons (HC's) in moving water; (b) a solution of organic acids in moving water, transformable to HC's at or enroute to the traps; (c) pre-formed, separate-phase, whole oil or gas mixtures moving through water, with water, or independently of water. The relative importance of each mode is expected to vary with local conditions. None is likely to operate exclusively.

HISTORY

According to history, flammable gases in the ascending waters of hot springs were known to Turkish, Grecian, and Roman bathers many centuries ago. Those folks weren't into the rock-oil business yet, but they knew gas and hot water went together. In this country, possibly the earliest scholarly treatment relating hot water to geology was by W. B. Rogers (105)[1] in 1842, who noted "a connection of thermal springs in Virginia with anticlinal axes and faults." Apparently the relation of oil and gas to anticlines and faults was then not yet known to the Western world.

The earliest association of oil and gas deposits with positive (high) temperature anomalies on anticlines seems to have been made by G. V. Abikh, who worked in the Baku area (USSR) from 1848 to 1866, as reported by Mekhtiev and Aliev (82). In this country, E. W. Evans (42) in 1866 reported the association of oil deposits with fractures and ground water movements on the Burning Springs Uplift in West Virginia. Modern observations of HC and hot water associations have become numerous (e.g.: 5, 11-40, 40, 54-56, 60, 61, 66, 68, 80-83, 92-94, 96, 100, 104, 114, 117, 118, 123-125, 129, 130, 140, 141).

In spite of the evidence that has been available for over 100 years, oil people have remained skeptical about the association of HC's and water and temperature. They have been unable to adopt or adapt the hydrothermal concepts which are familiar to people who work with ore deposits. Why, is a mystery. The reasons for the association of water and heat and hydrocarbons have now become quite clear, and implementation is overdue.

BASIC HYDROLOGY

It is important to agree on some gross basin hydrology. In young, compacting basins most of the original water is moving upward in relation to sinking stratigraphic levels. In older, pre-compacted basins, new waters are moving along some of the same routes, but for different reasons. Some of the waters are going down, in a recharge mode, while roughly equal volumes are coming up to discharge. It is essentially a material balance situation. In this paper it should become increasingly clear why most of the important oil in the world is found in areas of upward water movement which will be referred to as deep-water-discharge areas. And because those ascending waters are warmer than other waters, the reason for improved oil and gas prospects to be associated with warmer temperatures should also be clear.

Because of pressure and temperature differences found in every basin at almost every depth, we are assured there are not truly static basinal water systems, although rates of movement may vary from almost zero to rather high numbers. It is like a forced-draft system. The basinal water traffic is energized, guided, regulated, and controlled by geologic factors (which are not always well defined). In other words, the rock framework is an apparatus in which

[1] References appear at end of text.

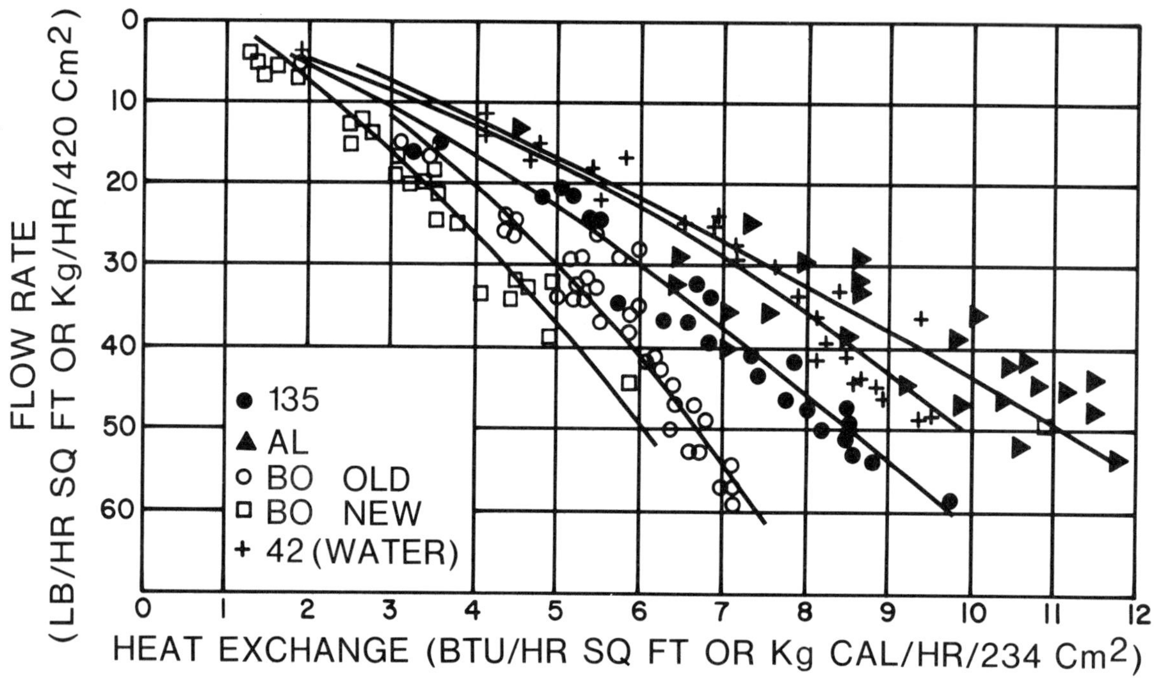

HEAT EXCHANGE FOR 10% SODIUM-CHLORIDE SOLUTION

FROM ADIVARAHAN ET AL, 1962

Fig. 1. Flow of brine dominates heat exchange in five samples of porous media (from Adivarahan et al., 1962).

things happen according to the laws of physics and chemistry and because energy is applied. Hydrologically speaking, the apparatus consists of aquifers and aquitards. In the geologic time frame, aquicludes don't seem to exist. In terms of heat exchange, the aquifers are the transmitters, and the aquitards are the insulators or filters.

HYDROTHERMICS

The essence of "hydrothermics" can be demonstrated by measuring the heat exchange across sedimentary material through which water is moving. In Figure 1, Adivarahan (1) and his coworkers at Northwestern University have graphed the results of flowing a 10% brine solution through five different porous media, three natural and two synthetic. It should be noted that the best fit curve for each medium approaches zero heat exchange for zero flow. The remaining heat exchange for zero flow is assumed to represent conductivity.

The flow rates for this bench work seem to be rather high in comparison to natural conditions. Field work by Bredehoeft and Popadopolous (18), however, shows that movement as slow as 1 mm per day, about 1 ft (36.5 cm) per year, can produce a significant temperature change. Others (e.g., 54, 140) have documented temperature anomalies to modest natural flow rates.

Figure 2 shows the results of a recent sea floor temperature survey (10) near Baku in the South Caspian. The data were obtained with a probe reaching 1.2 m into the bottom mud with temperature accuracy of 0.01°C. Here an oil-productive anticlinal closure shows a significant positive temperature anomaly in the bottom mud of only 0.5°C. This is precision work which can only be done in a low noise situation (e.g., under water or at depth underground). Temperature profiles crossing the

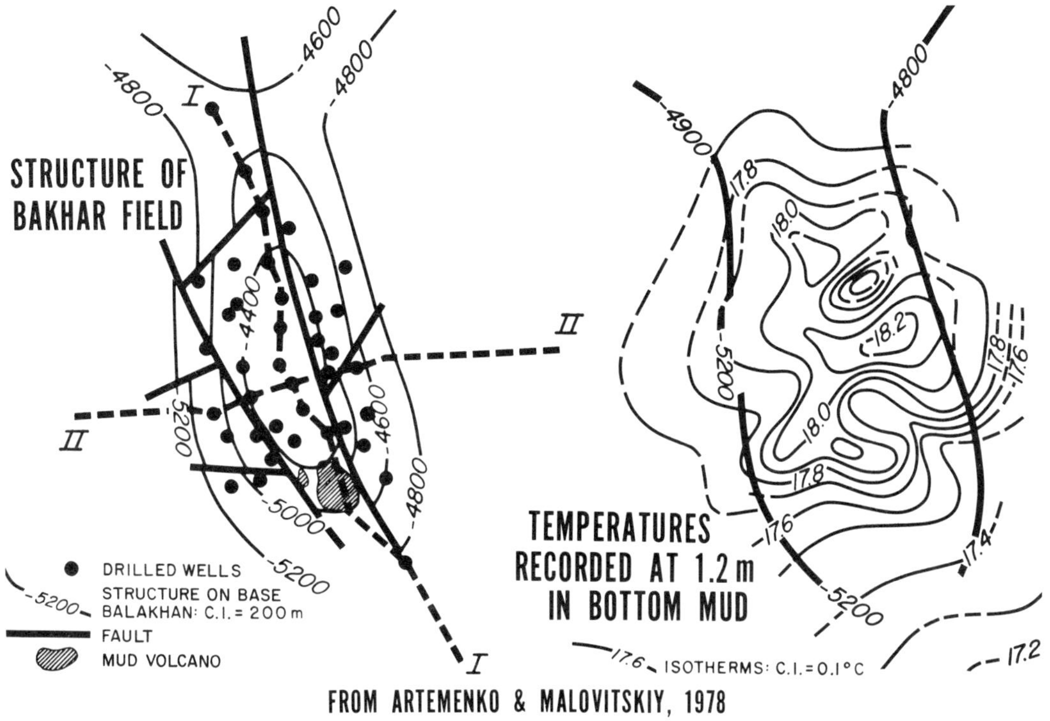

Fig. 2. Temperature anomaly mapped in bottom mud over Bakhar Oil Field in Caspian Sea near Baku (from Artemenko & Malovitskiy, 1978).

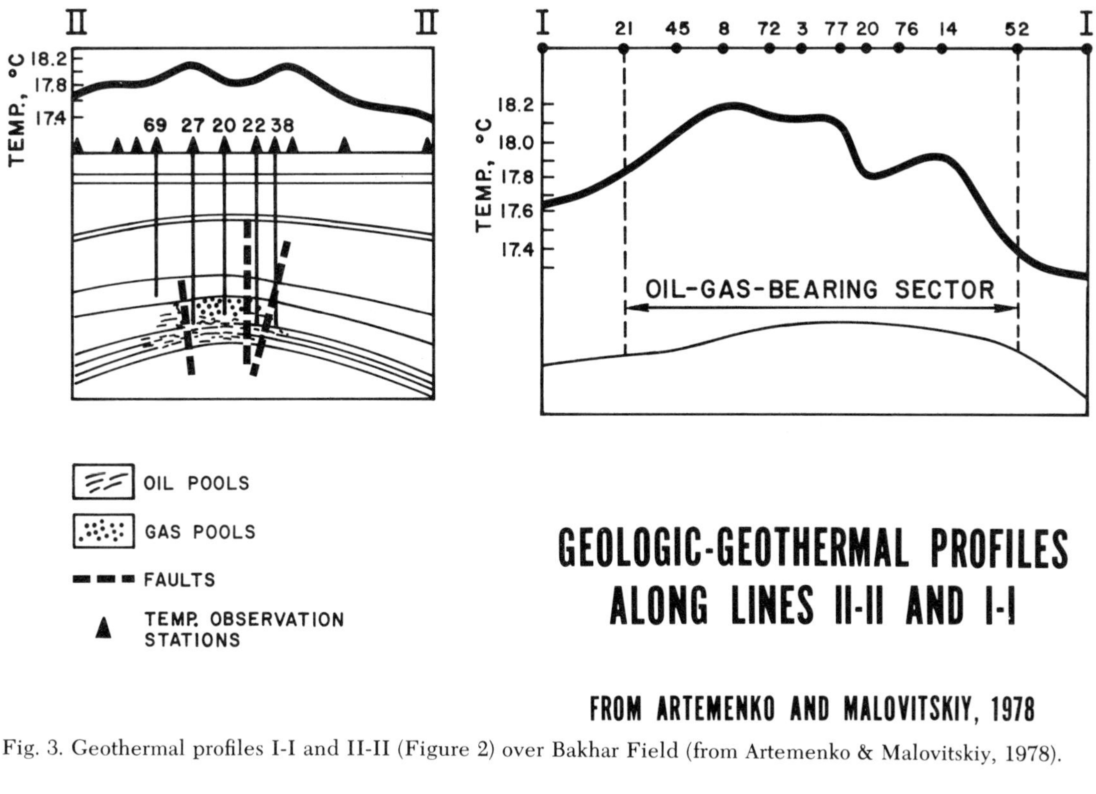

Fig. 3. Geothermal profiles I-I and II-II (Figure 2) over Bakhar Field (from Artemenko & Malovitskiy, 1978).

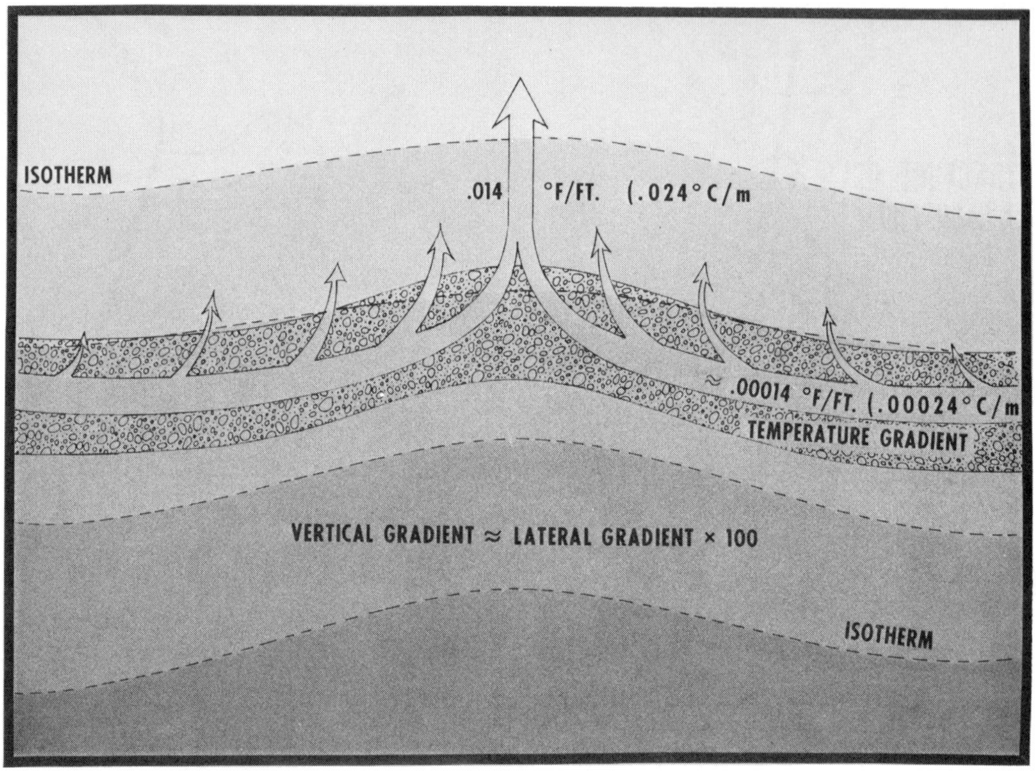

Fig. 4. Pressure gradients: vertical in relation to lateral flow through a trap.

long and short structural axes are presented in Figure 3. As will be seen later, other producing structures in the Baku area show that greater temperature relief (stronger anomalies) can be expected at greater depths.

TRAP FUNCTION

It is important to examine the functions of oil and gas entrapment in order to appreciate the importance of temperature in trapping HC's as well as in tracking the water that carries them.

Figures 4 and 5 show schematically the pressure and temperature conditions in an anticlinal aquifer. For a number of reasons (104), the anticline promotes and focuses water movement upward through the covering strata. This is what we have called "deep-water-discharge." Of special interest are the pressure and temperature environments of that water as it may carry HC's into or through the upward-converging reservoir space. As a result of the basic change in direction from lateral to upward movement, the water and whatever it carries will experience an abrupt change in pressure and pressure gradient, probably several hundredfold as shown in Figure 4. A similar change occurs in the temperature environments as shown in Figure 5, using rather conservative numbers. The lateral to vertical contract in temperature gradients could easily be a thousandfold.

The sudden reduction of pressure and temperature as the water passes through the trap can have a very pronounced effect on the capacity of the water to carry HC's along with it. Thus the HC's tend to collect and coalesce in the trap. One more picture of this trap is shown in Figure 6. The darker gray represents increased brine strength or salt in the water due to filtration by overlying membranes such as clays or silts and dirty sands. This brine is not connate water. It is a dynamic concentration (entrapment) of salt, which also works to exsolve the HC's from the feedstock waters as they pass through the trap. Ore deposit people call this "salting out."

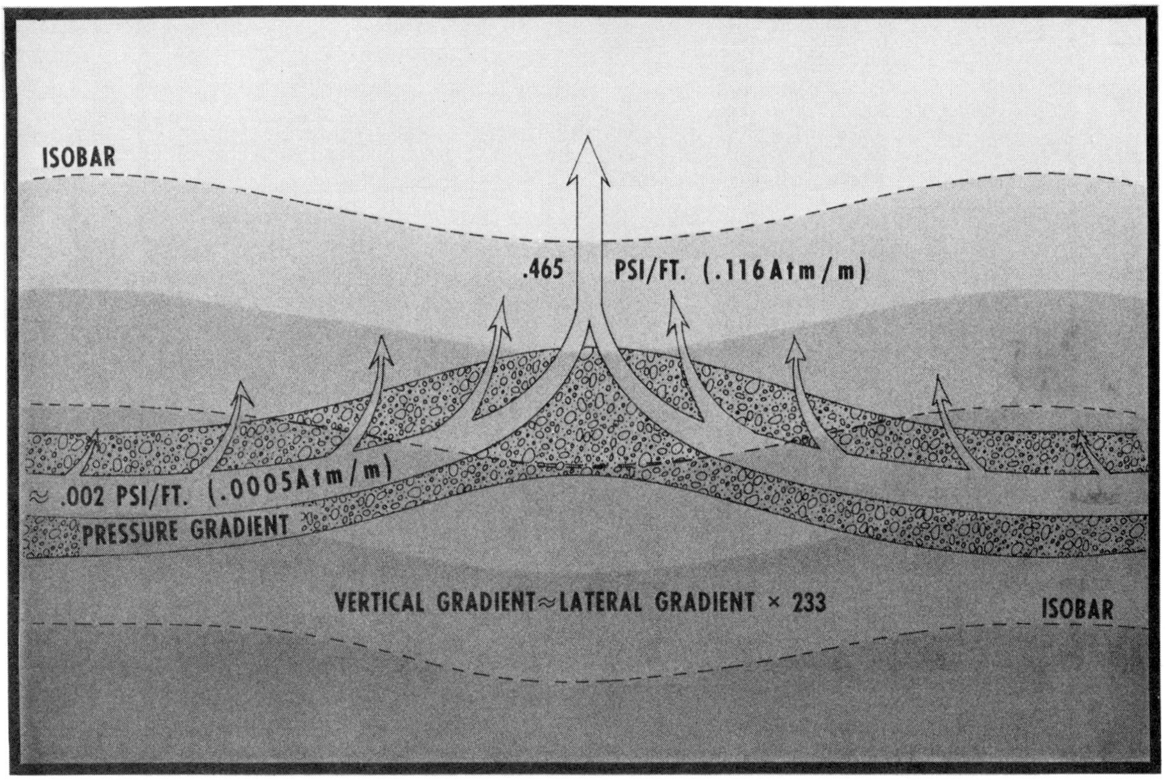

Fig. 5. Temperature gradients: vertical in relation to lateral flow through a trap.

TEMPERATURE DIFFERENCES

Although our earthly heat source is reasonably constant and the heat exchange is directed generally upward to the earth's surface, we are dealing with a three-dimensional, usually assymetric combination of intervening rocks and watercourses. The vertical heat flow rate varies both laterally and vertically (9, 132). Temperature depth gradients (TDG's) thus need to be examined on an interval basis, as we do with sonic velocities in seismic interpretation. They also need to be considered for both regional and local variation.

It has been observed (e.g., 137) that in mature basins point-to-point temperature differences decrease below depths of 5-7 km. The paucity of water and relative uniform density of rocks at such depths would logically forestall or neutralize temperature variation. It has also been observed (e.g., 140) that point-to-point temperature variations decrease at shallower depths, usually within 1,000 m of the surface. Probably this is because the shallow rocks are less capable of pore space confinement and flow channel restriction so that the more general freedom of water to move around allows equalization of temperatures.

Figure 7 compares the depth distribution of big oil volumes in giant fields with the depth distribution of greater and lesser point-to-point temperature differences. The big oil volumes have developed and/or stabilized in a modest depth range, not shallow and not deep, mainly between 1 and 4 km. The chances of finding large volumes of oil at significantly greater depths, extending this "giant oil" depth range, seem poor regardless of how many super-deep wells are drilled. There are some fairly sound physical and chemical arguments (104) as to why this should be so, but they are not within the scope of this paper.

On the right side of Figure 7 are plotted the maximum temperature differences found at common depths within regional (i.e., basinal) and local (i.e., oil field) areas. The depth scale

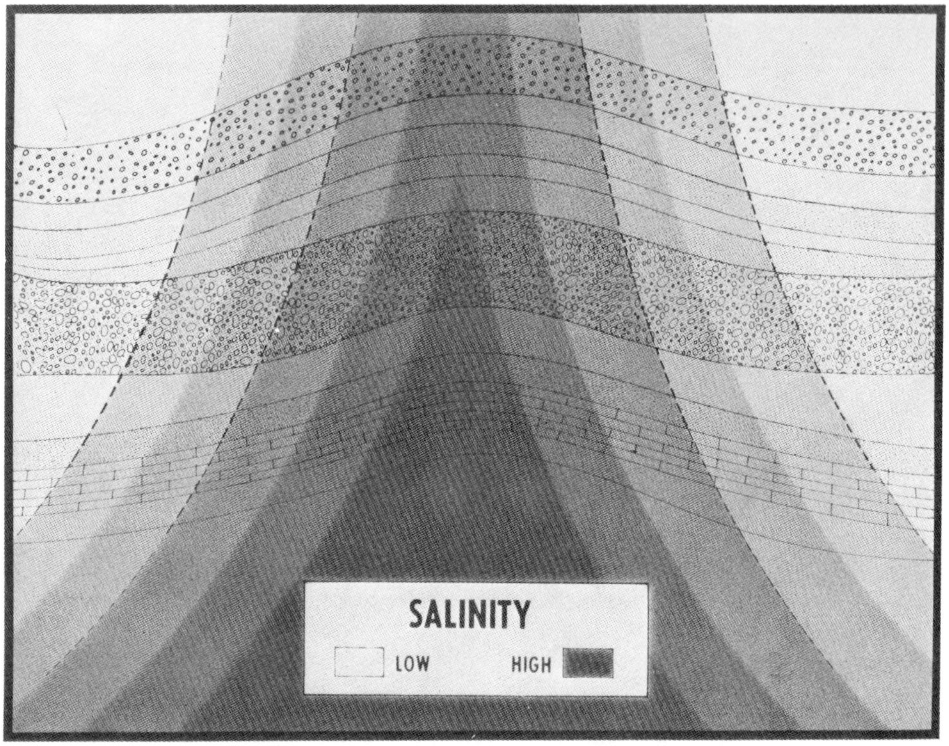

Fig. 6. Salinity increase by convergent ascending brine filtration.

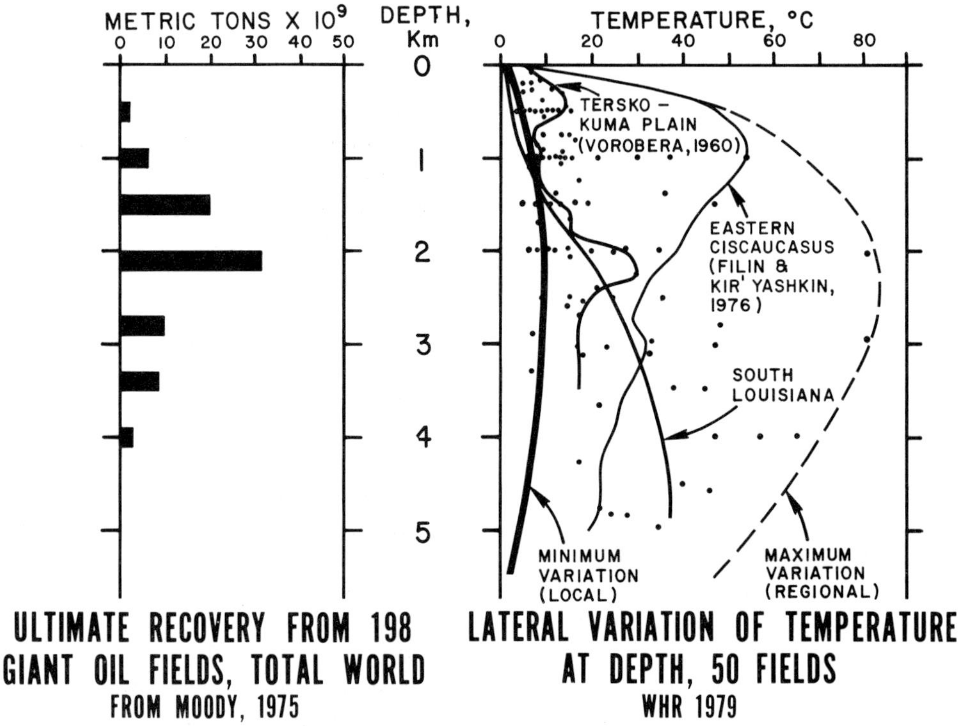

Fig. 7. Depth distribution of giant oil compared to temperature difference (in part from Moody, 1975).

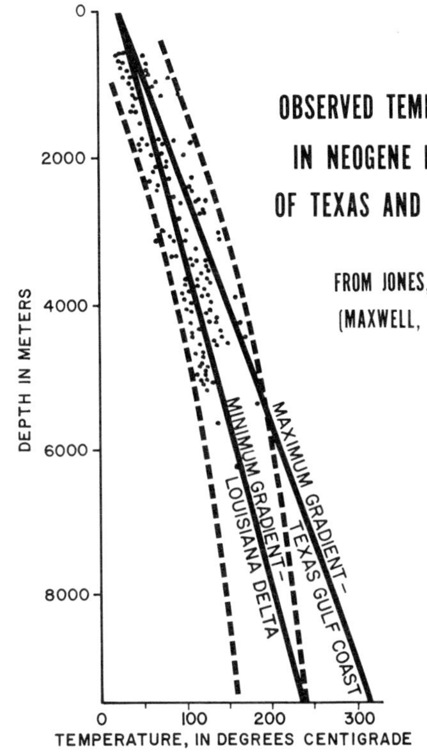

Fig. 8. Temperature-depth trends of the Gulf Coast (from Jones, 1975).

matches the "giant oil" graph to the left. Analysis of the reasons, geological and hydrological, for the temperature differences shown by these control points from over fifty oil and gas areas in different parts of the world would get pretty involved. The differences range up to about 15°C for local areas, usually on and off producing structures, and up to 70°C for regional areas containing major uplifts and depressions. Two boundary lines have been drawn showing the approximate limits of temperature difference within local areas (i.e., about 0-10°C) and regional areas (i.e., about 10-70°C). The loss of data density with increasing depth may impose some bias on these limits, but the general shape of the boundaries is believed to remain valid. There is an obvious swelling of temperature differences at intermediate depths, corresponding rather well in depth with the bulge for the "giant oil." Inasmuch as many of these data are taken from around oil and gas fields, we might surmise that the most effective depths for trapping HC's correspond closely to the depths showing the greatest variations in water movement as revealed by the differences in temperature.

Also on the temperature graph in Figure 7 are shown three irregular, subvertical lines which track the temperature differences obtained within three local areas. The typical irregularity of such plots is a measure of the geological diversity and the irregularity of water movement within a given area. The one for South Louisiana is believed to be typical of immature, subsiding basins, in which compaction waters are still very active and pronounced temperature differences persist to greater depth than would be expected in mature basins.

TEMPERATURE-DEPTH GRADIENTS

Temperatures recorded in relation to depths in drilled wells are useful for regional as well as detailed study of water movements and their effects on the geothermal regime. Presented in Figure 8 are some Gulf Coast temperature-depth data which show a certain conformity to a general trend, bounded by the dashed lines. Within the data set, however, it is possible to distinguish between a maximum temperature-depth gradient (TDG) recorded in the Texas coastal area and a minimum TDG recorded in the Louisiana delta area. The difference is thought to be caused mostly by increased subsidence, sedimentation, and the flow of juvenile compaction water which dominates deep-water-discharge (DWD) in the delta area.

To appreciate the comparative range of temperature-depth relations which can occur in different petroliferous regions, consider Figure 9. Here are four distinct temperature-depth systems. Obviously, the right-hand data group is reacting to a relatively strong heat source at depth. Note that the maximum temperature differences among the four regions at a common depth of 4,000 m is about 120°C, which is greater than the difference between freezing and boiling, at atmospheric pressure.

Two typical thermograms (of wells) illustrating the relation between TDG and lithology (more specifically, transmissibility) are shown in Figures 10 and 11. In general, the slope of the TDG line is steeper (less T-change per unit

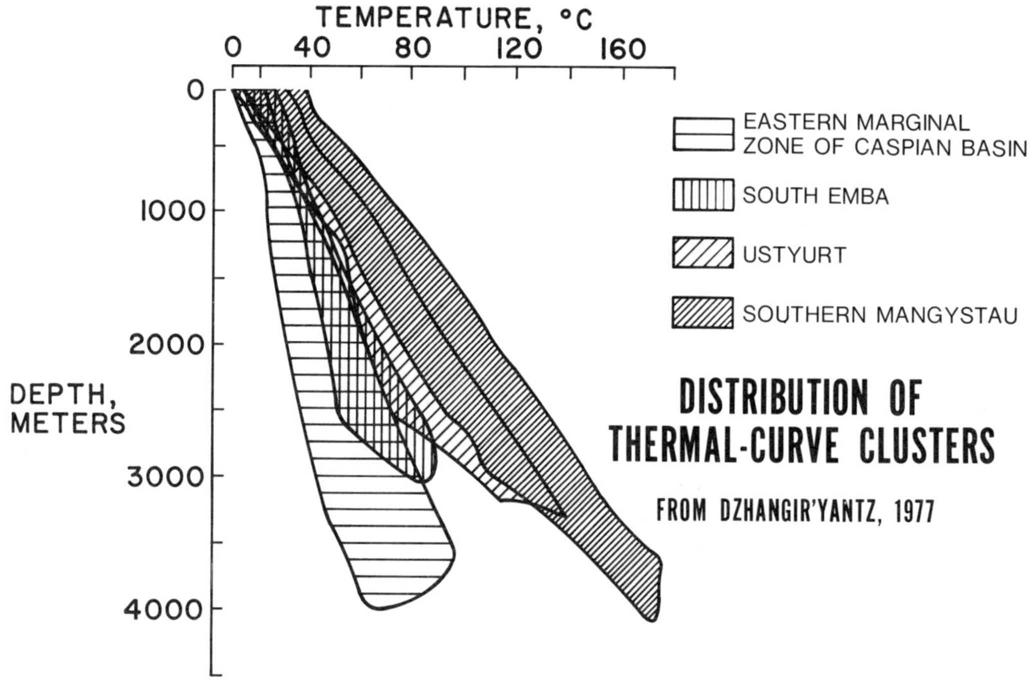

Fig. 9. Variant temperature depth trends of four different provinces (from Dzhangiryants, 1977).

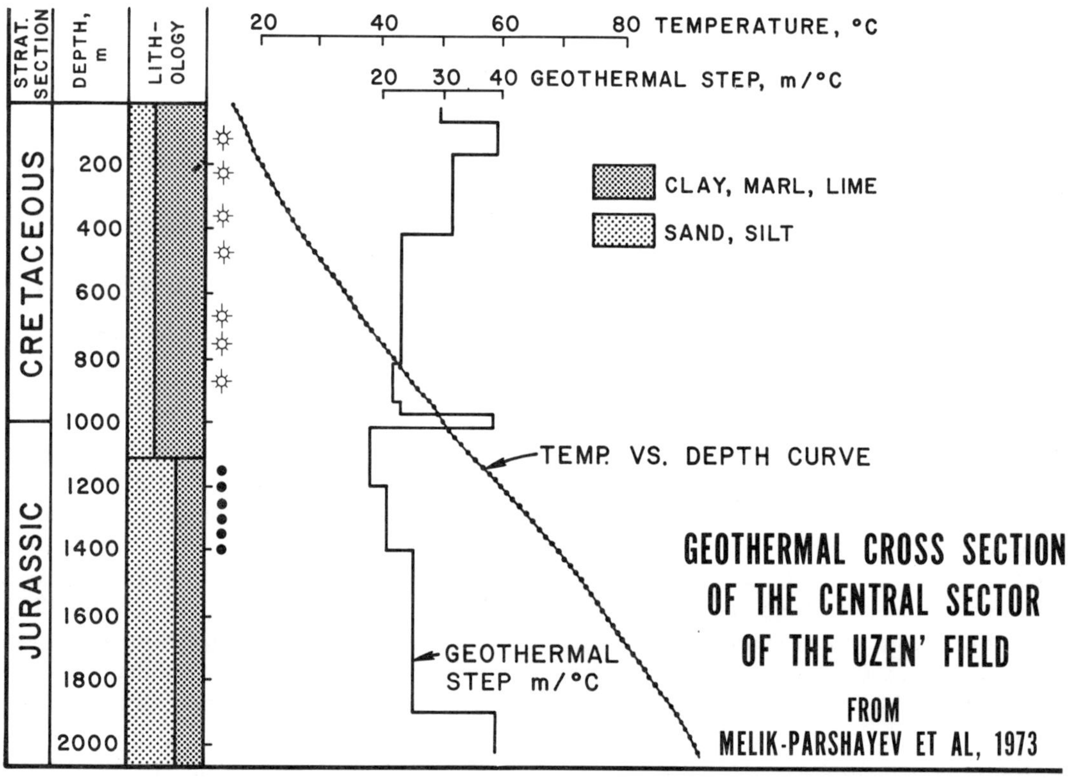

Fig. 10. Geothermal gradient and step in relation to lithology, Uzen Field, East Caspian region (from Melik-Pasheyev et al., 1973).

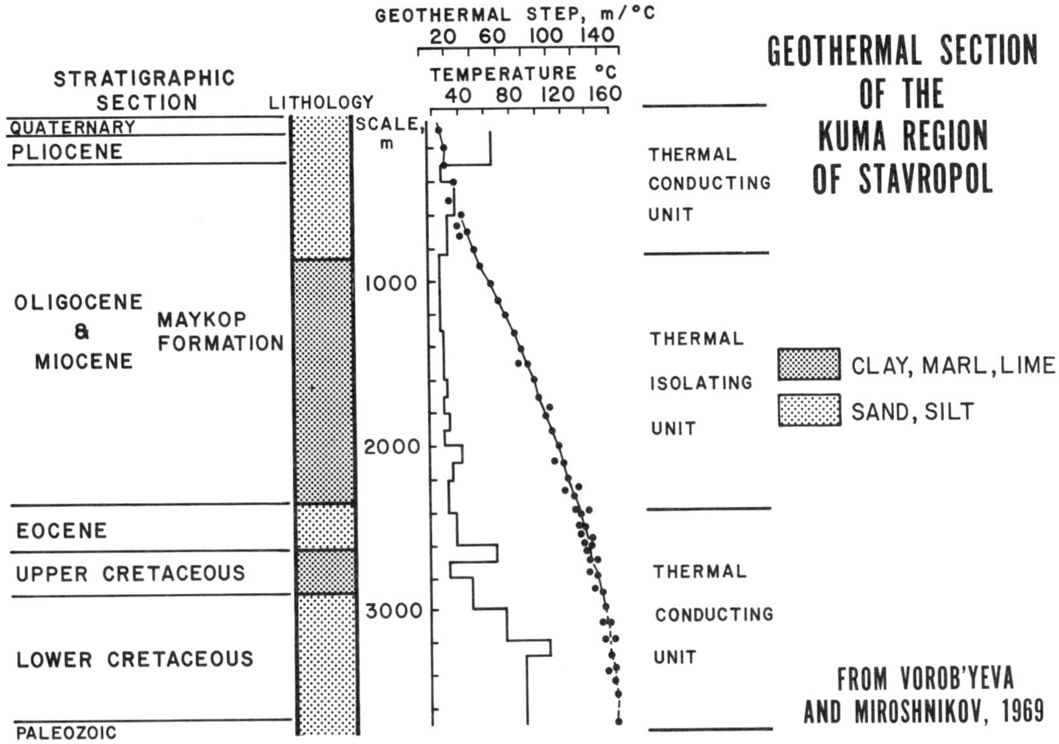

Fig. 11. Geothermal gradient and step in relation to lithology, Stavropol, Trans-Caucasus region (from Vorobyeva & Miroshnikov, 1969).

depth) for more transmissible strata and less steep (more T-change per unit depth) for less transmissible strata. Basically this means that temperatures are more equalized in the more transmissible (aquifer) strata where waters have more freedom to move around. Temperature equalization is also believed to be favored by a heat-bank effect, whereby upward heat exchange is impeded by an overlying aquitard—thus accumulating (pooling?) heat in the aquifer below.

Some rather important principles of TDG's are represented in Figure 12. There are three schematic temperature-depth plots: a, b, and c. The first simply compares the TDG's caused by weak and strong heat sources transmitting heat upward across the same homogeneous sedimentary section toward a common surface temperature. The next two plots show one TDG curve convex to the depth-axis and the other concave to the depth-axis. Case "b" represents a TDG decreasing (less T-change per unit depth) progressively upward. Case "c" represents a TDG increasing (more T-change per unit depth) upward. In case "b" the more transmissible (possible reservoir) strata are shallow, somewhat cooler, and probably oil-prone. In case "c" they are deep, somewhat warmer, and probably gas-prone.

The interpretation of the oil-prone and gas-prone conditions is less concerned with the influence of temperature on the generation of gas as compared to oil, but more concerned with the intensity of water activity required for assembling the heavier HC mixtures (i.e., oils) as compared to the lighter HC mixtures (i.e., gases). In case "c" the possible reservoir strata are "covered" by low transmissibility aquitards, thus reducing the upward component of water exchange and making it more difficult for the heavier HC's to be collected by the waters. The lighter HC's, being easier to come by and more easily accommodated in a limited water supply, may have a chance to collect and be held if conditions of entrapment (80, 104) exist. In case "b" the possible reser-

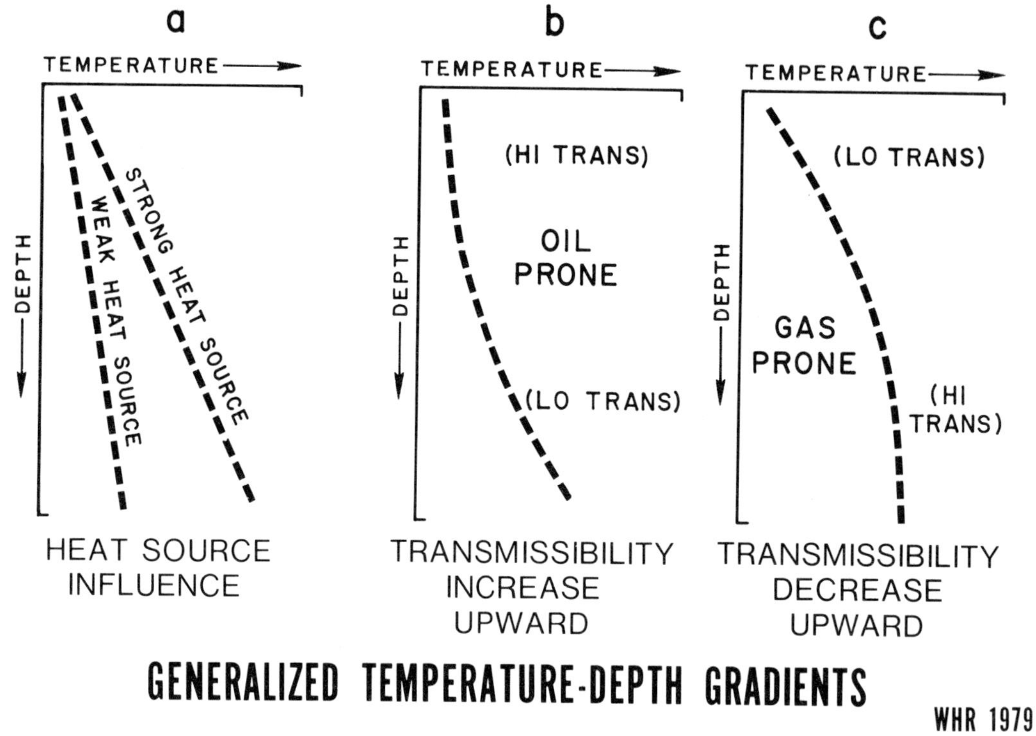

Fig. 12. Typical temperature-depth relations shown by schematic plots.

voir strata have a less dense cover, and there is considerably more freedom for upward water movement. If a variety of HC's is or has been available in those waters, and trap conditions exist (80, 104), a heavier mixture may be collected and held while the lighter HC's escape through the cover with the waters (a winnowing effect).

In the real geological world, of course, these schematic stereotypes don't exist. Rather, we find a complicated rock apparatus and confused fluid distributions. Some oils are deep, and some are shallow. Some gases are shallow, and some are deep. Still, it appears that as a general rule, oil-prone systems show more water activity *and* more temperature variation, regardless of depth, while gas-prone systems occur in comparatively less active water, regardless of depth.

TEMPERATURE RELIEF

As already stated, point-to-point temperature variations are always related to geologic variations, and it is chiefly the water that transmits the temperature signal. Basically, whatever alters the water traffic will alter the temperature. As a general rule, the amount of temperature relief will depend on relative rates of water movement. In other words, fast water usually causes high temperature relief (strong temperature anomalies), while slow water yields low relief. The latter, of course, requires more careful interpretation. This relationship is consistent with the schematic TDG curves in Figure 12, as interpreted. In most cases, increased temperature relief means increased probability of conditions favoring oil and gas deposits. It is important to look at this association in relation to depth as well as to lateral position. In other words, there are certain depth intervals as well as certain localities which are favored by temperature variation and the probability of hydrocarbon accumulation. The reasons are geological, both structural and stratigraphic.

RELATED ANOMALIES

In addition to oil and gas, there are many other water-related anomalous conditions

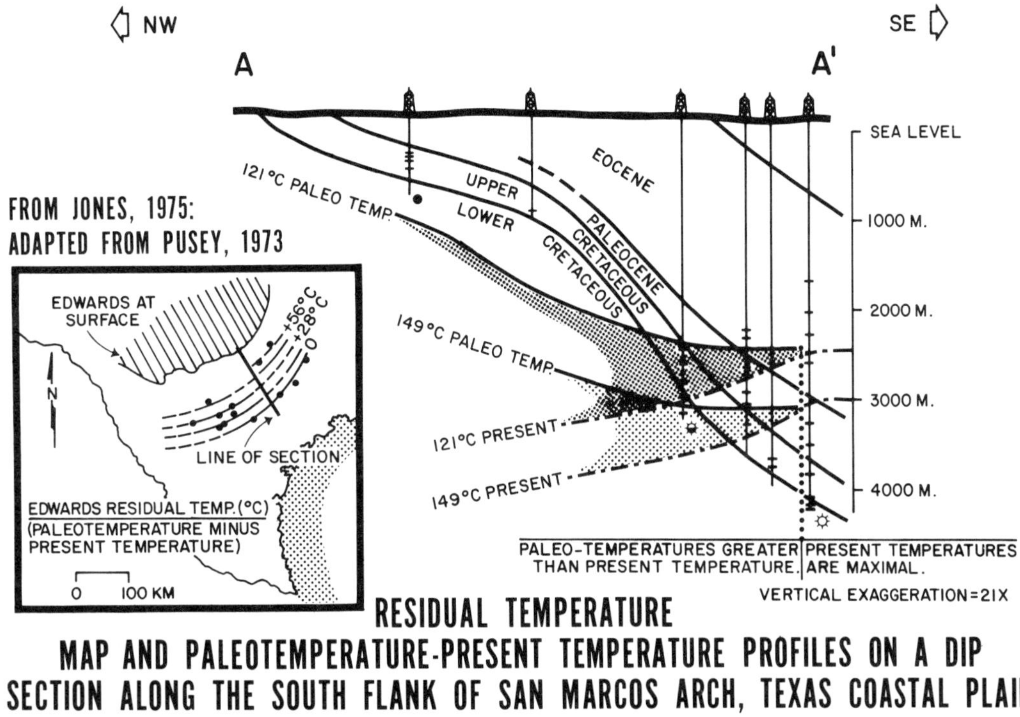

Fig. 13. South Texas dip section shows present temperature in relation to paleo-temperature estimated from kerogen ESR. Temperature difference (residual) shown on map insert (from Jones, [Pusey, 1973]).

which are likely to occur in company with geothermal anomalies (83, 84, 92). Included are pressure anomalies, saline water concentrations, hydrochemical plumes, reduced waters, bleached redbeds, salt solution and collapse, local diagenesis, mineralization, and caprock formation. Any and all of these phenomena could serve as diagnostic criteria of water activity possibly favoring oil and gas. None is a promise. Temperature and pressure are the only live, real time signals of water activity, and they don't promise oil and gas either. Properly interpreted, they can improve oil and gas *probability* significantly.

It should be pointed out that higher temperatures are related by both cause and effect to natural conditions which favor the collection of hydrocarbons into traps. They are the cause of much reduced water viscosity, increased solubility of salts, increased accommodation of hydrocarbons and other organic compounds in water, and the thermogenic transformation of residual organic source materials. It has been estimated (3) that a 10°C increase will cause a two- to threefold increase in rates of organic transformation. In their turn, higher temperatures are the effects of anomalous thermal conductivity (e.g., salt, basalt) and transmissibility (e.g., fracture zones, reefs, sand lenses, anticlinal convergence).

A curious ambiguity exists in the water-temperature relationship. High temperatures are caused in two different ways, one by fast water, and the other by slow water. Ascending waters carry the earth's central heat upward through a variety of sedimentary layers, subject to deviation, shunting, and impedence among the various materials. Where vertical transmissibility is high, flow is strong and temperatures are high through the transmissible section. Where vertical water flow is impeded by low-transmissibility or high-pressured overlying strata, the discharge of heat is restricted. Because the heat source is constant, the heat accumulates, and the temperatures below the aquitard will be high. At the same time, however, the temperature above the aquitard will be anomalously low. Obviously, it is important

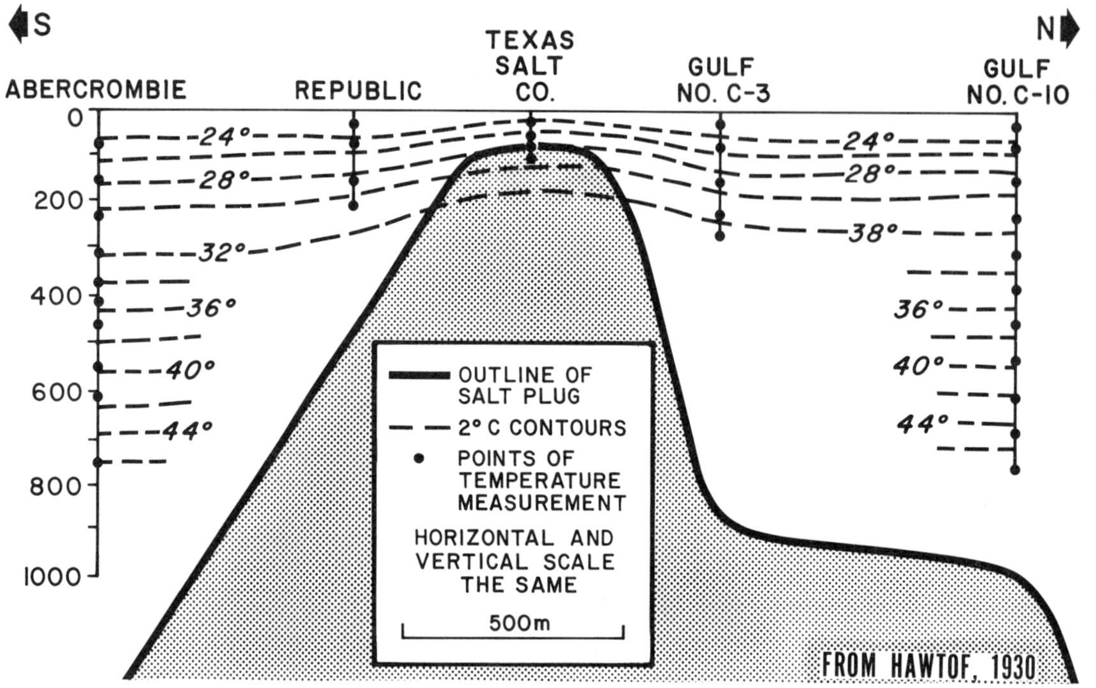

TEMPERATURE PROFILE OVER BLUE RIDGE SALT DOME, FORT BEND COUNTY, TEXAS

Fig. 14. Thermal anomaly over a Gulf Coast salt dome (from Hawtof, 1930).

to know the TDG's by intervals (55, 56, 68). For example, across a typical geopressured interval of the Gulf Coast, the TDG is above twice what it is above that interval. Moreover, an area of "leakage" (56) above a geopressured interval will be marked by a local thermal "high." We need to understand these relationships. An averaged TDG is seldom adequate to define the significant temperature activity.

The people who estimate maximum temperatures by vitronite reflectance and kerogen electron spin resonance (ESR) have given us something else to think about. Figure 13 shows the apparent relation of paleotemperature based on kerogen ESR to present temperature in South Texas. It appears that updip (NW) from a given strike line paleotemperatures are higher than present temperatures, while downdip from that line, the reverse may be true. According to the thermal maturation-generation argument, this could mean that HC generation is decreasing updip and increasing downdip at the present time.

FIELD EXAMPLES

Fifty years ago when drilling and logging and testing were less complicated and less expensive, and the engineers were more patient, there was a lot of interest in temperature work (e.g., 24, 25, 49, 51, 79, 96, 134). In this country and in Europe, some of the earliest known positive temperature anomalies were associated with salt domes or piercements (11-13, 29, 37, 38, 48, 51, 66, 79, 96, 99, 113, 119, 133, 144). This is a matter of interest to us because oil and gas are often found near salt bodies. A circa 1930 geothermal profile of such a body is presented in Figure 14. It has been commonly thought that the thermal conductivity of the salt per se, like a vertical calrod, is responsible for transmitting anomalous heat from depth. While this is no doubt true in some degree, there is clearly another influential source of heat which is often closely associated with salt domes (and other diapiric features). That is the water which ascends through the same zones of weakness that are invaded by the salt. Strong hot water

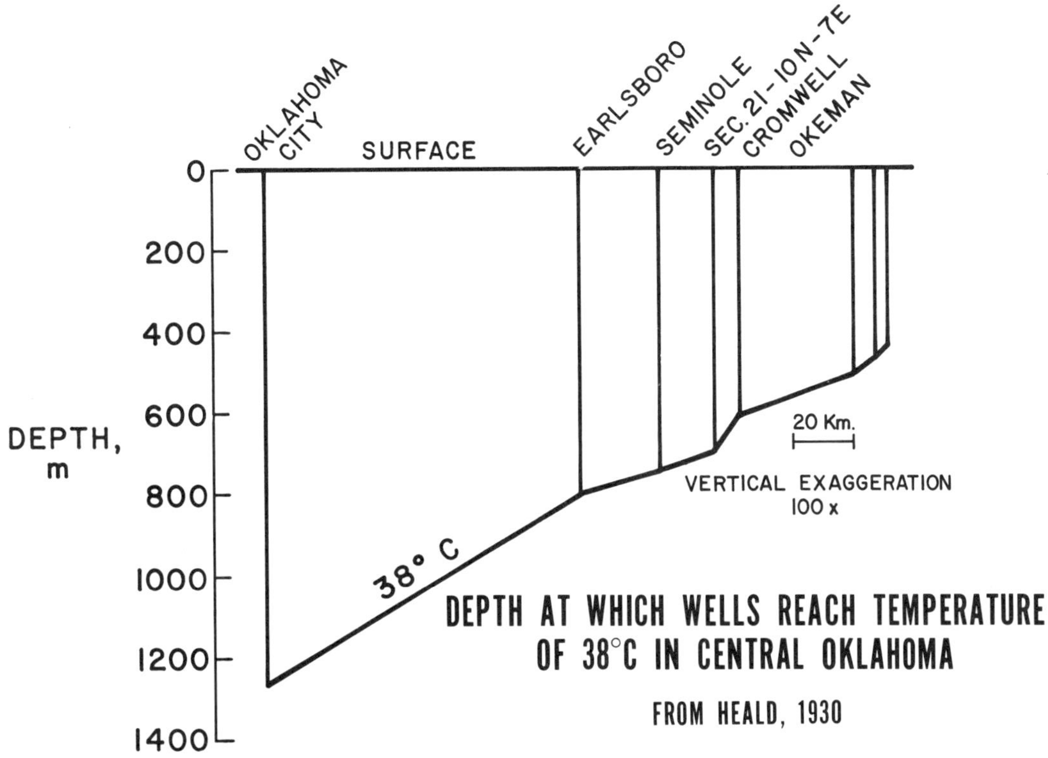

Fig. 15. Isothermal depth profile on north flank of Anadarko Basin, Oklahoma (from Herald, 1930).

flows have often been encountered in salt mining operations. In some cases the sediments immediately adjacent to or sheathing the salt appear to be sufficiently dilated (fluidized) to flow with the salt. This water from depth may well carry HC's as well as heat, thus accounting in part for some of the accumulations around salt domes. Where water-dilated sediments are present flanking a salt dome, the lower sonic velocity might account for some of the "rim syncline" effects seen on seismic profiles.

With regard to waters ascending near salt domes, we would not expect to find the same rate of water movement in a continuous ring around the sale. Temperature data, as well as ground water behavior, suggest an irregular pattern of rising waters. The irregularity of the water and the temperature patterns should be compared with the irregularity of oil and gas distribution around salt domes. It could be an important way of evaluating the probabilities of oil and gas entrapment at certain favored positions in relation to the salt body.

Figure 15 is an E-W profile of the 38°C (100°F) isotherm in Oklahoma. Going east, the depth decreases from 1,250 m to 460 m. That happens to be the direction of deep water movement—with an important upward component which contributes to surface drainage moving in the same direction. Figure 16 is a structure and isothermal (49°C) depth map of the Powell Field in Texas. Here the thermal effect of water ascending from depth in the Powell-Mexia fault zone is clear.

The Big Lake Field, Reagan County, Texas, is a good example of early temperature work (49). Figures 17 and 18, respectively, show structure on the pay sand, and a temperature map at a constant datum depth of 860 m. The relationship is clear. Another vintage temperature study treats the Eldorado Field of Kansas. In Figure 19, the subsea relief on the 38°C (100°F) surface is shown in relation to structure. The degree of coincidence shows typical variation, which is probably related to variations in vertical transmissibility.

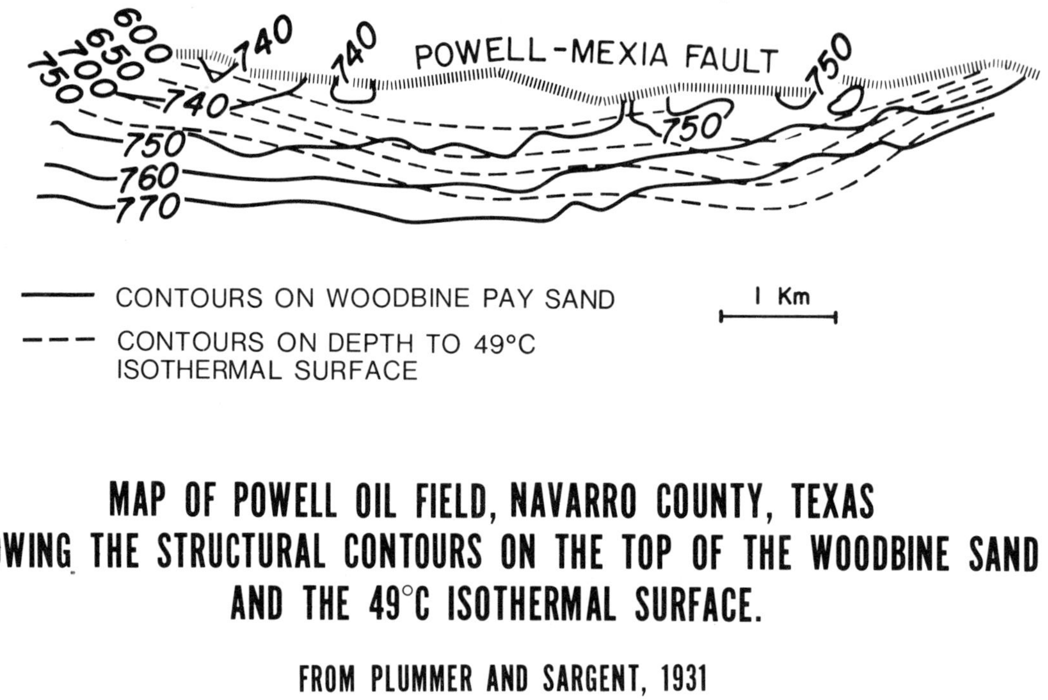

Fig. 16. Isothermal surface (49°C) in relation to structure on Woodbine sand, Powell Field, Texas. Note temperature maximum adjacent to Powell–Mexia Fault zone where vertical water movement is most active (from Plummer and Sargent, 1931).

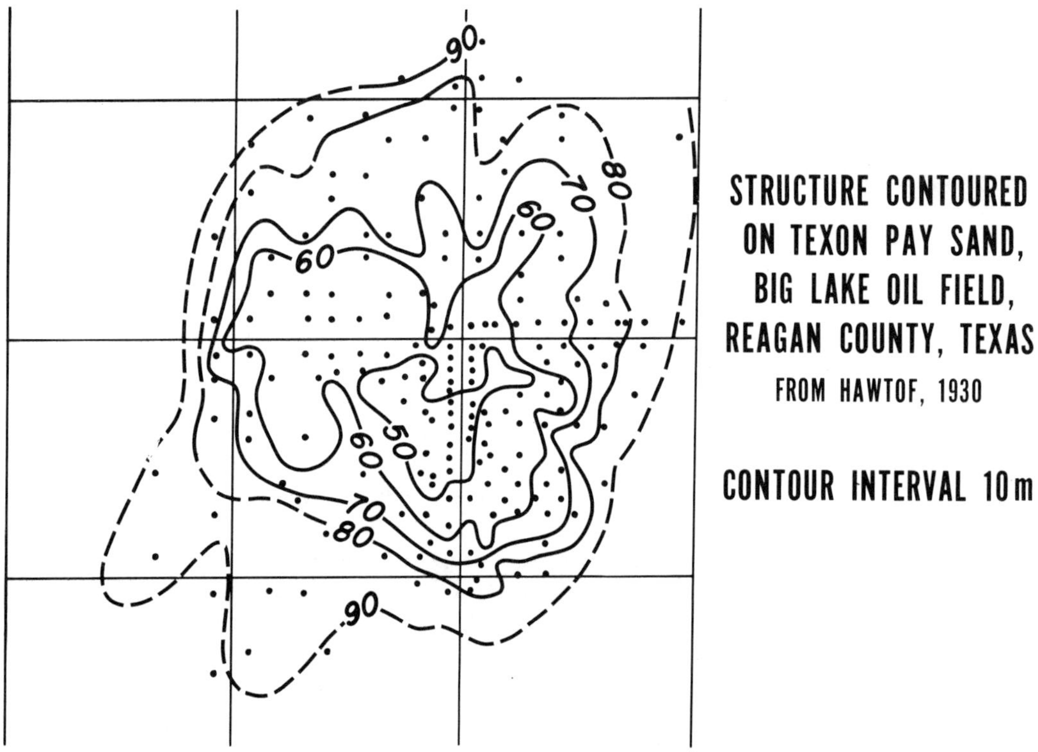

Fig. 17. Structure map of Big Lake Oil Field, Texas (from Hawtof, 1930).

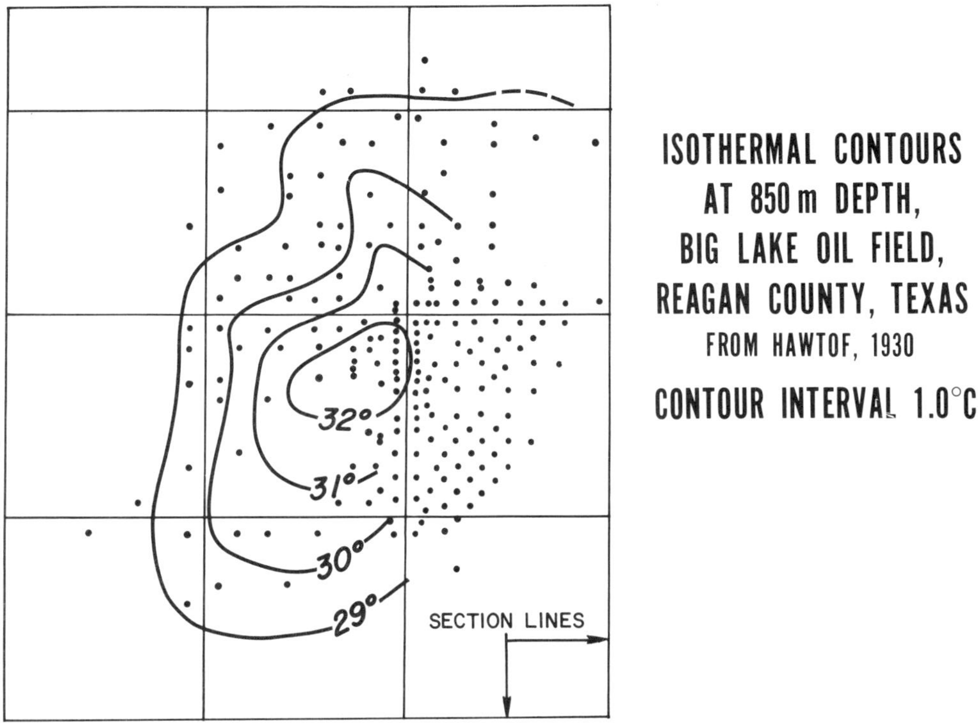

Fig. 18. Temperature map of Big Lake Oil Field, Texas, at 850 m depth (from Hawtof, 1930).

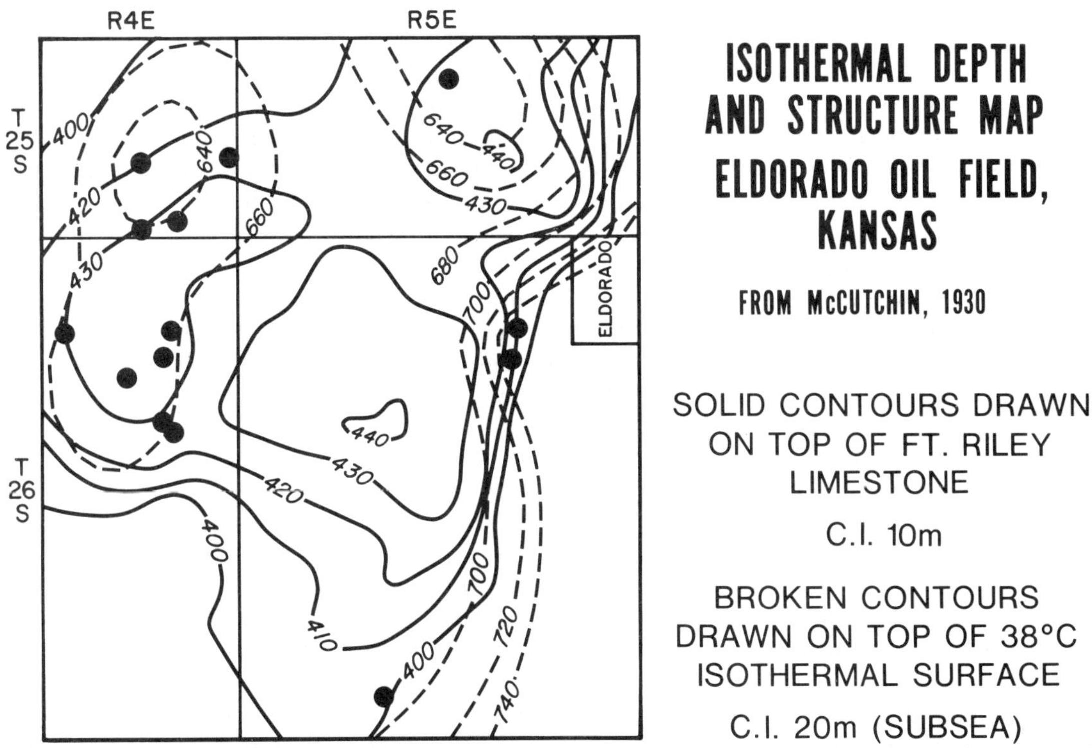

Fig. 19. Structure and 38°C isothermal depth map, Eldorado Oil Field, Kansas (from McCutchin, 1930).

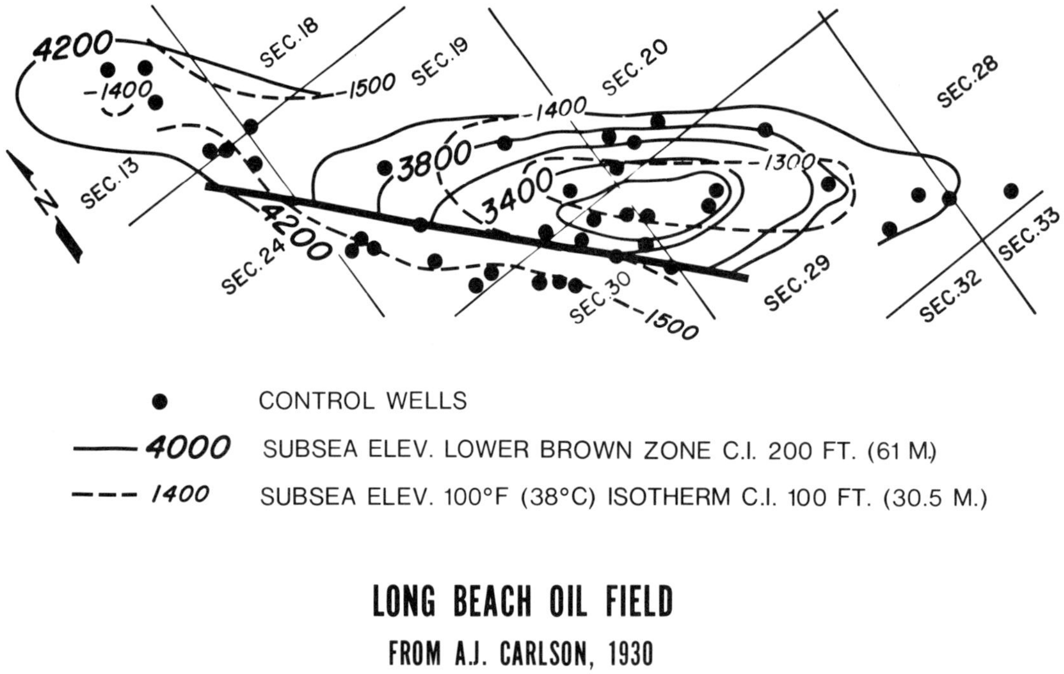

Fig. 20. Structure and 38°C isothermal depth map, Long Beach Field, California (from Carlson, 1930).

An early temperature study of the giant Long Beach Field is shown in Figure 20. In this illustration, a fault-controlled anticline with about 1,000 ft (300 m) of indicated closure on the Lower Brown Zone shows about 200 ft (60 m) of relief on the 100°F (38°C) isothermal surface, which is centered over the structural apex. The Long Beach Field has an impressive vertical sequence of pays, the deepest covering the largest area (the Christmas Tree model). The symmetry and coincidence of structure and temperature anomaly are consistent with a strong trap.

Another strong, multi-pay trap is the Santa Fe Springs Oil Field, shown in Figure 21. In this case, however, the temperature anomaly does not coincide with the structure mapped on the Lower Brown Zone (one of the deeper pays), but appears to be centered to the southeast, probably on the same structural axis. Deeper pays might be expected to be more fully developed in that direction which is a more active source of ascending water.

Having looked at several examples of early field data, we will now consider some contemporary data. The sequence will proceed from regional to local. Figure 22 shows a regional geothermal gradient map of the Gulf Coast. The control for this map is restricted to normally pressured depths, and the gradients represent an average between mean annual surface (zero depth) temperature and a maximum bottom hole reading at whatever total depth was reached in each well drilled. Without some datum depth control, such a map is not very useful for prospecting except to note generally warmer and cooler trends.

Figure 23 shows a somewhat more useful isothermal map of coastal Louisiana at a datum depth of 3,048 m (10,000 ft). Temperatures measured at other depths are extrapolated to the datum by means of averaged temperature-depth gradients. Although this involves some uncertainty, it's about the only way you can make a map with the available data. There are two warm trends on this map that deserve attention. The best-defined one follows the somewhat indistinct coastline, running about 15 to 25 km offshore most of the way. The other one almost parallels this one, about 55 km inland. The authors of this work noted these trends but did not explain them. A cursory view of the

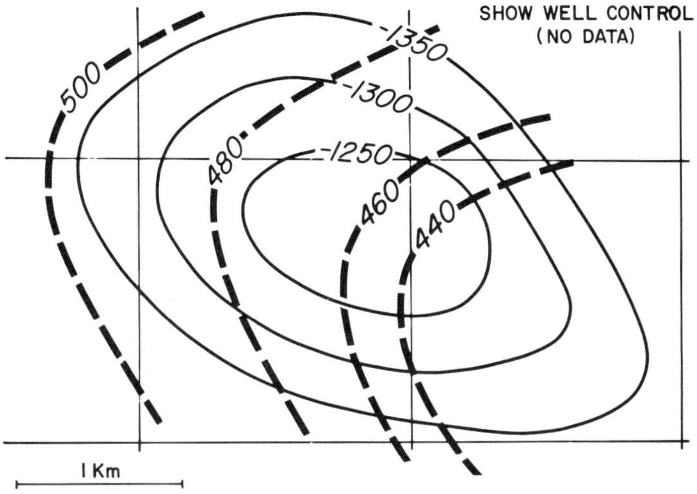

38°C ISOTHERMAL DEPTH WAS, SANTA FE SPRINGS OIL FIELD, CALIFORNIA.
C.I. 20 m (DASHED). SUBSEA STRUCTURE ON TOP OF MEYER ZONE. C.I. 50 m (SOLID).
FROM CARLSON, 1930

Fig. 21. Structure and 38°C isothermal depth map, Santa Fe Springs Field, California. Note: Geothermal apex appears to be southeast of structural apex (from Carlson, 1930).

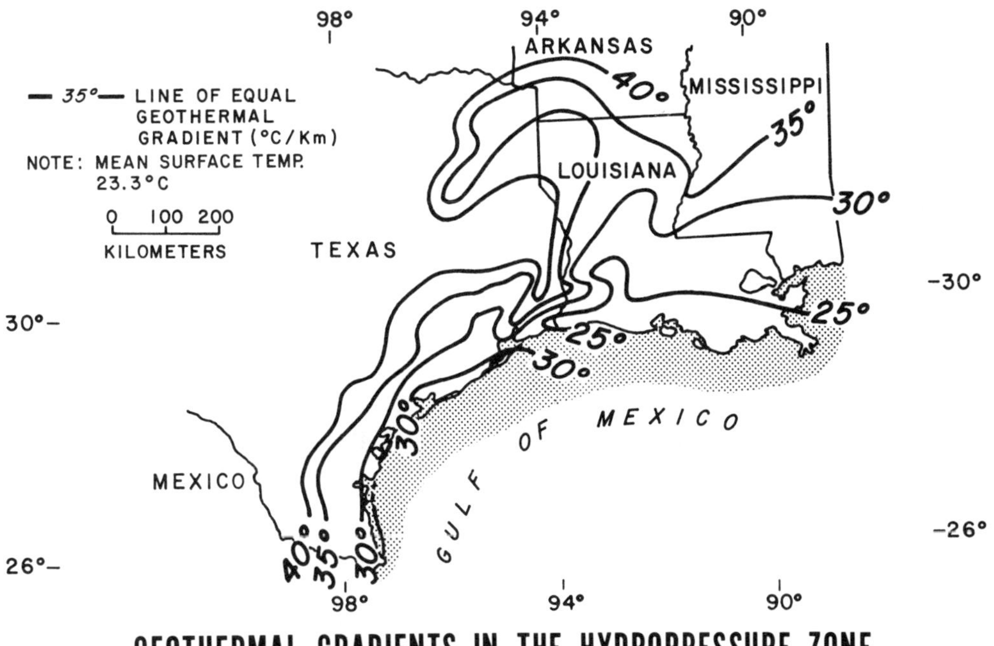

GEOTHERMAL GRADIENTS IN THE HYDROPRESSURE ZONE
IN THE NORTHERN GULF OF MEXICO BASIN
ADAPTED FROM JONES, 1975 · MOSES, 1961

Fig. 22. Gulf Coast regional geothermal gradients mapped at normal-pressure (hydropressure) depths (from Jones, 1975 [Moses, 1961]).

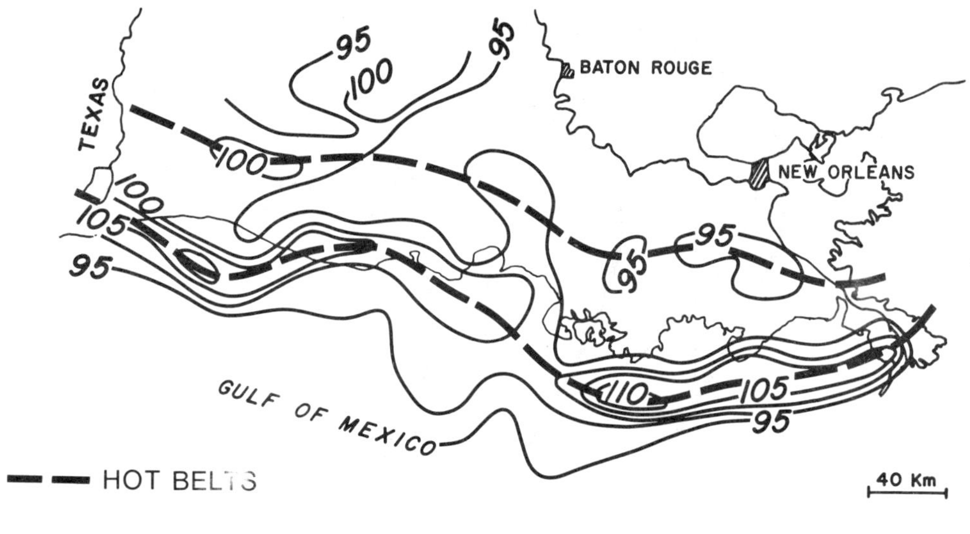

TEMPERATURES IN SOUTH LOUISIANA AT DEPTH OF 3,048 m. C.I. = 5°C.

FROM JAM L., DICKEY, AND TRYGGVASON, 1969.

Fig. 23. Geothermal map of South Louisiana at constant depth of 3,048 m. Note subparallel, coastwise hot belts (from Jam-L et al., 1969).

subsurface geology would seem to indicate that the trends are aligned with greater than average incidence of growth-faulting, geopressure, and diapirism. It seems quite probable that these conditions are accompanied by greater than average rates of ascending warm water.

Figure 24 is a geothermal gradient map and temperature-depth graph made fairly early in the exploration of the North Sea region. A number of important discoveries in the North Sea are related to the warm areas on this 1971 map. Temperature-depth data from eighty wells used for control are plotted in Figure 25. The data are somewhat biased, of course, because most of the wells were supposed to be on some kind of structure and therefore probably have above average temperatures. However that may be, the temperatures at a depth of 3.5 km (12,000 ft) can be seen to differ by as much as 50°C. It is generally understood that discovery wells are usually warmer and dry holes are usually cooler at equivalent depths under the North Sea.

At the south end of the North Sea play is the Groningen Gas Field of the Netherlands, where it all started about twenty years ago. Shown in Figure 26 is a temperature-depth graph for the top of the Zechstein Salt. The temperatures were recorded in wells which had been closed in for at least five weeks. At a given depth, the deviation of temperature from the average temperature-depth gradient is up to 8.3°C. The highest temperatures are found where the salt is domed, as salt conductivity would predict. The average TDG in the salt is 1.8°C/100 m compared to 3.1°C/100 m in the Rotliegend producing section.

A geothermal gradient map of the Alberta Basin is shown in Figure 27. The data are from many different stratigraphic levels, Devonian to Cretaceous, so the map is a mixture of averages and not very dependable for prospecting. Some qualitative observations can be made. There is a general increase of temperature northeastward in the direction of water movement toward the heavy oil region and the Canadian Shield. More locally, a warm trend follows the Red River for some distance. The North Saskatchewan and Athabasca rivers also flow through some warm areas. It is possible

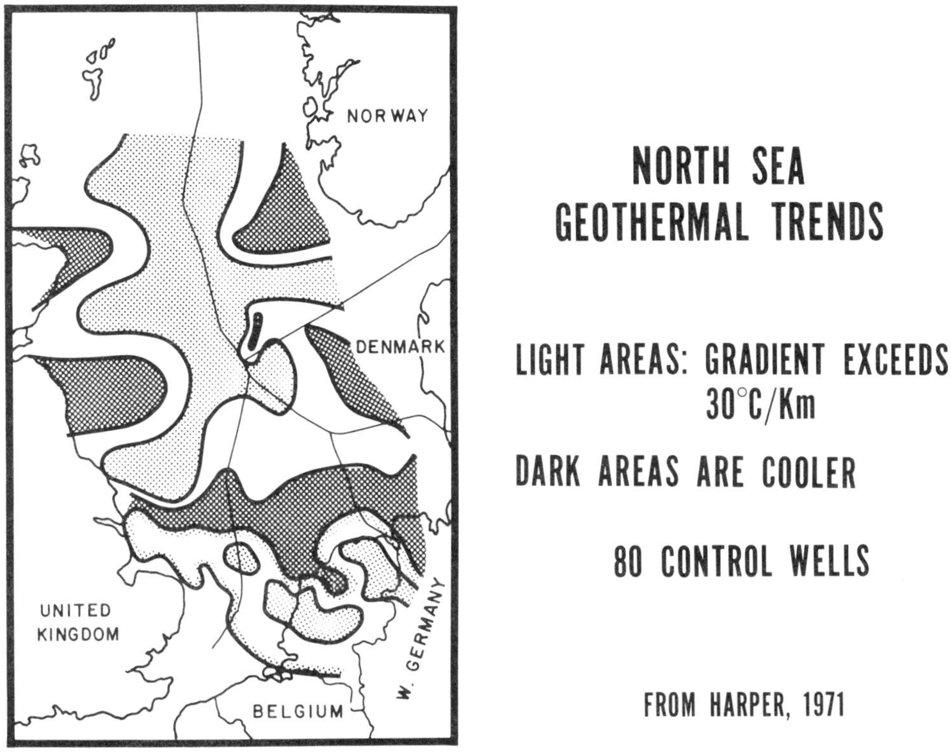

Fig. 24. Regional geothermal gradient pattern of the North Sea as mapped in 1971. For data see Figure 25 (from Harper, 1971).

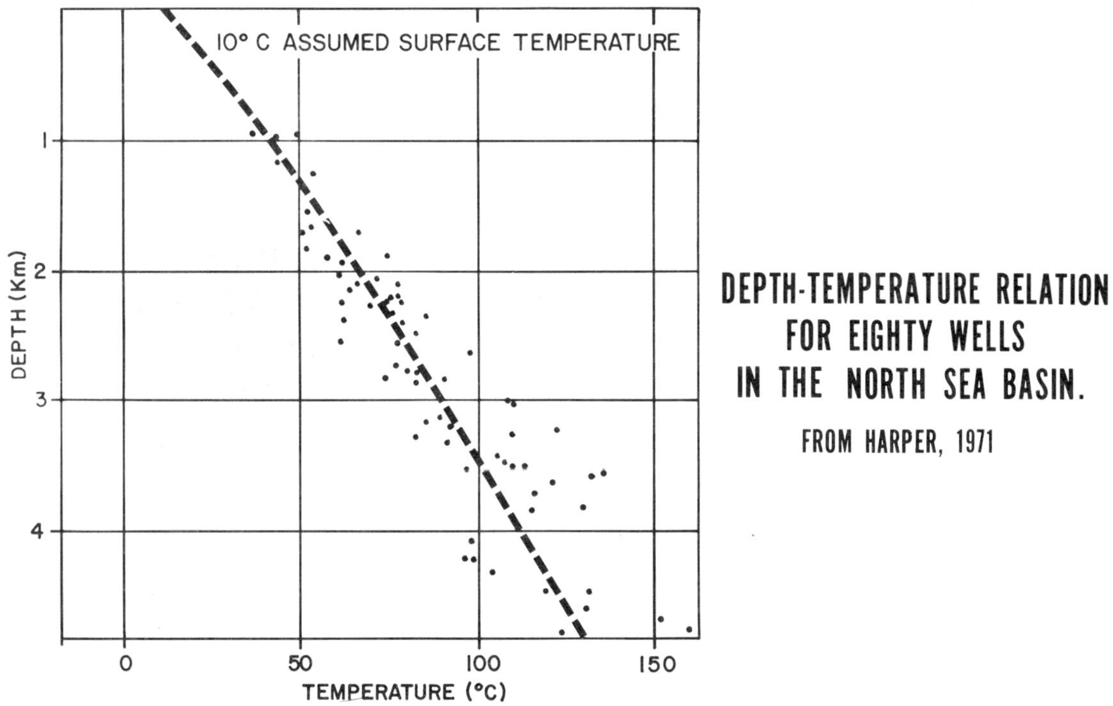

Fig. 25. Temperature-depth data for North Sea geothermal gradient map, Figure 24 (from Harper, 1971).

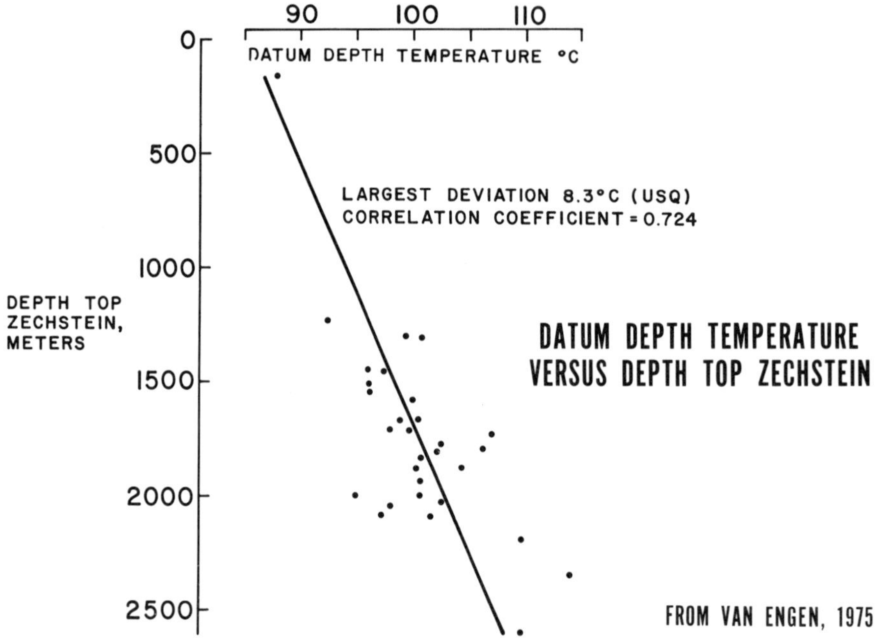

Fig. 26. Temperature-depth variation on top of Zechstein Salt, Groningen Gas Field, Netherlands (from Van Engen, 1975).

that these rivers are fed in part by ascending waters. Production is found in most of the warm areas and in some of the cooler areas.

It is now of interest to follow through some regional temperature work in the Trans-Caucasus plain to some detail in the Baku area

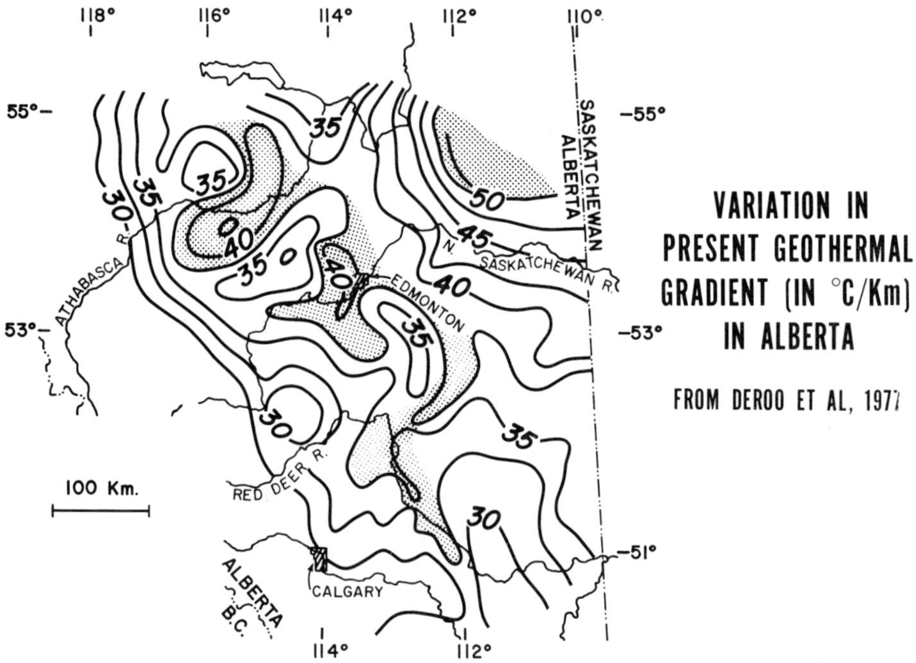

Fig. 27. Geothermal gradient map of the Alberta Basin (from Deroo et al., 1977).

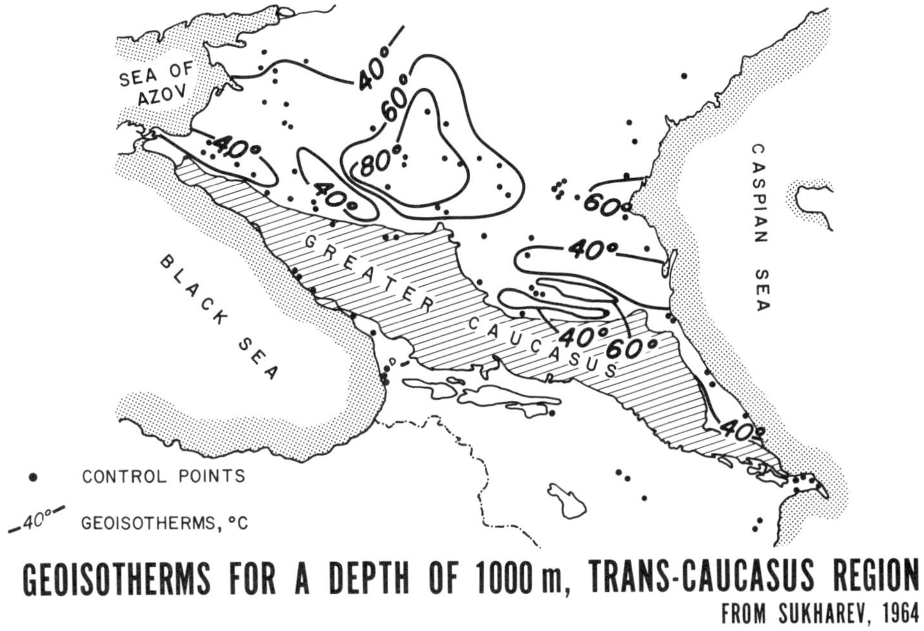

Fig. 28. Geothermal map of Trans-Caucasus region at 1,000 m depth (from Sukharev, 1964).

on the south Caspian shore. Figures 28 and 29, respectively, show isothermal maps at datum depths of 1,000 m and 2,000 m in the region between the Black Sea and the Caspian. On the 1,000 m map, the maximum temperature difference from on structure to off structure is 40°C. On the 2,000 m map it is 70°C. Prospects for entrapment appear to be favored at the 2,000 m

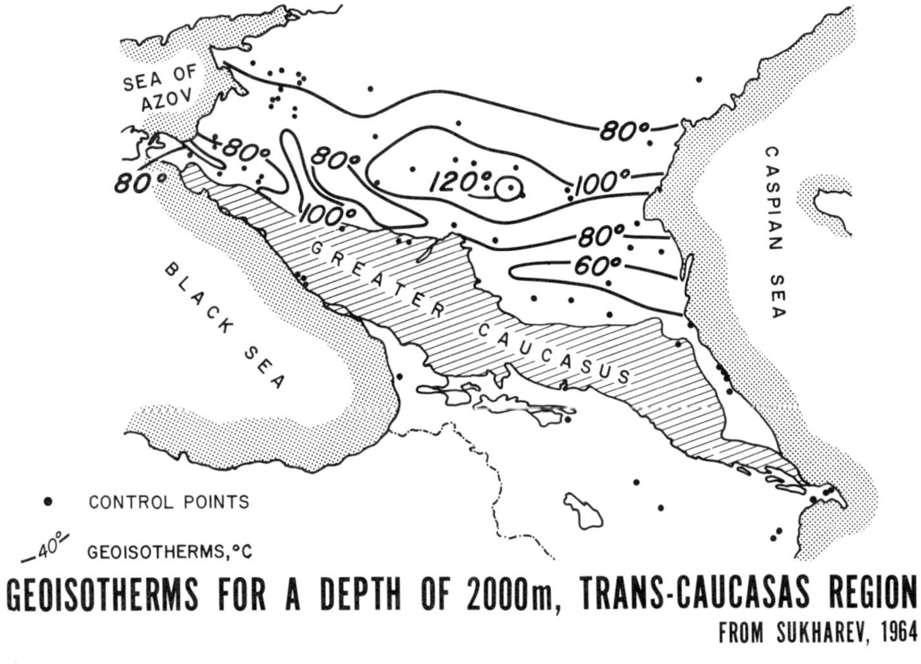

Fig. 29. Geothermal map of Trans-Caucasus region at 2,000 m depth (from Sukharev, 1964).

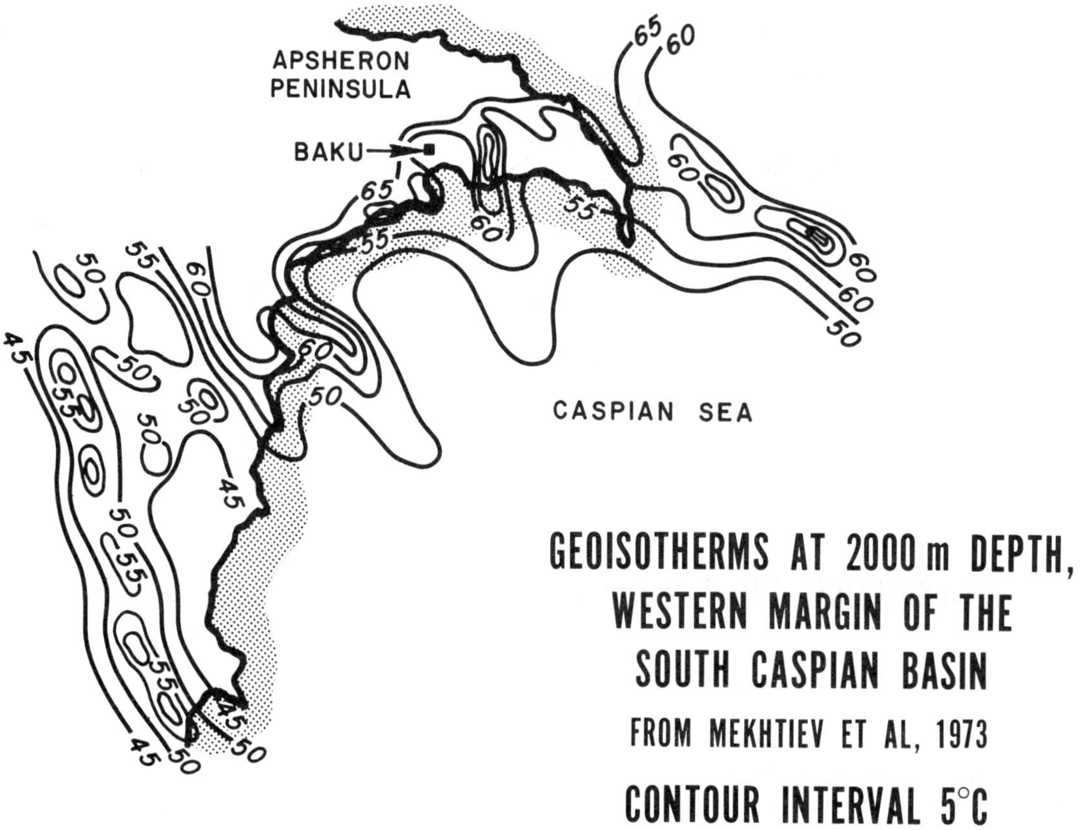

Fig. 30. Geothermal map of western margin, South Caspian Basin, at 2,000 m depth (from Mekhtiyev et al., 1973).

level, where variations of confinement are undoubtedly responsible for the greater differences of temperature. This is in fact proven by the better production developed below 2,000 m in the area.

In the southeast corner of these regional maps is the Apsheron Peninsula pointing into the Caspian Sea. Figure 30 shows an enlargement of this area around the famous oil town of Baku. On this map, the isotherms at the 2,000 m level show a maximum temperature difference of 20°C on and off structure in this local coastal area. Attention is called to the north-south thermal trend just east of Baku and plunging offshore to the south. Figure 31 details the geothermal picture on that trend.

On Figure 31, the isothermal map at 500 m depth shows several thermal anomalies closely associated with as many structural anomalies (i.e., closures, terraces). This map is based on over 2,300 temperature measurements made in nearly 300 wells, all of which had been shut in for at least 30 days. Maximum temperature relief at 500 m in depth is 14°C. Isothermal surfaces at 40°, 50°, and 80°C (maps not included here) show depth relief of 400 m, 700 m, and 1,000 m respectively. The greater relief at greater depth suggests improved prospects of entrapment.

Attention is called to the anomaly just onshore, the Zykh Field, which appears in Figure 32. This map shows local relief of about 300 m on the 40°C surface. It also shows important evidence of ascending water; i.e., a lake and a large mud volcano, which happen to be at the intersection of two structural trends—one plunging south, the other running east-west near the south shore of the peninsula. Water chemistry in the Zykh Field, supported by over 5,000 samples, also confirms deep water discharge (104, 115).

It is of interest to note that the Bakhar Field,

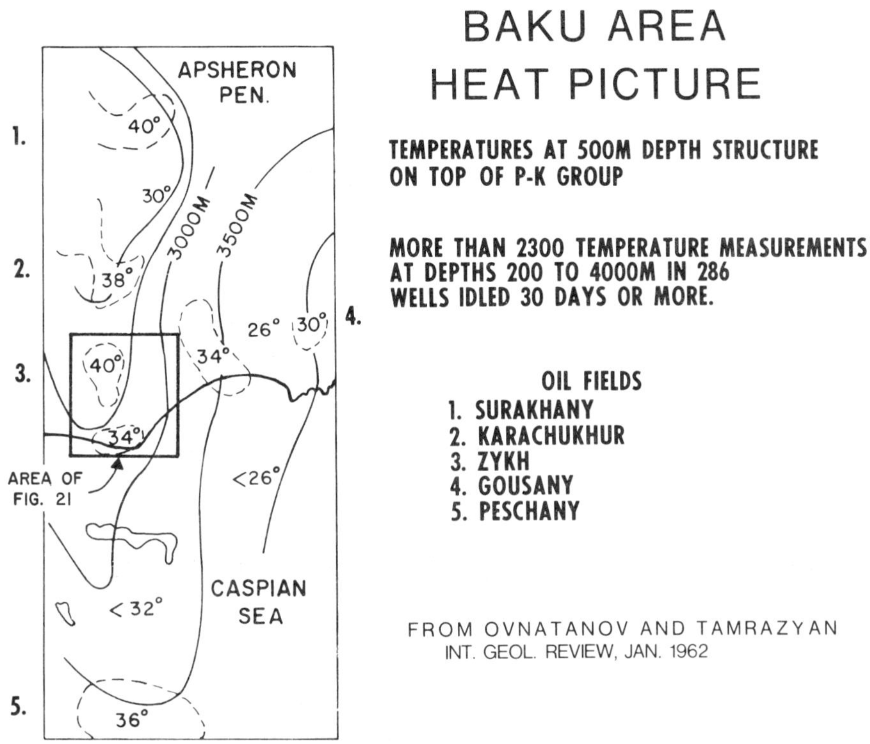

Fig. 31. Geothermal map of Baku area at 500 m depth (from Ovnatanov and Tamrazyan, 1962).

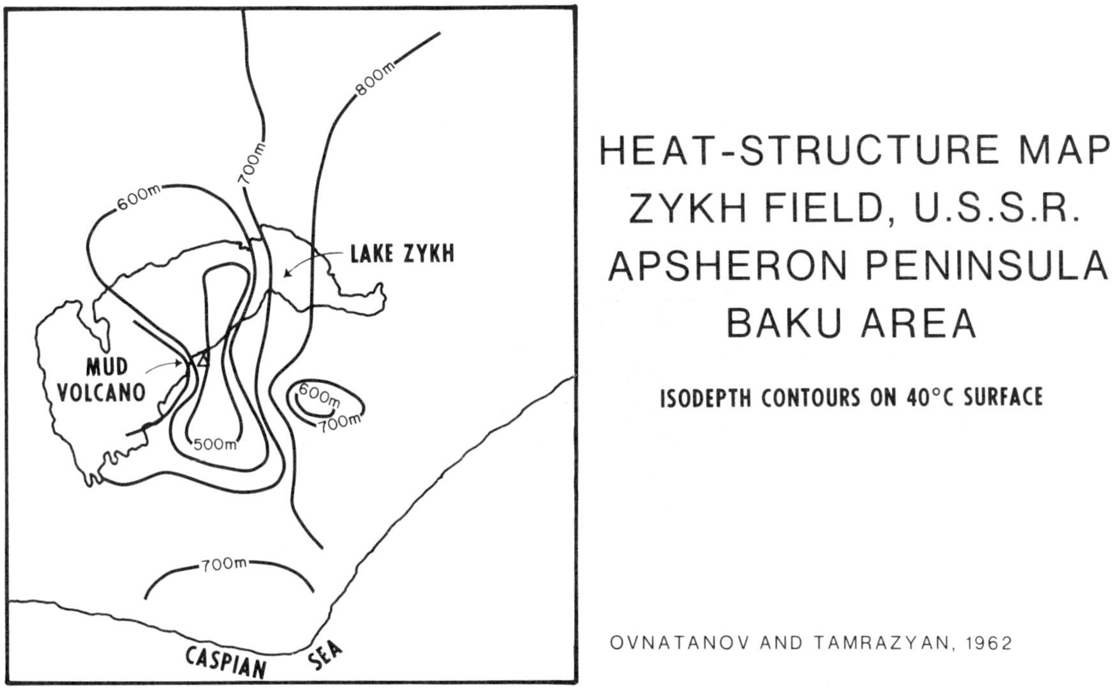

Fig. 32. 40°C isothermal depth map of Zykh Field, Baku area (from Ovnatanov and Tamrazyan, 1962).

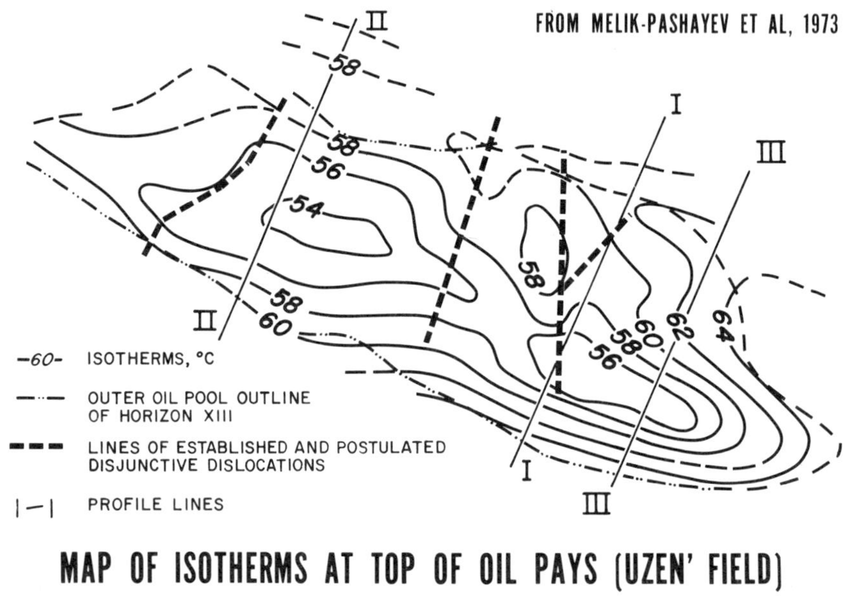

Fig. 33. Geothermal map on top of producing zone, Uzen Field, East Caspian region (from Melik-Pashayev et al., 1973).

which was mapped in the shallow mud offshore (Figures 2, 3), is farther out on the south plunge of this same trend.

The next area of interest is the large Uzen Field on the Mangyshlak Peninsula of the East Caspian area. Figure 33 shows isotherms mapped on top of the Jurassic oil pay. The relief is 10°C from crest to flank for a 300 m difference in depth. Also shown are the outer pool outline, several cross-faults, and three lines of cross-section. Figure 34 shows cross-section I-I with a gentle thermal upwelling over the structure.

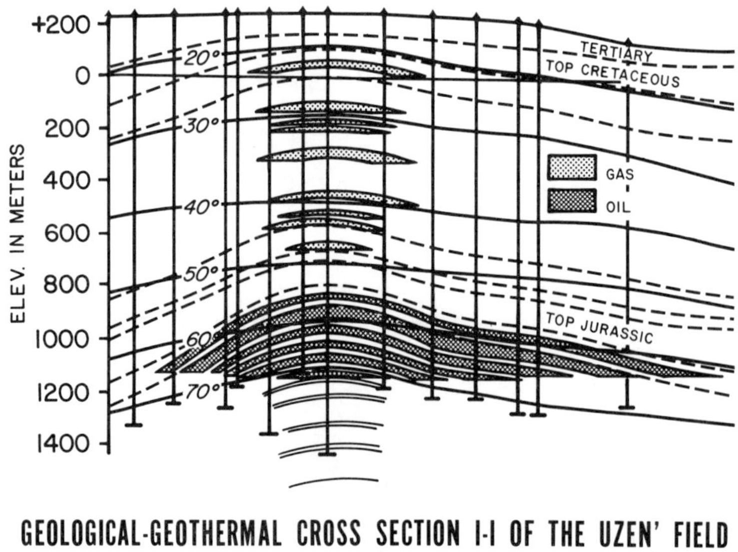

Fig. 34. Geothermal profile (I-I, Figure 33) of producing zones, Uzen Field, East Caspian region (from Melik-Pashayev et al., 1973).

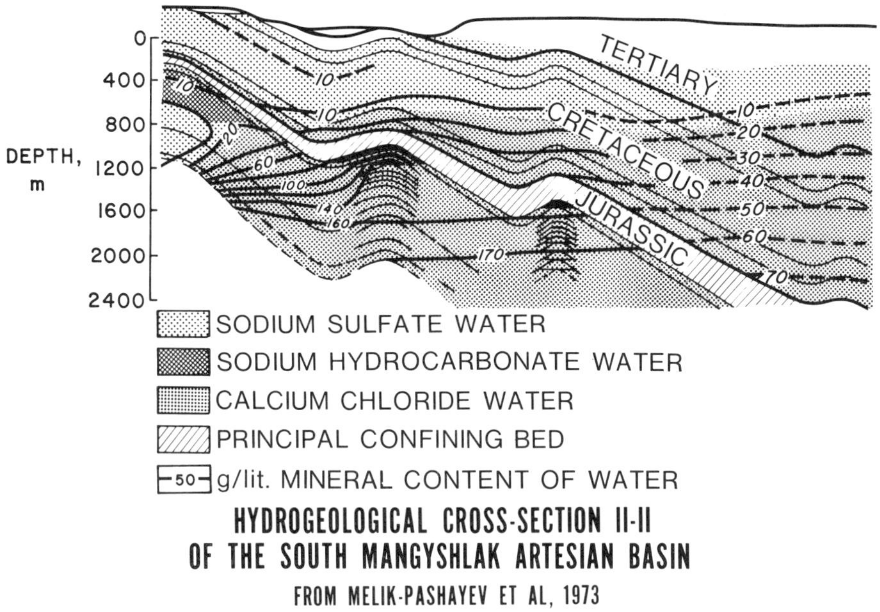

Fig. 35. Hydrogeological profile (II-II, Figure 33) of Uzen Field, East Caspian region (from Melik-Pashayev et al., 1973).

Section II-II is in Figure 35. This is a hydrochemical picture in which the numbered lines show mineral content of the water in grams per liter. Of special interest is the mineral concentration in close to the Uzen structure and below the covering aquitard. Fresher waters are obviously invading from the left (southwest), but they do not intrude the salt concentration in the Uzen trap. As indicated, the waters below 800 m are dominated by calcium chloride. From surface to 800 m, the sulfates are prominent, and there is a minor invasion by

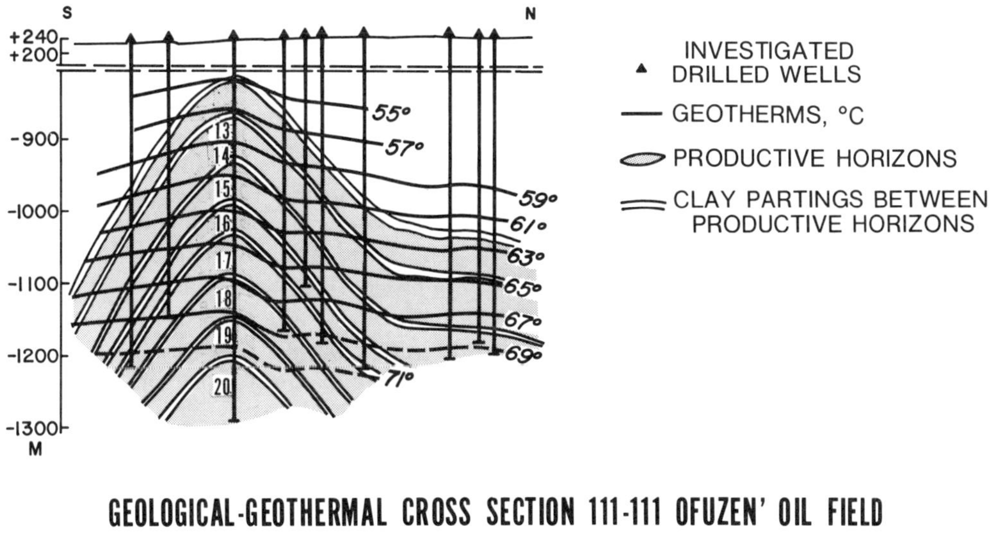

Fig. 36. Geothermal profile (III-III, Figure 33) of producing section, Uzen Field, East Caspian region (from Melik-Pashayev et al., 1973).

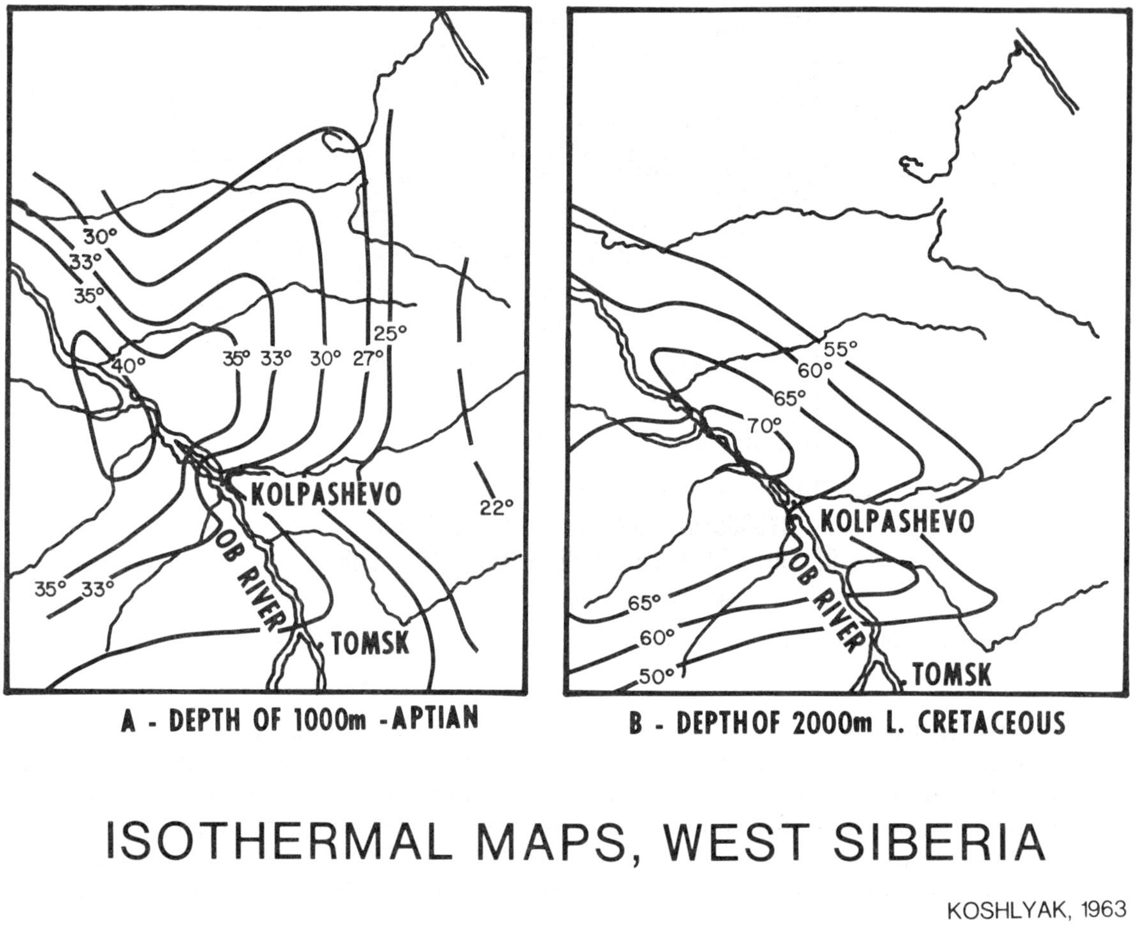

Fig. 37. Geothermal maps at 1,000 m and 2,000 m depths on a producing structure, West Siberia Basin (from Koshlyak, 1963).

sodium hydrocarbonate from updip southwest.

Figure 36 is the isothermal section on line III-III. It shows temperature details in the Jurassic producing zone. There seems to be relief of about 75 cm from crest to flank on each isotherm. This is basically a Middle Jurassic oil pay sequence lying below an Upper Jurassic aquitard (low permeability membrane).

A hydrothermally favored oil field on the Ob River in the West Siberian Basin is shown in Figure 37, mapped at two different levels. The deeper map has less control to the northeast, but in the central area near the river the temperature relief is obviously stronger at the 2,000 m level than at 1,000 m. A profile of the area is shown in Figure 38. The Soviet author attributes most of the thermal anomaly to a buried igneous ridge as a heat source. While this may be valid in part, the structural draping of Mesozoic strata over the ridge probably promotes deep-water-discharge into the overlying Ob River surface drainage, thereby making a hydrothermal contribution to the anomaly.

Another local, oil-productive temperature anomaly is shown in Figure 39. The Osa Field, in the Voga-Ural Basin, shows an apex 3°C hotter than the flank at 80 m greater depth, whereas under normal conditions it should be about 3°C cooler.

A Soviet field study of special interest is shown in Figure 40. The Don Medvedica Uplift appears to be another river-related hydrothermal anomaly. Paralleling a straight section of the Volga River, it is probably associated with deep, sub-vertical faulting. The gross uplift is outlined by dip symbols, and some

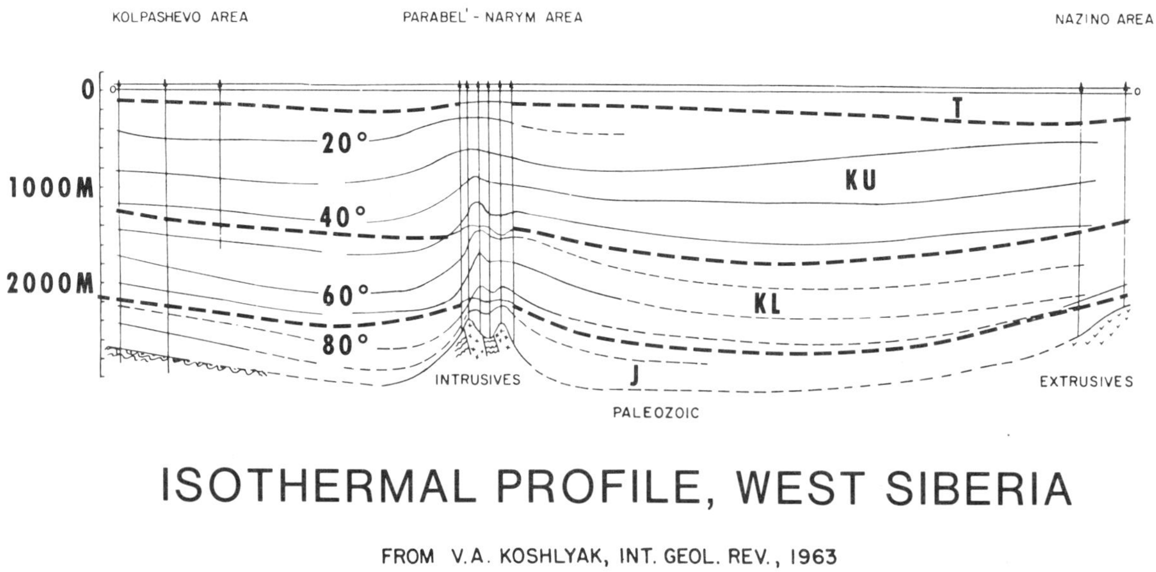

Fig. 38. Geothermal profile of anomaly mapped Figure 37 (from Koshlyak, 1963).

eighteen producing fields occur on it. Temperature relief at 2,000 m depth is at least 15°C. Believed to be very significant are the surface discharges of chemically reduced waters, represented by crosses on the map. The present author presumes the reduced condition to be characteristic of oil and gas deposits and a reliable criterion of deep-water discharge. The reduced condition is believed not to be inherited from a depositional environment but to be imposed on the trap by ascending, oxygen-poor waters.

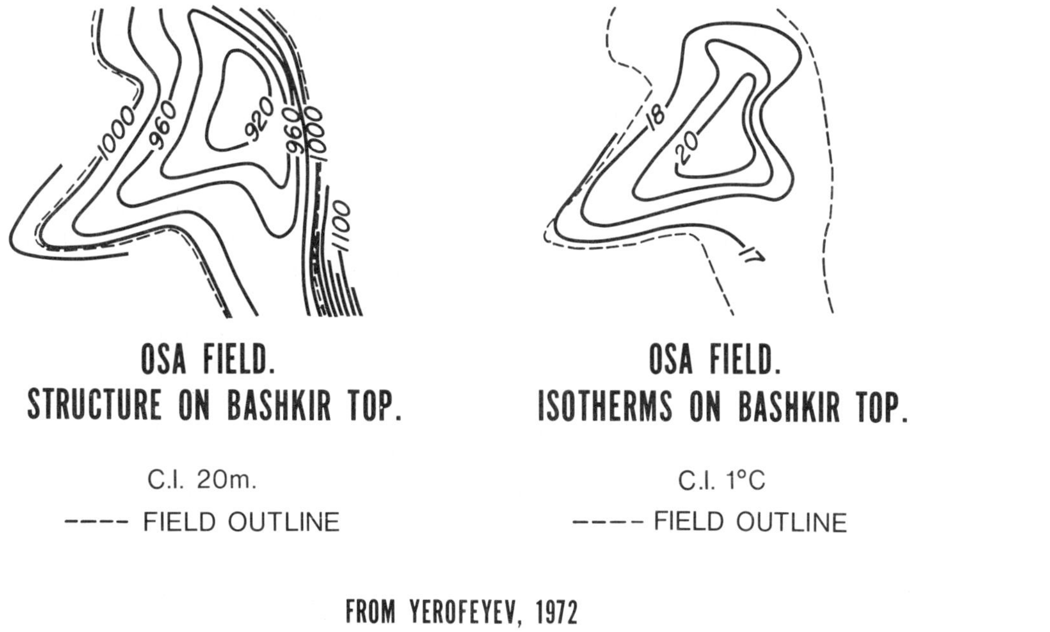

Fig. 39. Structure and isothermal maps on top of Bashkir interval in the Osa Field, Volga-Ural region (from Yerofeyev, 1972).

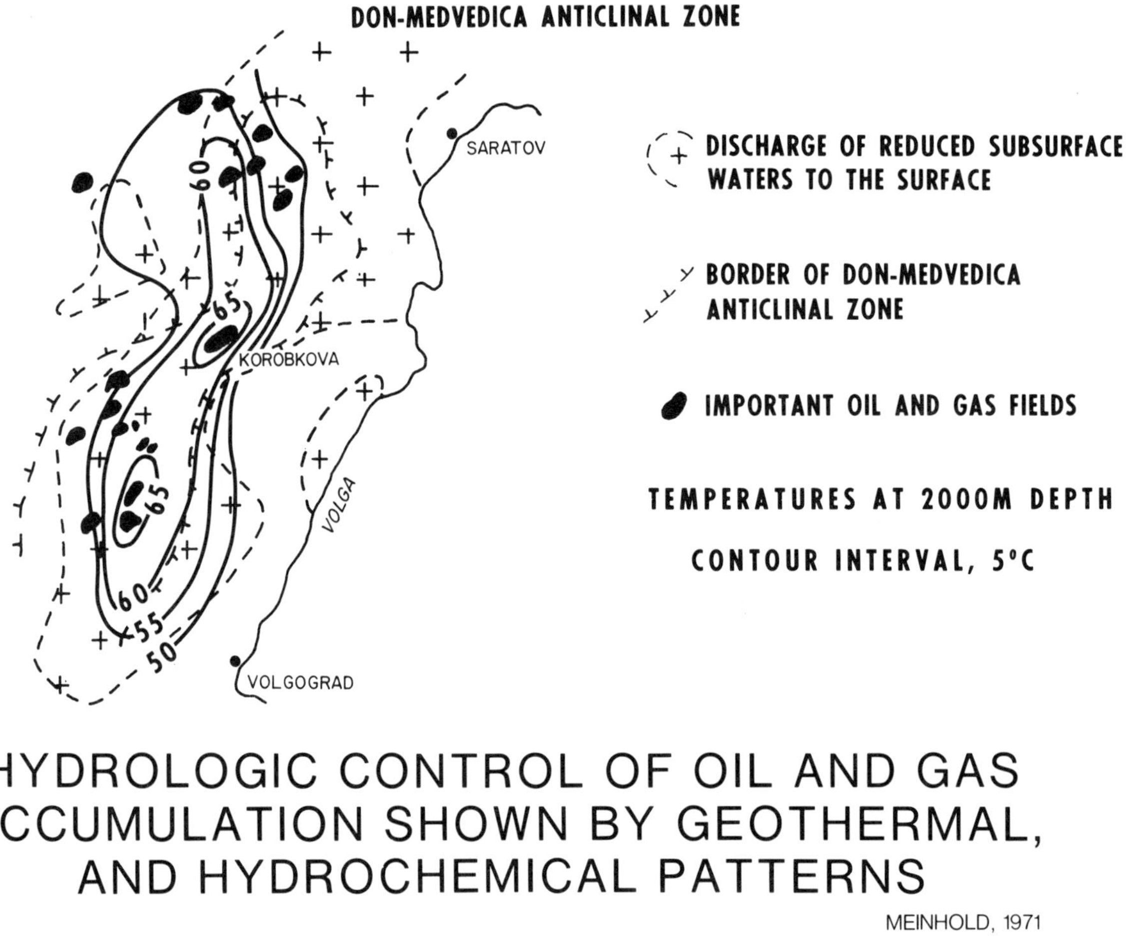

Fig. 40. Structure, temperature, and water discharge on Don Medvedica oil-producing uplift, North Caspian region (from Meinhold, 1971).

Some very interesting pressure-temperature relations to structure and grow-faulting have been defined (66) in the Bayou Carlin—Lake Sake area of coastal Louisiana. Figure 41 shows subsea depth of the top of the geopressured (under compacted) interval. There are three highs. One almost totally covers the Lake Sand field just south of a prominent growth fault; one covers the east end of the bayou Carlin field which lies on a structural plateau between two growth faults; and one lies on that same plateau, just west of the Bayou Carlin Field.

Figure 42 is an isothermal map of that same top of geopressure shown by depth in Figure 41. In this case the thermal highs are slightly offset from the producing fields, but the coincidence is still significant. The high area of geopressure west of the Bayou Carlin field, however, shows distinctly low temperature.

Figures 43 and 44, respectively, show isothermal maps of two structural horizons—one in the hydropressure regime at about 3,000 m depth and the other in the geopressure regime at about 4,000 m depth. Contour intervals are approximately 10.0°F(5.6°C). The geothermal highs are more nearly coincident with the producing fields at the deeper structural horizon than at the shallow one.

Isothermal maps at constant depths of 3,048 m and 3,658 m (both in hydro pressure) are shown in Figures 45 and 46. At both levels the thermal highs show notable coincidence with growth faults and with the producing fields.

Figures 47 and 48, respectively, are isother-

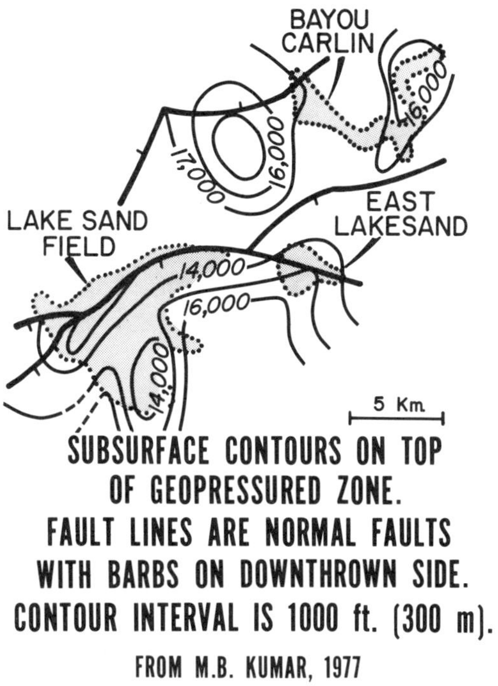

Fig. 41. Depth to top of geopressured (overpressured) zone, Bayou Carlin–Lake Sand Area (BCLSA), coastal Louisiana (from Kumar, 1977).

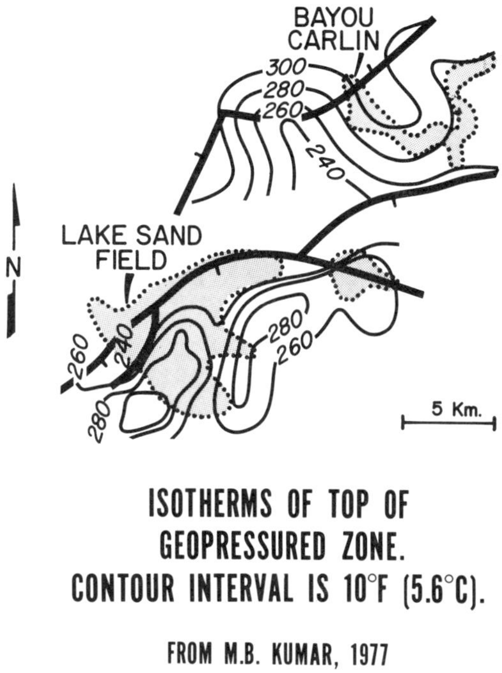

Fig. 42. Isotherms on top of geopressured (overpressured) zone, BCLSA (from Kumar, 1977).

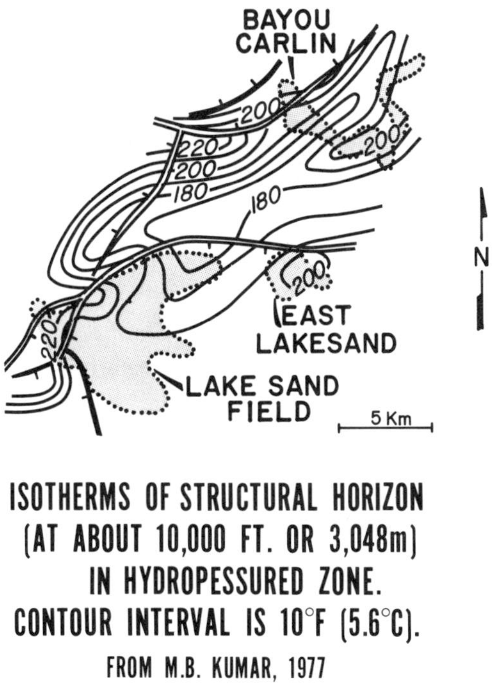

Fig. 43. Isotherms in hydropressured (normal hydrostatic) zone on a structural surface at about 10,000 ft (3,048 m), BCLSA (from Kumar, 1977).

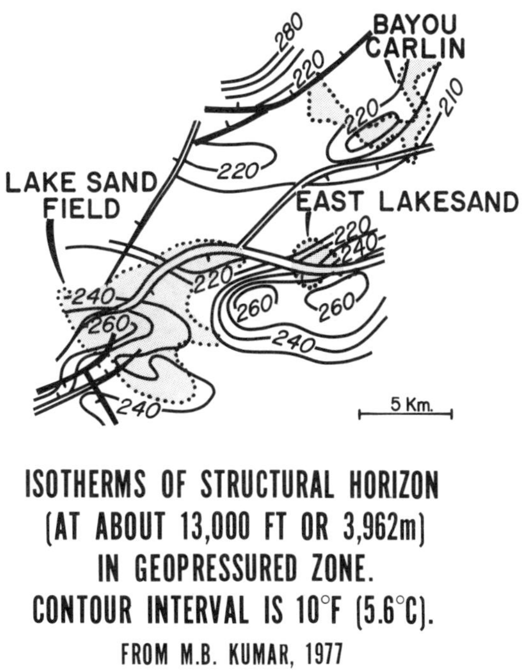

Fig. 44. Isotherms in geopressured (supra-hydrostatic) zone on a structural surface at about 13,000 ft (3,962 m), BCLSA (from Kumar, 1977).

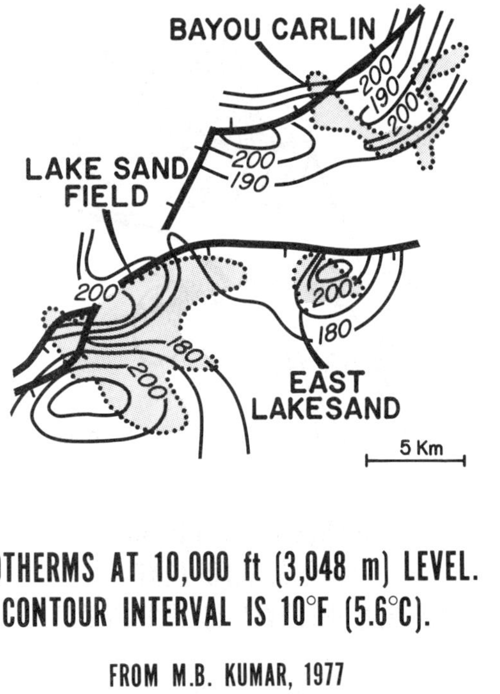

Fig. 45. Isotherms at level, 10,000 ft (3,048 m) datum, BCLSA (from Kumar, 1977).

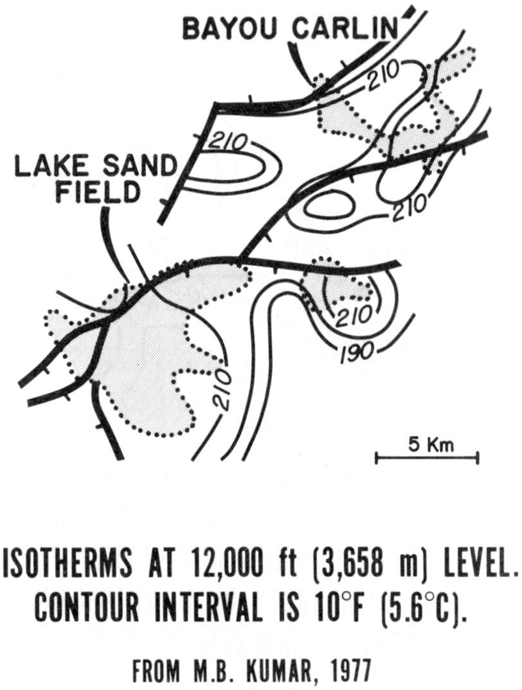

Fig. 46. Isotherms at level, 12,000 ft (3,658 m) datum, BCLSA (from Kumar, 1977).

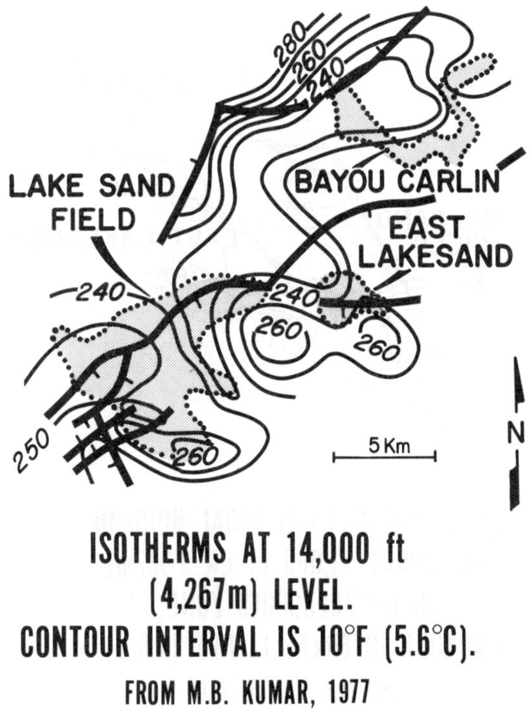

Fig. 47. Isotherms at level, 14,000 ft (4,267 m) datum, BCLSA (from Kumar, 1977).

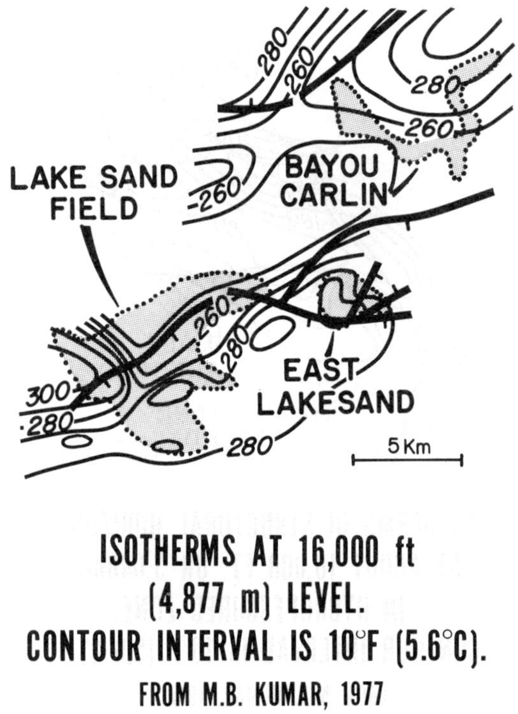

Fig. 48. Isotherms at level, 16,000 ft (4,877 m) datum, BCLSA (from Kumar, 1977).

VARIATIONS OF TEMPERATURE AT GIVEN DEPTHS, EMBA REGION

DEPTH, m	TEMPERATURE, °C MAXIMUM	MINIMUM	RANGE, °C
ZONE OF SALT DOMES			
500	31.9	21.2	10.7
750	42.1	27.0	15.1
1000	52.4	31.5	20.9
ZONE OF SOUTH EMBA UPLIFT			
500	31.3	22.2	8.1
750	37.2	27.7	9.5
1000	42.9	33.7	9.6

FROM DZHANGIR'YANTZ, 1965

Table 1. Local temperature differences on salt domes and uplifts (from Dzhangiryants, 1965).

mal maps at 4,267 m and 4,877 m depths (both in the geopressure regime). The coincidence with growth faults is present at these depths, but the coincidence with productive traps is less distinct.

It appears from this study of the Bayou Carlin—Lake Sand area that a relationship exists between growth faulting, structure, geopressure, thermal anomalies, and hydrocarbon accumulations. The relationship seems to persist from the geopressure regime upward through the transition zone into the hydropressure regime. This apparent vertical continuity, or perpetuation, of a deeper pressure-temperature situation upward into a shallower zone suggests the importance of an ascending water system which is associated with the upward collection of hydrocarbons and which also carries a significant temperature signal.

Some local temperature differences around salt domes in comparison to those around an uplift are presented in Table 1. As expected, the differences for the salt domes are greater than for the uplift (e.g., at 1,000 m depth it is a twofold difference).

The foregoing examples illustrate the implementation of temperature data in defining or predicting oil and gas productivity. It seems that productive situations are almost invariably warmer than average. A clear exception to that rule has not been found. Some of the known petroleum-related thermal anomalies not covered in this paper are the following:

HOT FIELDS		HOT BASINS
Wilmington, CA		Gipsland, Australia
Prudhoe Bay, AK	Minas, Sumatra	Tampico, Mexico
Salt Creek, WY	Hassi Massaoud, Alg.	East Texas
Normal Wells, CAN	Gawar, Saudi Arabia	

Higher temperatures may favor HC mobility by reducing the viscosity of water. The effects of temperature on the real viscosity of waters with varied mineral content are shown in Figure 49. For a water of given mineral content, a

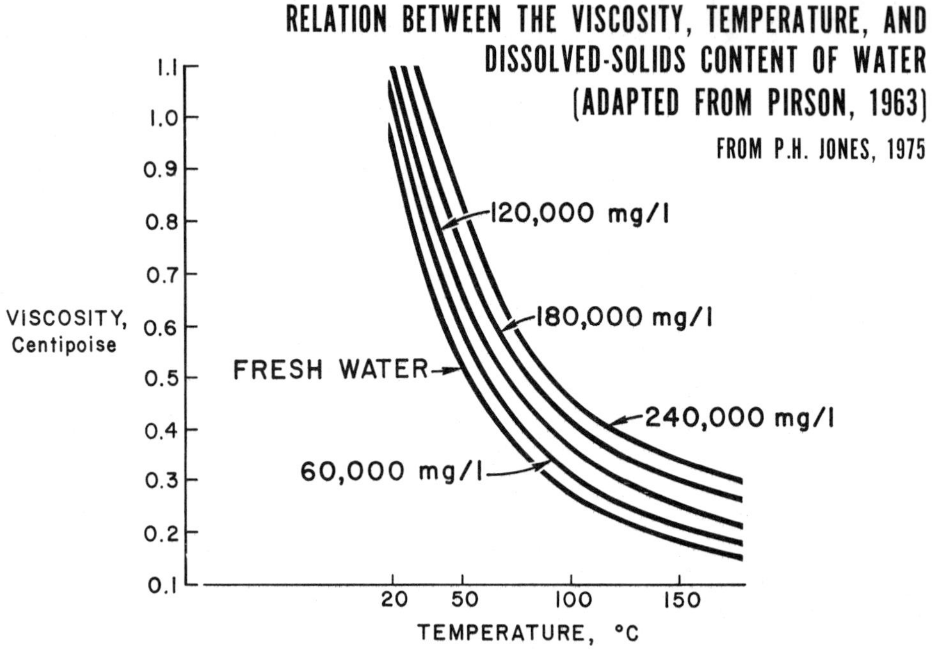

Fig. 49. Variation of water viscosity with temperature and dissolved solids (from Jones, 1975).

temperature increase of about 33°C reduces the viscosity by 50%. Not only does that allow water to move more easily through low-permeability media, but the increased temperature also increases the capacity of the water to accommodate hydrocarbons and other organic compounds.

It has been pointed out by Klemme (61) that

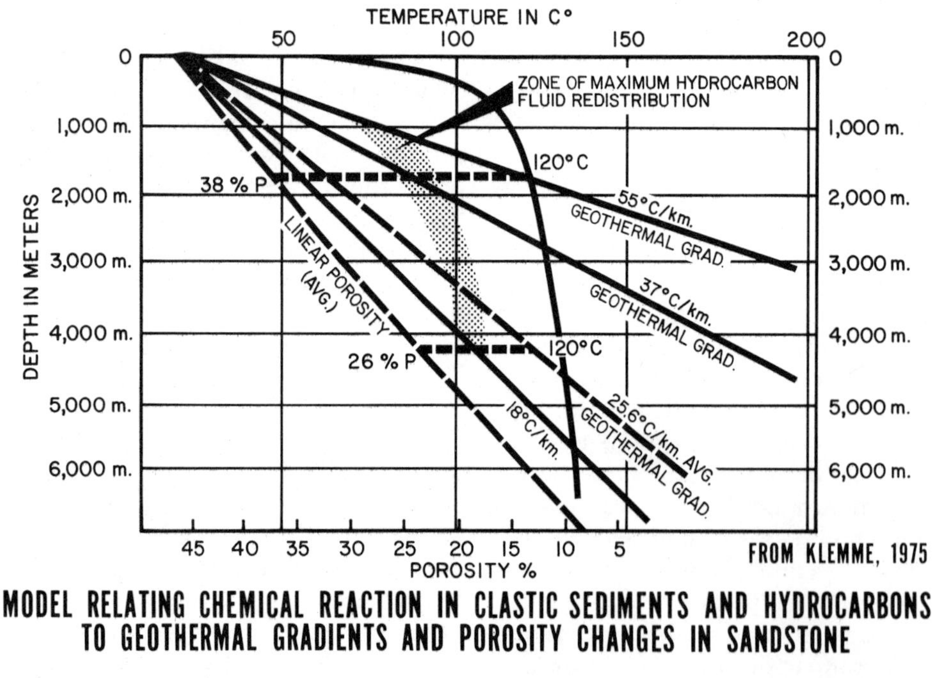

Fig. 50. Effects of porosity and temperature on chemical reactivity in clastic sediments (from Klemme, 1975).

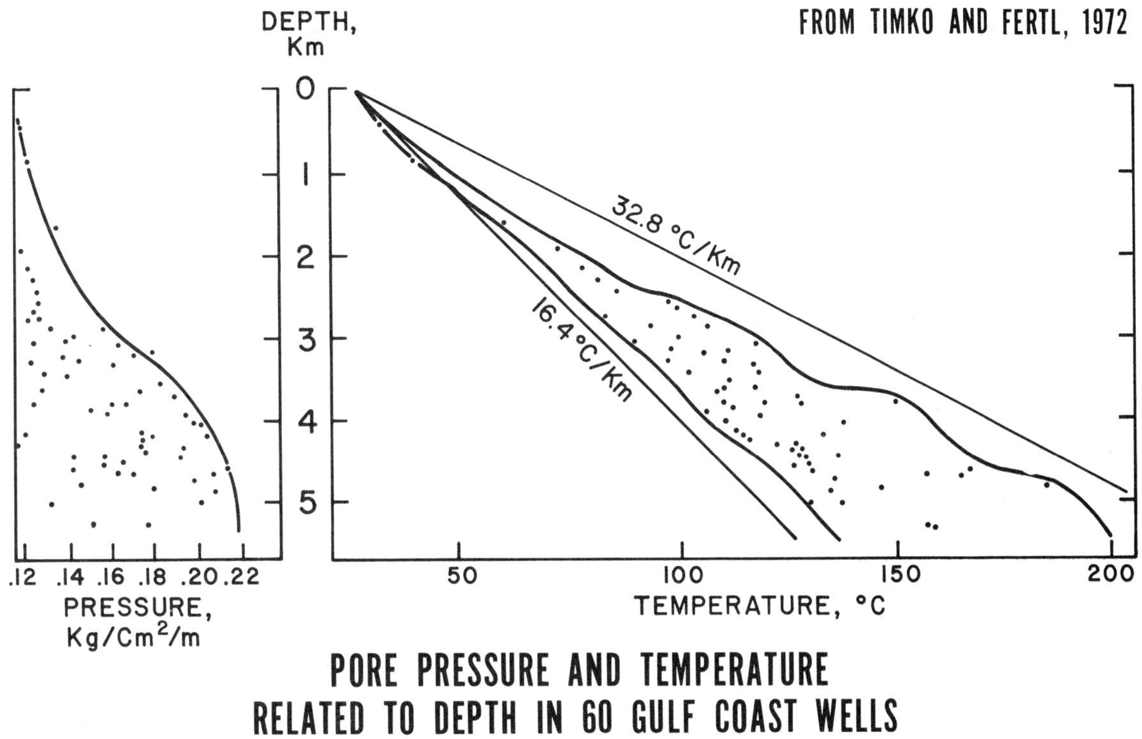

Fig. 51. Pressure and temperature variation with depth in Gulf Coast wells (from Timko and Fertl, 1972).

higher temperatures in compactable clastic sediments favor hydrocarbon generation and accumulation at shallower depths because of increased porosity and chemical reactivity. Figure 50 diagrams those relationships. It shows, for example, that with a high TDG of 5.5°C/100 m, at 121°C an average sandstone might have 38% porosity at a depth of about 1,750 m. With a lower average TDG of 2.5°C/100 m, the same sandstone would have to be buried about 4,250 m to achieve the same temperature (121°C), and the porosity would be only 24%.

The curved line in the diagram starting on the temperature scale at about 60°C and extending asymtotically to about 130°C is intended to represent the exponential increase of chemical reactivity downward on an imaginary vertical scale in response to increased temperature. The heavy band marked "zone of maximum hydrocarbon fluid redistribution" is bounded by the high TDG (5.5°C/100 m) and the average TDG (2.5°C/100 m) and covers a temperature range from about 65° to 110°. Within this range, a brine of 50g/l mineral content would have a viscosity of about 0.2 - 0.5 centipoise (see Figure 49). Hydrocarbons supposedly would be generated from organic-rich shales, and average sandstone porosities would range from about 30% down to about 17%. The general conclusion from these estimated relationships has to be that favorable conditions are accompanied by higher TDGs. To what extent the favored conditions are dependent upon the higher temperatures per se may remain a moot question.

Although somewhat loosely controlled and loosely defined, the fluids in the subsurface (which are in the order of 99.995% water) behave as a PVT system under material balance control. Pressures and temperatures tend to go up and down together. Figure 51 shows companion plots of pressure and temperature data from 60 representative Gulf Coast wells. The spread of the data below 2,500 m in this highly productive region certainly shows that neither pressure nor temperature gradients are linear.

The relation of temperature to pressure has

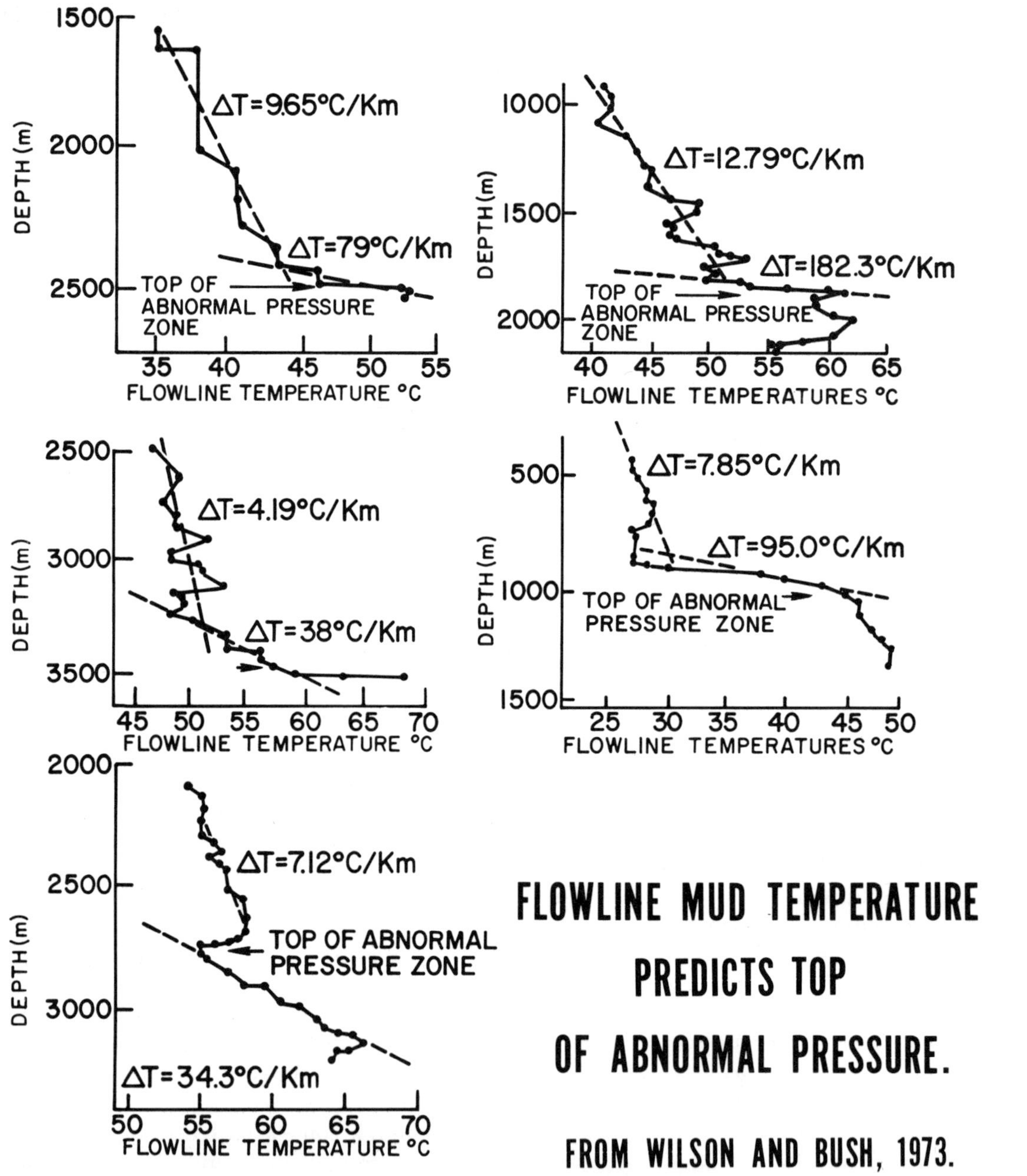

Fig. 52. Flowline mud temperature plotted against depth predict abnormal pressure in five different areas (from Wilson and Bush, 1973).

been put to practical use in the prediction of geopressure zones in drilling wells by monitoring the temperature of the circulating mud. Figure 52 shows flow line temperatures plotted against depth from five wells which encountered geopressures in different parts of the world.

DATA ACQUISITION

The acquisition of temperature data from well logs and production tests has been well covered in the literature (7, 15, 21, 22, 24, 25, 28, 33, 44, 49, 50, 53, 87, 98, 108-111, 116, 126, 127, 130, 133, 134, 139). The principle of acquiring stabilized temperature data is shown

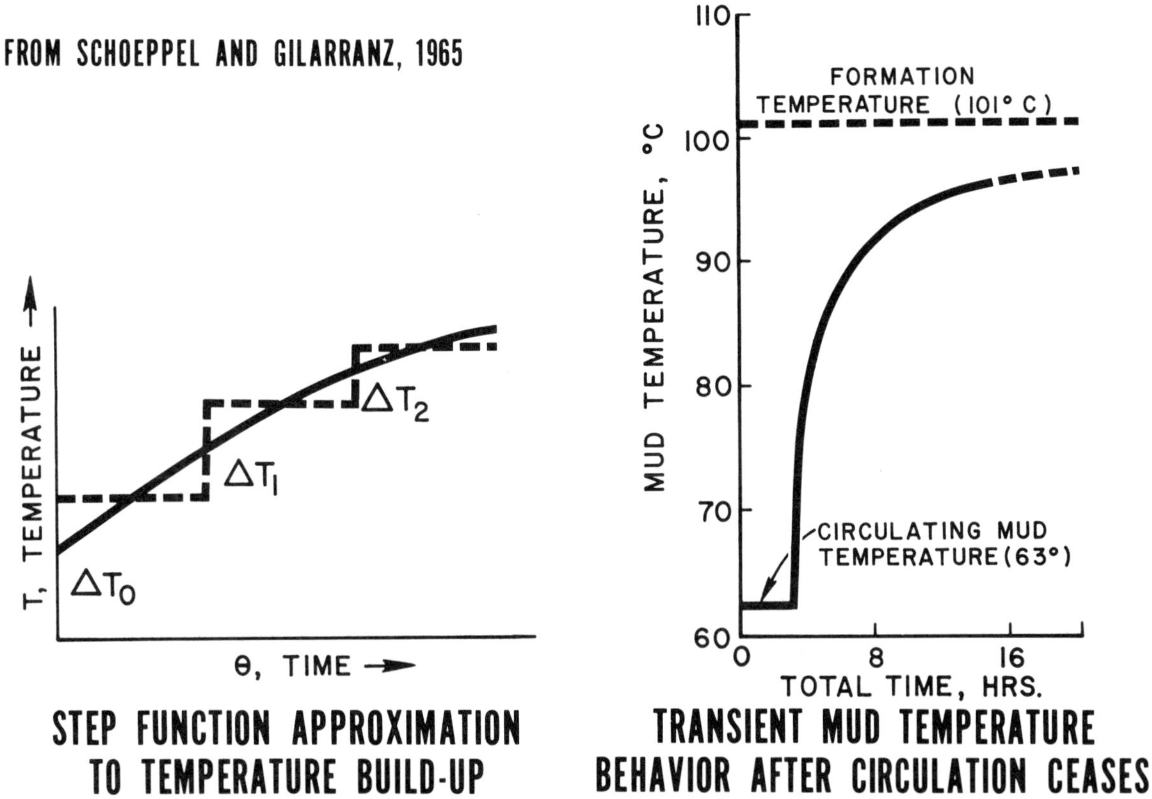

Fig. 53. Projection of well-bore temperature true formation temperature by incremental readings (from Schoppel and Gilarranz, 1965).

by two simple diagrams in Figure 53 illustrating the projection of well-bore temperatures toward true formation temperatures after infinite time. It has been shown by a number of investigators (e.g., 30, 66, 111, 116, 126) that temperatures recorded at two or more time intervals after well circulation has ceased can be projected to within about 10% of true formation temperature. With such data, it is possible to make useful regional maps. The main problem, however, is getting enough temperature detail to evaluate the vertical and lateral differences in gradients and to coordinate the temperature data with hydropressure and geopressure regimes.

Special mention should be made of the joint ASPG-USGS project to assemble all available temperature data from drilled wells. The geothermal gradient map of North America shows many interesting regional features. Since it is based on averaged TDG's, some oil and gas regions appear hot and others cold. This type of data compilation does not permit depth interval discrimination. Such a two-dimensional display is, however, useful for reconnaissance (116).

Like any other exploration technique, geothermics has a noise problem. The most obvious natural sources of noise are surface weathering and solar effects, both of which are less troublesome in water-covered areas (e.g., see Figures 2, 3). In some onshore areas (35), significant noise-free anomalies can be found below 20-40 m, in others (33, 155), only below 100 m, sometimes 1,000 m. Thermochemical (58, 140) and radioactive temperature effects are difficult to evaluate. Tectonic compression and the friction of fault movements (7) also have poorly known thermal effects.

Artificial sources of temperature noise include drilling, circulating, cementing, treating, and producing operations—especially where gas expansion occurs. Empirical techniques have been worked out (14, 20, 30, 38, 111,

126, 127) in local areas to compensate for these disturbances.

Even under the best conditions, most determinations of subsurface temperatures made in wells drilled for oil and gas in the western countries will be semiquantitative. Only in the occasional idled producer or dry hole will temperature be monitored long enough to approach a static, undisturbed value—and then only if someone has a special reason to want that temperature. As a matter of fact, in a produced well the natural thermal condition may never be restored.

We need to devise techniques for obtaining and using large amounts of semiquantitative temperature data from drilled wells. A possibly important source seems to be at hand. Referring back to Figure 52, we see several TDG plots of flowline mud. Such temperatures are recorded for engineering purposes (139). These plots have not been adjusted for drilling variables such as circulation rate, rotary speed, penetration rate, material additions, and tripping. The known variable effects can probably be averaged empirically to permit more accurate temperature definition. The data would thus be more useful in exploration as well as in enhancing the safety and the efficiency of the drilling operation.

The acquisition of semiquantitative thermal data is not limited to subsurface measurement in drilled wells. Inasmuch as the earth's heat is directed upward to the earth's surface, it should be possible under favorable conditions to map the effects of that emanation at or near the surface. Efforts to use infrared remote sensing for generalized earth surface temperature evaluation have, however, been disappointing. One of the indirect and semiquantitative methods of obtaining temperature information involves looking at the effects of ascending waters in areas which are subject to freezing and thawing. Basically, it is a matter of interpreting the patterns of freezing and thawing as modified by anomalous hydrothermal activity. The patterns may be seen in such media as snow melt, frost, lake ice, permafrost thickness, etc. Some arctic lakes are late to freeze and early to thaw because of hydrothermal support from below. Permafrost is thin or absent in some oil productive areas of the Arctic, suggesting hydrothermal activity.

CONCLUSION

The temperature signal continuously transmits geological and hydrogeological information to anyone who wants to listen. Of course, no one will listen unless he has a reason. And he won't have a valid reason unless he understands where the signal comes from and how it relates to something of value. It is important to understand the behavior of the water which is doing so much of the work in putting oil and gas accumulations together and transmitting the interesting signals. It is most unfortunate that the average oil-seeker is so averse to water. He doesn't want water in his well, and this is understandable. The value of the water-borne temperature signal for finding additional major reserves of oil and gas should, however, make a working knowledge of water worthwhile.

REFERENCES

1. Adivarahan, P., 1962, Heat transfer in porous rocks through which single-phase fluids are flowing: Society of Petroleum Engineering Journal, Sept. 1962, p. 290-296.
2. Adolphson, D. G., and LeRous, E. F., 1968, Temperature variations of deep flowing (water) wells in South Dakota: U.S. Geological Survey Professional Paper 600-D, p. DC60-62.
3. Afanasyev, T. P., et al., 1975, Geothermal conditions in oil-gas regions: International Geology Review, v. 17, no. 2, p. 164-166.
4. Aliyev, S. A., et al., 1977, Heat flow in the central part of the Kura depression: Petroleum Geology (U.S.S.R.), v. 14, no. 2, Feb. 1977, p. 45-47.
5. ———, and Rastamor, R. I., 1977, Geothermal characteristics of the Azerbaydzhan part of the Kura depression: Petroleum Geology (U.S.S.R.), v. 14, no. 11, Nov. 1977, p. 478-479.
6. ———, et al., 1977, Heat flow in the central part of the Kura depression: Petroleum Geology (U.S.S.R.), v. 14, no. 2, Feb. 1977, p. 45-47.
7. Anglin, F. M., and Beck, A. E., 1965, Regional heat flow pattern in western Canada: Canadian Journal of Earth Sciences, v. 2, no. 3, p. 176-182.
8. Arabadzhi, M. S., et al., 1974, Role of regional basement faults in forming geothermal field of young platforms: International Geology Review, v. 16, no. 2, Feb. 1974, p. 180-187.
9. Arbolishvili, D. Z., 1958, The geothermal regime of some areas in the Georgian SSR: Petroleum Geology (U.S.S.R.), v. 2, no. 12B, p. 1034-1038.

10. Artemenko, V. I., and Malovitskiy, Ya. P., 1977, Results of a bottom geothermal survey in the Bakhar oil-gas field: Petroleum Geology (U.S.S.R.), v. 15, no. 4, Apr. 1978, p. 150-152.

11. Beskopylnyy, V. N., 1974, Dependence of geothermal field on geologic structure and the oil potential of the northeast part of the Pripyat depression: Petroleum Geology (U.S.S.R.), v. 11, no. 11, Dec. 1974, p. 510-515.

12. ———, 1975, Geothermal zonality and distribution of oil field in the Pripyat Basin: Petroleum Geology (U.S.S.R.), v. 11, no. 2, Jan. 1975, p. 542-545.

13. ———, 1976, Oil formation in the inter-salt sediments of Pripyat Basin: Petroleum Geology (U.S.S.R.), Aug. 1976, p. 359-361.

14. Beskrovny, N. S., et al., 1973, Presence of oil in hydrothermal systems associated with volcanism: International Geology Review, v. 15, No. 4, Apr. 1973, p. 384-393.

15. Birch, F., 1954, The present state of geothermal investigations: Geophysics, v. 19, no. 4, Oct. 1954, p. 645-659.

16. Bobelyuk, V. P., 1965, Some results of temperature measurement of oil-bearing strata of the Nebit-Dag field: Petroleum Geology (U.S.S.R.), v. 5, no. 11, p. 605-607.

17. Boldizar, T., 1964, Heat flow in the Hungarian Basin: Nature, v. 202, June 27, 1964, p. 1278-1280.

18. Bredehoeft, J. D., and Popadopolous, I. S., 1968, Rates of vertical groundwater movement estimated from the earth's thermal profile: Water Resources Research Journal, v. 1, no. 2, p. 325-328.

19. Buckley, S. E., et al., 1958, Distribution of dissolved hydrocarbons in subsurface water, *in* Habitat of oil, American Association of Petroleum Geologists, p. 850-882.

20. Bullard, E. C., and Niblet, E. R., 1951, Terrestrial heat flow in England (emphasis on DWD@ Eakring anticline): Monthly Notices, Royal Astronomical Society, Geophysical Supplement, no. 6, p. 222-238.

21. Bullard, H. M., et al., 1964, Production logging as applied to post-primary production, pt. 2: Producers Monthly, v. 28, no. 10, Oct. 1964, p. 22-32.

22. Bunter, M. A. G., 1974, How flowline mud temperature plotting can help predict (pressure) transition zones: Proceedings of SEAPLEX Meeting, Jan. 3, 1973, v. 1, p. 95-113.

23. Byerly, P. E., 1965, Geothermal surveys in sedimentary rocks near Grants and Laguna, New Mexico: Geophysics, v. 30, no. 3, June 1965, p. 396-402.

24. Carlson, A. J., 1930, Geothermal conditions in oil producing areas of California: American Production Industry Production Bulletin 205, p. 109-139.

25. ———, 1931, Geothermal variations in Coalinga area, Fresno County, California: American Association of Petroleum Geologists Bulletin, v. 15, p. 829-836.

26. Cartwright, Keros, 1971, Redistribution of geothermal heat by a shallow aquifer: Geological Society of America Bulletin, v. 82, Nov. 1971, p. 3197-3200.

27. Clark, Sydney P., Jr., 1963, Temperatures in the earth's crust: American Physical Society Transactions, v. 3, no. 1, p. 769-790.

28. Cronoble, James M., 1969, Geothermal gradients and their application to petroleum geology: The Mountain Geologist, v. 6, no. 1, p. 27-39.

29. Dalyan, I. B., et al., 1977, Geothermal conditions in the sub-salt rocks of the eastern part of the Peri-Caspian depression and prospects for oil and gas: Petroleum Geology (U.S.S.R.), v. 14, no. 11, Nov. 1977, p. 488-489.

30. Da Silva Carvalho, H., and Vacquier, Victor, 1977, Method for determining terrestrial heat flow in oil fields: Geophysics, v. 42, no. 3, Apr. 1977, p. 584-593.

31. Deroo, G., et al., 1977, The origin and migration of petroleum in the western Canadian sedimentary basin: Alberta Geological Survey of Canada Bulletin 262.

32. Domenico, P. A., and Paliauskas, V. V., 1973, Theoretical analysis of forced convective heat transfer in regional ground-water flow: Geological Society of America Bulletin, v. 84, Dec. 1973, p. 3803-3814.

33. Donaldson, I. G., 1962, Temperature gradients in the upper layer of the earth's crust due to convective water flows: Journal of Geophysical Research, v. 67, no. 9, Aug. 1962, p. 3449-3459.

34. Druzhinin, A. V., 1961, Relationship between geothermal conditions of the sedimentary cover and the structure of the crystalline basement: Petroleum Geology (U.S.S.R.), v. 5, no. 3, p. 158-162.

35. D'yakonov, D. I., 1961, Problem of determination of basic geothermal parameters: Petroleum geology (U.S.S.R.), v. 5, no. 9, p. 500-503.

36.———, 1963, Geothermy in regional geology and for study of subsurface tectonics: Petroleum Geology (U.S.S.R.), v. 4, no. 11B, July 1963, p. 657-661.

37. Dzhangiryants, D. A., 1965 [1970], Geothermal characteristics of the Emba region: Petroleum Geology (U.S.S.R.), v. 9, no. 1, English translation Dec. 1970, p. 55-61.

38. ———, 1977, Types of geothermal curves and their interpretation: International Geology Review, v. 19, no. 2, Feb. 1977, p. 245-248.

39. Dzhibuti, S. S., 1960, Some data on the geothermy of the West Turkmen artesian basin: Akademiya Nauk SSSR Laboratoriia Gidrogeologicheskikh Problem, Trudy 30, p. 104-109.

40. ———, 1961, Geothermal conditions of the West Turkmen artesian basin (applicable to the solution of some problems of hydrogeology and oil and gas possibilities): Akademiya Nauk SSSR Izvestiya Seriya Geologicheskaya, no. 5, p. 95-100.

41. Eckstein, Y., and Simmons, G., 1978, Measurement

and interpretation of terrestrial heat flow in Israel: Geothermics, v. 6, p. 117-142.

42. Evans, E. W., 1866, On the oil producing uplift of West Virginia (Burning Springs): American Journal of Science, v. 42, p. 334-343.

43. Filin, A. S., and Kiryashkin, V. M., 1976, Geothermal conditions of deeply subsided horizons of the eastern Cis-Caucasus in connection with their oil-gas productivity: Petroleum Geology (U.S.S.R.), v. 13, no. 7, July 1976, p. 300-301.

44. Florida Division of Resource Management, 1977, Geothermal nature of the Floridan Plateau: Kohut, F. A., Henry, H. R., and Banks, J. E.; Griffin, G. M., Reel, D. A., and Pratt, K. W.; Sproul, C. R., Smith, D. L., and Fuller, U. R., F.D.R.M. Spec. Pub. #21.

45. Grechishnikov, N. P., 1976, Paleogeothermal unconformities: International Geological Review, v. 18, no. 7, July 1976, p. 804-806.

46. Griffin, G. M., et al., 1969, Geothermal gradients in Florida and Southern Georgia: Gulf Coast Association of Geological Societies Transactions, 19th Annual Meeting, p. 189-193.

47. Hanshaw, B., 1965, Radiocarbon determinations for estimating ground-water flow velocities in central Florida: Science, v. 148, Apr. 23, 1965, p. 494-495.

48. Harper, M. L., 1971, Approximate geothermal gradients in the North Sea Basin: Nature, v. 230, Mar. 26, 1971, p. 235-236.

49. Hawtof, E. M., 1930, Results of deep well temperature measurements in Texas: American Petroleum Industry Production Bulletin no. 205, p. 62-108.

50. Head, H. J., et al., 1978, Maps showing formation temperatures and configurations of the tops of the Minnelusa formation and the Madison Limestone, Powder River Basin, Wyo., Mont., and adjacent areas: U.S. Geological Survey Open File Report 78-905, 12 p.

51. Heald, K. C., 1930, The study of earth temperatures in oil fields on anticlinal structure: American Petroleum Industry Production Bulletin no. 205, Oct. 1930, p. 1-8.

52. Hopper, M. L., 1971, Approximate geothermal gradients in the North Sea Basin: Nature, v. 230, Mar. 26, 1971, p. 235-236.

53. Jam-L, P., et al., 1969, Subsurface temperature in South Louisiana: American Association of Petroleum Geologists Bulletin, v. 53, no. 10, Oct. 1969, p. 2141-2149.

54. Jones, P. H., 1968, Geochemical hydrodynamics—a possible key to the hydrology of certain aquifer systems in the northern part of the Gulf of Mexico Basin: U.S. Geological Survey Open File Report, 20 p.

55. ———, 1975, Geothermal and hydrodynamic regimes in the northern Gulf of Mexico Basin: United Nations Development and Use of Geothermal Resources Symposium Proceedings, 2d, San Francisco, May 20, 1975, p. 429-440.

56. ———, 1978, Interacting dynamics of pressure, temperature, and water salinity: American Association of Petroleum Geologists Research Symposium, Problems of Petroleum Migration, Oklahoma City, [in press].

57. Kappelmeyer, O., 1957, The use of near-surface temperature measurements for discovering anomalies due to causes at depth: Geophysical Prospecting, v. 5, p. 239-258.

58. Kissin, I. G., and Kazintsev, Ye. A., 1965, Temperature conditions of the oil-gas regions of central and eastern Ciscaucasus: Petroleum Geology (U.S.S.R.), v. 6, no. 2., Oct. 1965, p. 95-98.

59. Klemme, H. D., 1972, Heat influences size of oil giants: Geothermal Gradients: Oil and Gas Journal, pt. 1, July 17, 1972, p. 136-144; pt. 2, July 24, 1972, p. 76-78.

60. ———, 1975, Geothermal gradients, heat flow and hydrocarbon recovery, *in* Petroleum and global tectonics, Princeton, N. J., Princeton University Press, p. 251-304.

61. Kortsenshtein, V. N., 1964, On the deep geothermal regime of the South Mangyshlak trough: Doklady Akademiya Nauk SSSR, v. 159, no. 2, p. 336-339.

62. Koshlyak, V. A., 1961, Thermal properties of Mesozoic and Cenozoic sediments on the eastern part of the West Siberian Lowland: International Geology Review, v. 3, p. 1011-1018.

63. Krebtov, A. I., 1961, Internal heat of oil and gas areas: Akademiya Nauk SSSR Doklady, v. 136, no. 5, p. 1069-1072.

64. Kruglikov, N. M. K., 1963, On the problem of the geothermal role of ground water movement: (abstract only) Geologiya-Razvedka Institut Trudy no. 220; Geologiya Sbornik, no. 8, p. 260-272.

65. Kruzhinin, A. V., 1968, Geothermal characteristics of the rocks of Samara Luka: Petroleum Geology (U.S.S.R.), v. 7, no. 6, May 1968, p. 343-348.

66. Kumar, M. B., 1977, Geothermal and geopressure patterns of Bayou Carlin–Lake Sand area, south Louisiana: implications: American Association of Petroleum Geologists Bulletin, v. 61, no. 1, p. 65-78.

67. Kutuzov, O. N., and Solovyev, Ye. A., 1976, Results of geothermal investigations in near-surface sediments in the Tarasov and Rybachyey areas of the Kaliningrad region: Petroleum Geology (U.S.S.R.), v. 13, no. 1, p. 13-14.

68. Landes, K. K., 1966, Eometamorphism can determine oil floor: Oil and Gas Journal, v. 64, no. 18, May 2, 1966, p. 172-177.

69. Larskaya, E. S., and Zhabrev, D. V., 1964, The effect of temperature and pressure of strata on the composition of dispersed organic matter (as exemplified by Meso-Cenozoic deposits of West Ciscaucasia): Doklady A. N. SSSR, v. 157, no. 4, p. 897-900.

70. Lasky, B. H., 1963, Surface alteration studies map subsurface structure: World Oil, June 1963, p. 75-79, 93.

71. Levorsen, A. I., 1967, Geology of Petroleum (2d ed.):

W. H. Freeman & Co., San Francisco, p. 173, 195-196, 389, 415-421, 427-429.

72. Levy, A. G., 1977, Oil and gas maturation zones in the Jurassic and Cretaceous trends of northwest Florida and adjacent parts of Georgia and Alabama: M.S. Thesis, University of Florida.

73. Lewis, C. R., and Rose, S. C., 1970, A theory relating high temperatures and overpressures: Journal of Petroleum Technology, Jan. 1970, p. 11-16.

74. Makarenko, F. A., and Sergiyenko, S. I., 1975, Abyssal heat flow in local oil and gas-bearing continental structures: International Geology Review, v. 17, no. 3, p. 249-255.

75. Maksimov, S. P., et al., 1976, Geothermal conditions of development of the cycle of oil-gas formation: Petroleum Geology (U.S.S.R.), v. 13, no. 11, p. 496-498.

76. Matviyenko, V. N., 1974, Lithologically uniform complexes in study of heat flow of oil-gas regions: Petroleum Geology (U.S.S.R.), v. 11, no. 11, Dec. 1974, p. 499-502.

77. ———, 1976, Comparative characteristics of geothermal conditions in some fields of West Siberia: Petroleum Geology (U.S.S.R.), v. 13, no. 2, Feb. 1976, p. 53-54.

78. ———, and Sergiyenko, S. I., 1977, The heat field of petroliferous regions of Ciscaucasia: International Geology Review, v. 19, no. 8, p. 987-992.

79. McCutchin, J. A., 1930, Determination of geothermal gradients in oil fields on anticlinal structures in Oklahoma: American Petroleum Industry Production Bulletin no. 205, p. 19-61.

80. Meinhold, R., 1971, Hydrodynamic control of oil and gas accumulation as indicated by geothermal, geochemical, and hydrological distribution patterns: Proceedings of the World Petroleum Congress, 8th, Moscow, v. 2, p. 55-66.

81. Mekhtiev, Sh. F., and Aliev, S. A., 1962, Factors affecting the geothermal step in the oil fields of Azerbaydzhan: Petroleum Geology (U.S.S.R.), v. 4, no. 3A, June 1962, p. 151-154.

82. ———, and Aliev, S. A., 1968, On source problems of geothermy in the Caucasus: International Geological Congress, 23rd, v. 5, p. 103-110.

83. ———, et al., 1973, Distribution of oil and gas banks in South Caspian basin, as influenced by thermodynamic parameters of the depths: International Geological Review, v. 15, no. 8, Aug. 1973, p. 960-970.

84. Melik-Pashayev, V. S., et al., 1973, Geothermal conditions of Uzen Oil Field: International Geological Review, v. 15, no. 12, Dec. 1973, p. 1365-1372.

85. Miroshnikov, M. V., 1958, Geothermal nature of the section of the Meso-Cenozoic sediments of the Stavropol Uplift (shales, lower conductivity. Ss, Ls, schists, high):Neft i Gaz, no. 5, p. 21-27.

86. Moody, J. D., 1975, Distribution and geological characteristics of giant oil fields, *in* Petroleum and Global Tectonics, Princeton, N.J., Princeton University Press, p. 307-320.

87. Moses, P. L., 1961, Geothermal gradients now known in greater detail: World Oil, May 1961, p. 79-82.

88. Neruchev, S. G., 1964, Variations in the composition of autochthonous bitumoids (sic) of organic matter of argillaceous rocks acted upon by factors of metamorphism and oil yield: Doklady A. N. SSSR., v. 157, no. 8, p. 901-903.

89. Nwachukwu, S. O., 1976, Approximate geothermal gradients in Niger Delta sedimentary basin: American Association of Petroleum Geologists Bulletin, v. 60, no. 7, p. 1073-1077.

90. Osadchiy, V. G., and Filyus, R. I., 1975, Geothermal conditions of the upper strata of the sedimentary complex of the Ciscarpathian foreland downwarp: Petroleum Geology (U.S.S.R.), v. 12, no. 5, May 1975, p. 262-264.

91. Oswald, G. K. A., and Robin, G. de Q., 1973, Lakes beneath the Antarctic Ice Sheet: Nature, v. 245, Oct. 5, 1973, p. 251-254.

92. Ovnatanov, S. T., and Tamrazyan, G. P., 1962, Thermal conditions in the anticlinal zone of Surakhany-Karachukhur-Zykh-Peschany (Apsheron Peninsula): International Geology Review, v. 4, no. 1, p. 79-88.

93. ———, 1970, Thermal studies in subsurface structural investigations, Apsheron Peninsula, Azerbaijan, USSR: American Association of Petroleum Geologists Bulletin, v. 54, no. 9, p. 1677-1685.

94. ———, 1970, Thermal studies as a technique in subsurface structural investigations: U.N. Symposium on Development and Utilization of Geothermal Resources, Pisa, 1970, v. 2, pt. 2 (Geothermics Special Issue 2).

95. Plewa, S., 1963, Results of experimental investigations of geothermal grades and gradients: Nafta, v. 19, no. 9, Sept. 1963, p. 205-209 [in Polish].

96. Plummer, F. B., and Sargent, E. C., 1931, Underground waters and subsurface temperatures of the Woodbine sand in northeast Texas: University of Texas Bulletin, no. 3138, Oct. 8, 1931, 175 p.

97. Pokrovskiy, V. A., 1960, Thermal waters of the European part of the U.S.S.R.: Akademiya Nauk SSSR Laboratoriia Gidrogeologicheskikh Problem, Trudy 30, p. 99-103.

98. Pollack, H. N., 1976, Thermal profile of the Michigan Basin: Invited Paper T40A, Society of Exploration Geophysicists Annual Meeting, 1976.

99. Pomarnatskiy, M. A., 1963, Geothermal features of the salt dome region of the Caspian Basin: Geologiya Razvedka Institut Trudy, no. 220; Geologiya Sbornik no. 8, p. 248-259.

100. Price, L. C., 1973, Solubility of petroleum in water as function of temperature and salinity and its significance in primary petroleum migration: Annual AAPG-SEPM Meeting, Abstract, American Association of Petroleum Geologists Bulletin, v. 57, no. 4, p. 801.

101. Pusey, W. C. III, 1973, Paleotemperatures in the Gulf Coast using the ESR-Kerogen method: Gulf

Coast Association of Geological Societies, Transactions, v. 23, p. 195-202.

102. Rayevskiy, V. Yu., 1974, Use of trend analysis for geothermal investigations in oil-gas regions (as in Crimean Plain and northern Black Sea region): International Geology Review, v. 16, no. 3, Mar. 1974, p. 263-264.

103. Roberts, W. H. III, 1964, Thermal exploration for ascending waters associated with petroleum entrapment: Unpublished file note, Gulf Research & Development Co., Sept. 7, 1964, 3 p.

104. Roberts, W. H., 1978, The design and function of oil and gas traps: Proceedings of American Association of Petroleum Geologists Research Symposium on Problems of Petroleum Migration, Oklahoma City, [in press].

105. Rogers, W. B., 1842, On the connection of thermal springs, in Virginia with anticlinal axes and faults: Association of American Geologists and Naturalists, Transactions, 1840-1842, p. 323-347.

106. Rogers, Wm. B., 1842 (professor of Natural Philosophy, University of Virginia), Observations of subterranean temperature in the coal mines of eastern Virginia: Association of American Geologists and Naturalists, Transactions, 1840-1842, p. 532-538.

107. Rovnin, L. L., et al., 1978, Effectiveness of the use of thermal exploration in the Orenburg region: Petroleum Geology (U.S.S.R.), v. 15, no. 2, 4 p.

108. Ryabukhin, G. E., and Nesterov, I. I., 1960, Geothermal observations in deep wells of the Omsk area (West Siberian Lowland): Sovetskaya Geologiya, no. 1, p. 129-134.

109. Sass, J. H., 1964, Heat-flow values from eastern Australia: Journal of Geophysical Research, v. 69, no. 18, Sept. 15, 1964, p. 3889-3893.

110. ———, et al., 1971, Heat-flow in the Western United States: Journal of Geophysical Research, v. 76, Sept. 10, 1971, p. 6376-6413.

111. Schoppel, R. J., and Gilarranz, S., 1965, Use of well log temperatures as a criterion for the evaluation of regional geothermal gradients; test case, Oklahoma: AIME Fall Meeting, 40th, Oct. 1965, SPE no. 1297, 12 p.

112. Schoff, S. L., Salso, J. H., and Garcia, J., 1964, Source of heat in a deep artesian aquifer, Bahia Blanca, Argentina: U.S. Geological Survey, Professional Paper 501-D, p. D153-D157.

113. Selig, F., and Wallick, G. C., 1964, Temperature distribution in salt domes and surrounding sediments: Abstracts of Programs of Geological Society of America, annual meeting, 1964, p. 181.

114. Sergiyenko, S. I., 1978, Influence of the thermal field on oil-gas content of Mesozoic rocks of the West Siberian Platform: International Geology Review, v. 20, no. 9, p. 1102-1110.

115. Sevostyanov, K. M., and Vilenskiy, M. A., 1976, The thermal field and the presence of oil and gas in the far east: International Geology Review, v. 18, no. 5, May 1976, p. 563-567.

116. Shelton, J. W., et al., 1974, U. S. geothermal, geologic patterns compared: Oil & Gas Journal, Nov. 25, 1974, p. 173-174.

117. Simkhayev, V. Z., and Stepanova, I. Z., 1977, Geothermal aspects of exploration for hydrocarbons in the lower Kura depression: Petroleum Geology (U.S.S.R.), v. 14, no. 1, p. 7.

118. Sokolov, V. A., et al., 1963, Migration process of gas and oil, their intensity and directionality: World Petroleum Congress, 6th, Frankfort, Proceedings, sec. 1, paper 47, PD2, p. 493-505.

119. Spicer, H. C., 1964, Geothermal gradients and heat flow in the Salt Valley Anticline, Utah: Bollettino di Geofisica Teorica ed Applicta, v. 6, no. 23, Sept. 1964, p. 263-282.

120. Stallman, R. W., 1965, Steady one-dimensional fluid flow in a semi-infinite porous medium with sinusoidal surface temperature: Journal of Geophysical Research, v. 70, no. 12, June 15, 1965, p. 2821-2827.

121. Storer, T. C., 1965, Smackover drilling problems in South Texas: World Oil, June 1965, p. 108-109.

122. Straus, Joe M., and Schubert, G., 1977, Thermal convection of water in a porous medium: effects of temperature-and-pressure-dependent thermodynamic and transport properties: Journal of Geophysical Research, v. 82, no. 2, Jan. 10, 1977, p. 325-333.

123. Sukharev, G. M., Taranukha, Yu. K., and Vlasov, S. P., 1964, Geothermal features of Caucasian oil and gas deposits: International Geology Review, v. 6, no. 9, Sept. 1964, p. 1541-1547.

124. ———, and Bartsev, O. B., 1965, Some data on the thermophysical properties of rocks, composing the sections of oil and gas deposits of the northern part of Sakhalin Island: Doklady Akadamiia Nauk SSSR, v. 162, no. 1, p. 64-165.

125. Sultanov, A. B., et al., 1975, Thermal regime of the Sangachaly-More-Duvanny Island–Bulla Island Field: Petroleum Geology (U.S.S.R.), v. 12, no. 9, Sept. 1975, p. 449-450.

126. Summers, W. K., 1972, Approximation of thermal gradient in southeastern New Mexico using bottom-hole temperatures from electric logs: American Association of Petroleum Geologists Bulletin, v. 56, no. 10, p. 2072-2074.

127. Timko, D. J., and Fertl, W. H., 1972, How downhole temperatures, pressures affect drilling. Part 6 — correlating geopressure gradients with hydrocarbon accumulations: World Oil, Nov. 1972, p. 79-82.

128. Toth, J., 1978, Cross-formational gravity flow of groundwater: a mechanism of the transport and accumulation of petroleum: Proceedings of the American Association of Petroleum Geologists Research Symposium on Problems of Petroleum Migration, Oklahoma City [in press].

129. Tsaturyants, A. B., et al., 1962, New data on geothermal investigations throughout Azerbaijan: Akademiya Nauk Azerbaijan SSR. Doklady, v. 18, no. 4, p. 45-48.

130. ———, and Gadzhieva, T. A., 1963, Geothermal step in the oil-gas regions of Azerbaydzhan: Petroleum Geology (U.S.S.R.), v. 4, no. 7-B, Feb. 1963, p. 433-436.

131. Tyvand, Peter A., 1977, Heat dispersion effect on thermal convection in an isotropic porous media: Journal of Hydrology, v. 34, p. 335-342.

132. Uyeda, S., and Watanabe, T., 1970, Preliminary report of terrestrial heat flow study in the South American continent; distribution of geothermal gradients: Tectonophysics, v. 10, p. 235-242.

133. Van Engen, H., 1975, An interpretation of Groningen subsurface temperature data: Geologie en Mijnbouw, v. 54, no. 3-4, p. 177-183.

134. Van Orstrand, C. E., 1930, Description of apparatus for measurement of temperatures in deep wells (<3500′): American Petroleum Industry Production Bulletin, no. 205, p. 9-18.

135. Vlasova, G. M., Sukharev, G. M., and Taranukha, Yu. K., 1964, Geothermal features of Meso-Cenozoic deposits in the eastern Forecaucasus: Izvestiia Vysshikh Uchebnykh Zavedenii, Geologie I. Razvedka, no. 2, Feb. 1964, p. 3-12 [in Russian].

136. Vorobeva, K. I., 1960, Geothermal features of the Ozek-Suat oil field and of other regions of the Tersko-kuma Plain: Petroleum Geology (U.S.S.R.), v. 4, no. 6, p. 359-362.

137. ———, and Miroshnikov, M. V., 1969, Main regularities in the distribution of heat at depth in the East Stavropol area: Petroleum Geology (U.S.S.R.), v. 8, no. 6, Dec. 1969, p. 348-354.

138. Ward, P. L., and Jacob, K. H., 1971, Microearthquakes in the Ahuachapan geothermal field, El Salvador, Central America: Science, v. 173, July 23, 1971, p. 328-330.

139. Wilson, G. J., and Bush, R. E., 1973, Pressure prediction with flowline temperature gradients: Journal of Petroleum Technology, Feb. 1973, 135-142.

140. Yerofeyev, V. F., 1969, Nature of thermal anomalies of Volga-Ural oil-gas basin: International Geology Review, v. 11, no. 12, Dec. 1969, p. 1382-1389.

141. ———, 1972, Geothermal activity at depth and distribution of deposits of hydrocarbons: International Geology Review, v. 14, no. 1, p. 49-53.

142. Zaydan, S., 1963, Geothermal characteristics of the Tuymazy oil field: Petroleum Geology (U.S.S.R.), v. 7, no. 12, p. 693-699.

143. Zinger, A. S., et al., 1977, Experience of a seismic and geothermal program of investigation under the conditions in the lower Volga region in connection with exploration for oil-gas structures: Petroleum Geology (U.S.S.R.), v. 14, no. 3, Mar. 1977, p. 133-134.

144. ———, 1977, Geotemperature analysis of conditions of the main phase of oil formation in the supra-salt section of West Peri-Caspia: Petroleum Geology (U.S.S.R.), v. 14, no. 8, p. 333-334.

HYDROGEOLOGY OF PETROLEUM AND NATURAL GAS

Paul H. Jones

P. H. JONES HYDROGEOLOGY, INC.

INTRODUCTION

The cardinal test of any theory regarding the origin, migration, and accumulation of petroleum and natural gas is its compatibility with subsurface conditions in *all* known occurrences. To be acceptable, a theory must be capable of universal application. Formulation of an acceptable theory was not possible until recent years because of deficiencies in knowledge regarding certain physical and chemical properties of petroleum and natural gas in subsurface environments, the conditions in which were also poorly known and little understood. Principal among the unknowns were (1) reaction kinetics of kerogen and smectite diagenesis; (2) chemical products of, and physical changes in, source rock resulting from kerogen and smectite diagenesis; (3) aqueous solubility of petroleum hydrocarbons at elevated temperatures and pressures; (4) hydrodynamic and structural deformation mechanisms associated with source-rock mineral transformations; (5) role of geopressure in the fluid hydrocarbon regime; (6) constraints on migration of petroleum in the liquid phase and gas in the vapor phase; and (7) trapping mechanisms, and the role of repeated trap failure in the accumulation of the vast amounts of oil and gas that occur in giant fields.

Most of these unknowns have received intensive study during the past twenty years, in both the laboratory and the field. Results of laboratory studies have provided the necessary basis for analysis and interpretation of subsurface conditions explored by deep drilling in "hostile" environments, such as geopressure zones, making possible the formulation of a new theory in which deep sedimentary basin hydrology defines the key factors.

Principal elements of this theory relate to the aqueous solubility of petroleum hydrocarbons in subsurface environments; diagenetic changes in sediments and interstitial fluids; hydrodynamic and buoyancy forces in fluid hydrocarbon migration and entrapment; the hydrogeology of sedimentary basins as it relates to migration paths, geothermal regimes, sedimentary tectonics, hydrochemical facies distribution, and geopressure; exsolution and entrapment in the petroleum window; and the modes of accumulation of natural gas.

PETROLEUM HYDROGEOLOGY

Until recently, petroleum geologists have given relatively little attention to the hydrology of deep sedimentary basins. White's paper on saline waters in sedimentary rocks (1965) identifies and describes diagenetic processes and molecular forces that may explain mappable changes in the composition of saline formation waters and account for their regional movement. White's paper lays the basis for the science of deep sedimentary basin hydrology; its great importance, in terms of petroleum hydrogeology, is partly the result of earlier work by Weaver (1960), who examined the possible uses of clay minerals in the search for oil, noting a remarkable correlation between oil production, age of reservoir rock, and the relative abundance of smectite (see Figure 1); and of Burst's later (1969) work on the diagenesis of Gulf Coast clayey sediments, and its possible relation to petroleum migration and accumulation (Figure 2). It is no coincidence that the median depth of occurrence of 5,368 oil reservoirs in the Gulf Coast is about 1,500 ft above the depth of the 221°F (105°C) isotherm (Burst's 2nd stage smectite dehydration level). The reaction rates, and the temperature ranges in which both smectite dehydration and kerogen conversion occur as a function of increasing depth, temperature, and fluid pressure, are indicated in Figure 3.

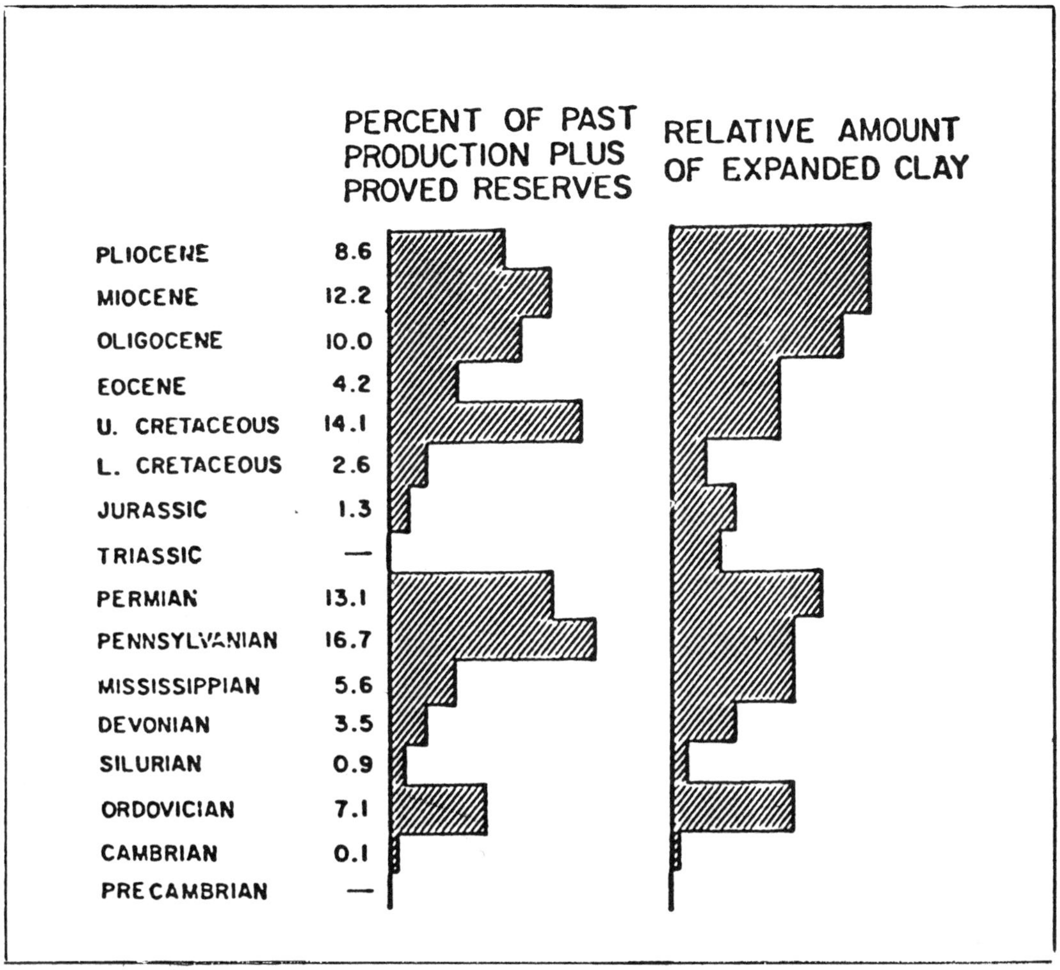

Fig. 1.

In an analysis of the hydrology of Neogene deposits in the northern Gulf of Mexico basin, Jones (the author, 1969) linked together sediment facies distribution, geologic structure, clay mineral type, salinity of sand-bed formation waters, geotemperature, and formation fluid pressure in a three-dimensional model. It appears certain that interrelations of these five fundamental parameters explain known petroleum and natural gas occurrences in deep sedimentary basins of Cenozoic age. In this and later studies, Jones (1975; 1977) has demonstrated that the most important physical interface in young petroliferous basins, from the standpoint of petroleum exploration, is the top of the geopressure zone (Stuart, 1970). More specifically, it is the zone of pressure transition, lying between the depth at which interstitial fluid pressure reflects a gradient of 0.70 psi/ft, and the depth at which that pressure reflects a gradient of 0.50 psi/ft. In a discussion of the hydrodynamics of geopressure, Jones (1969b) describes the behavior of water in geopressured systems in the northern Gulf of Mexico basin as non-Newtonian; hydrodynamic forces are gravity-related, but the dominating forces such as osmosis, diagenesis, and terrestrial heat flow—are molecular. The pressure gradient in the transition zone may be steep or gradual, as shown in Figure 4; pressure changes with depth are accompanied by correlative changes in temperature (Figure 5), in the salinity of water in sand-bed aquifers (Figure 6), and in the porosity of both shale and sand beds (Figure 7). Foster and Custard (1979) consider the smectite-illite transformation that

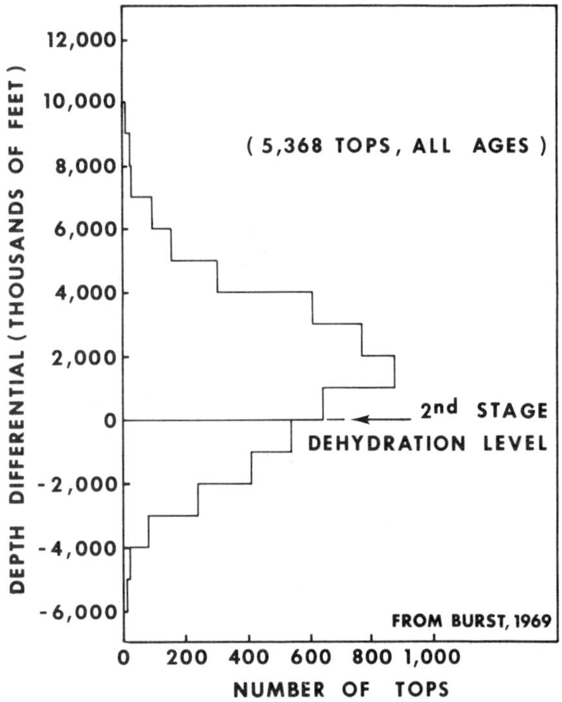

Fig. 2.

occurs in the pressure transition zone (Figure 3) to be "a necessary but insufficient criterion for the generation of such pressure regimes."

Price's discoveries (1973) regarding the aqueous solubility of petroleum hydrocarbons (and whole petroleum crude) at elevated temperatures (Figure 8), coupled with new concepts in deep sedimentary basin hydrology, enabled Jones (1975) to propose a new theory regarding the origin and migration of oil and gas; they are dissolved in and move with formation waters where temperatures and pressures are sufficiently high; where temperature and pressures are lowered sufficiently, they exsolve; thus, their occurrence is defined by the hydrogeology of the sedimentary basin. Oil and gas accumulations may be expected to occur in the traps nearest to the place where the petroleum hydrocarbons come out of solution in response to pressure and/or temperature drop, or by "salting out" of solution (Price, 1976a) where water salinity is increased. Myers (1968)

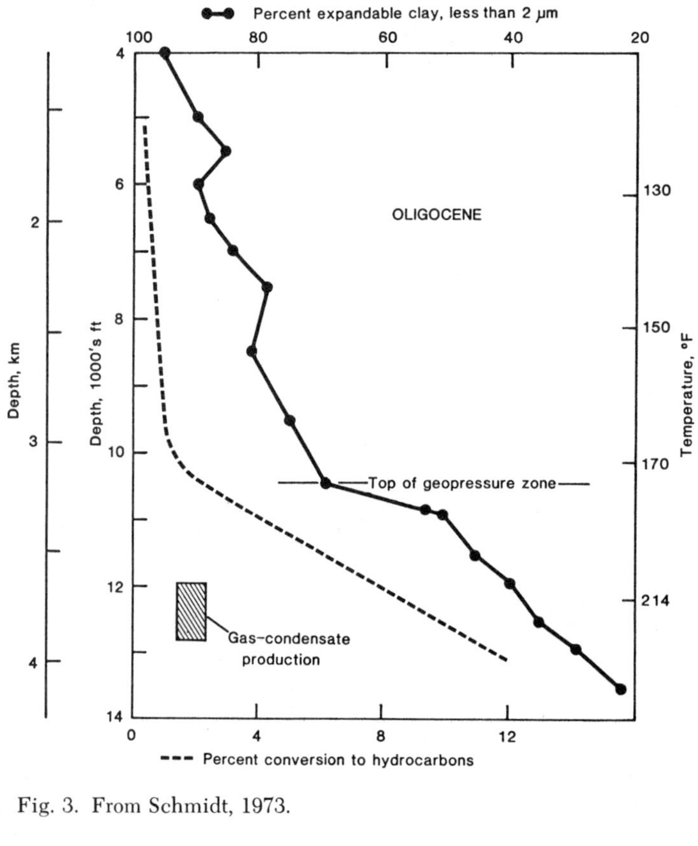

Fig. 3. From Schmidt, 1973.

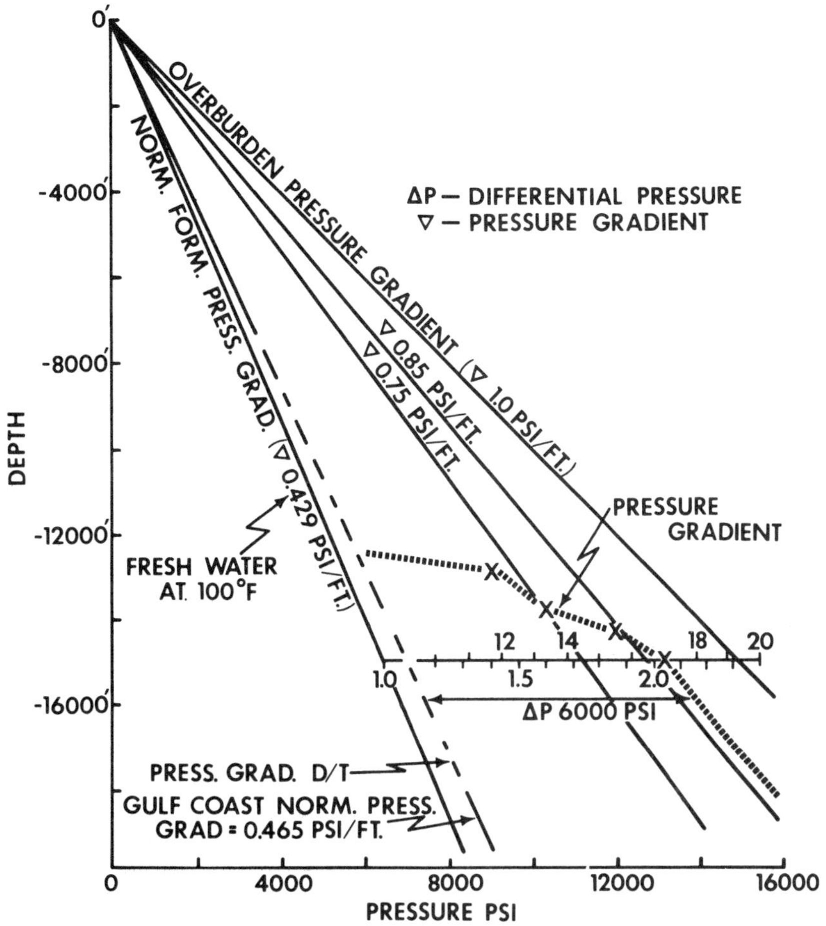

Fig. 4.

had recognized earlier the significance of pressure differential as an oil and gas "trapping" mechanism (Figure 9).

Fertl and Timko (1973), citing the results of a statistical study, reported that 90% of the commercial oil found in the northern Gulf of Mexico basin occurs in reservoirs having original pressures reflecting depth/pressure gradients less than 0.68 psi/ft. These authors also reported that 99% of the commercial oil was found in reservoirs having initial temperatures less than 300°F (150°C). (See Figure 10.) These reports predate publication of Price's (1973) results regarding the aqueous solubility of petroleum.

The importance and significance of the depth-related pressure and temperature constraints on the occurrence of commercial oil reported by Fertl and Timko (1973) are now apparent: they identify a phase-change interface, or boundary, below which whole oil occurs only in aqueous solution (Jones, 1978).

Price (1976, p. 237-238) summarized the fluid hydrocarbon regime defined by sedimentary basin hydrogeology in the following terms:

1. An increase in temperature causes an increase in the aqueous solubilities of petroleum-forming hydrocarbons; the increase is gradual to 100°C (212°F) where a more drastic increase takes place because of a change in the solution mechanism.

2. The most insoluble (in water) compounds of petroleum have their aqueous solubilities increased most strongly with increases in temperature.

3. Increases in salinity cause a decrease in the aqueous solubilities of petroleum-forming hydrocarbons. Brine concentrations in excess of 150,000 ppm salts cause pronounced decreases in aqueous-hydrocarbon solubilities.

4. Increases in salinity and decreases in temperature would cause the exsolution of hydrocarbons from the aqueous phase in natural systems.

5. Undercompacted, high-porosity shales and

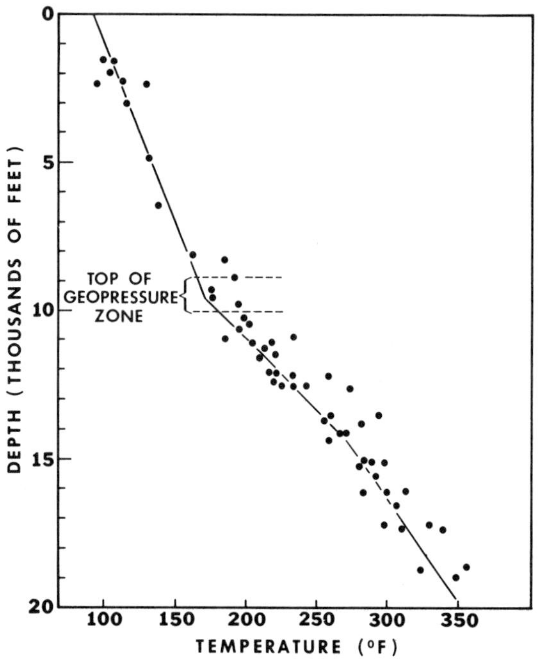

Fig. 5.

abnormal-fluid pressures in many basins of the world are evidence of a deep source of water which is able to mobilize and transport large amounts of hydrocarbons because of the increased solubilities from the very high temperatures in the deep-basin environment.

6. As deep-basin compaction proceeds to completion, the deep hot waters carrying large amounts of hydrocarbons are focused in fault planes and move vertically to shallower basin depths where the hydrocarbons are pooled in a reservoir.

7. Hydrocarbons in the Gulf Coast and many other basins thus would be nonindigenous and may have originated from depths in excess of 20,000 ft (6.10 km). Temperatures of at least 225°C (437°F) and possibly as much as 400°C (752°F) are involved.

8. Mass-balance calculations support the model presented in this work (Appendix).

9. This model predicts that Tertiary deltas with thick shale sequences, which have abnormal fluid pressures and a high content of swelling clays, will be productive.

10. The upward movement along faults of hot fresh waters bearing petroleum from depth in a basin to shallower depths is amenable to the development of powerful new petroleum-exploration techniques.

The fluid hydrocarbon generation-migration model described by Price has been seriously questioned by others (McAuliffe, 1979, p. 761-781); but experimental results recently reported by Bray and Foster (1979, p. 697-698) support it. According to Bray and Foster:

> Experimental data support the evolution of carbon dioxide and hydrocarbon gas concurrently with oil generation and demonstrate effective accommodation of oil in water saturated with carbon dioxide and hydrocarbon gas at the temperatures and pressures which are present at depth in source rocks. Oil carried by gas-saturated water migrating from source rocks can be unloaded in and near reservoir traps or enroute to a reservoir by removal of carbon dioxide from the water. This unloading is accomplished by reaction of carbon dioxide with "carbon dioxide-starved" or unconditioned sediments contacted by the oil-bearing water as it moves up faults and permeable strata or by coming out of solution because of low pressure at shallow depths. Carbon dioxide reactions with the source rock matrix have previously been saturated with the large amount of carbon dioxide generated which also provided a surplus to saturate the pore water. After the pore water is saturated with carbon dioxide, the mobilized oil can leave the source rock with water expelled by compaction. In noncompacting situations the hydrocarbons can diffuse over reasonable distances from the source rocks into adjacent permeable beds.
>
> The observed capacity of gas-saturated water to carry oil enables reasonable and available volumes of migrating water to transport and unload enough oil to fill reservoir traps.

The driving forces in the secondary migration of petroleum are buoyancy and hydrodynamics. Buoyancy force results from differences in fluid density; it is measured as the difference in pressure between the water phase and the oil phase at any point above the free water level in a reservoir. The free water level is the level at which water would stand in a large open well screened in the reservoir. The buoyancy gradient—the rate of buoyant pres-

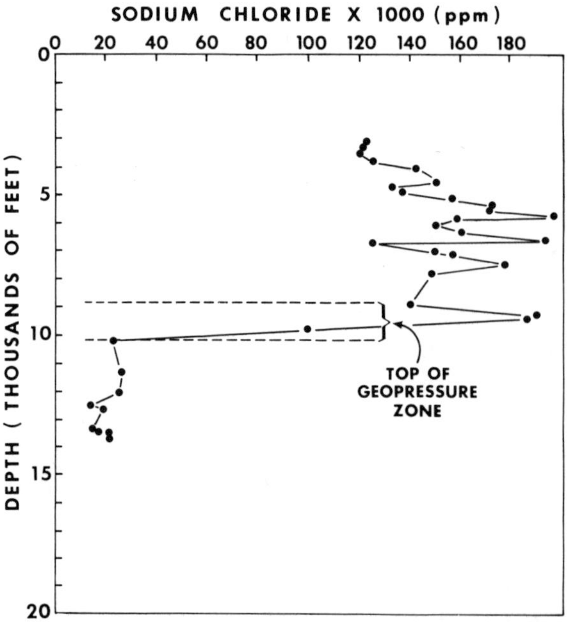

Fig. 6. From Schmidt, 1973.

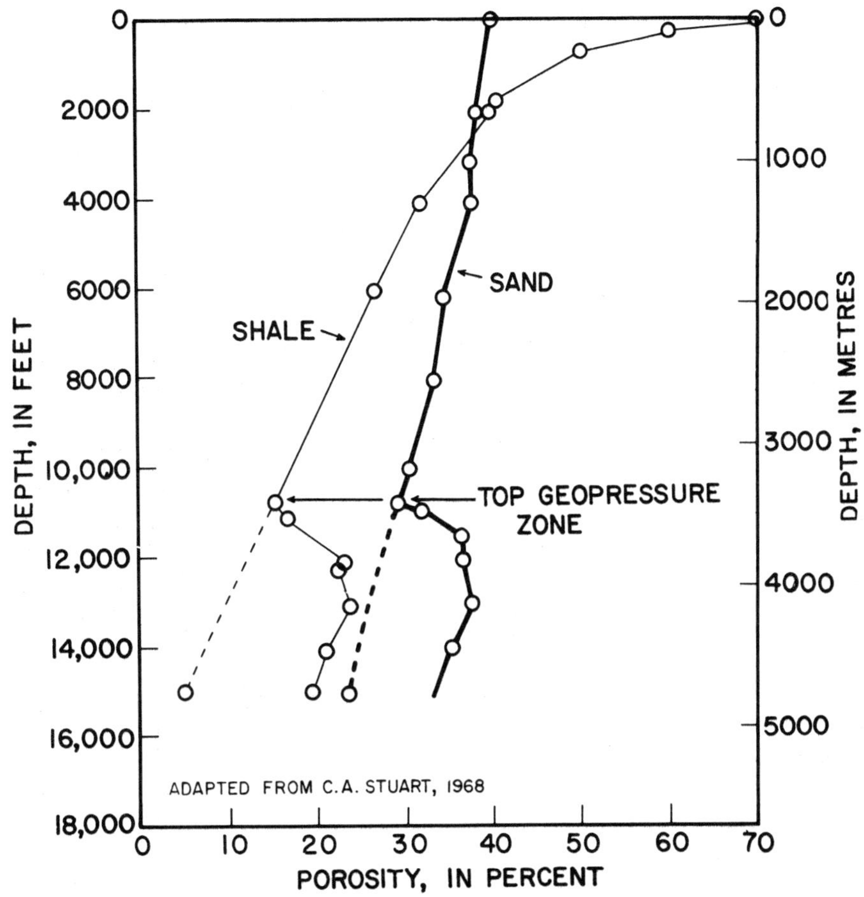

Fig. 7.

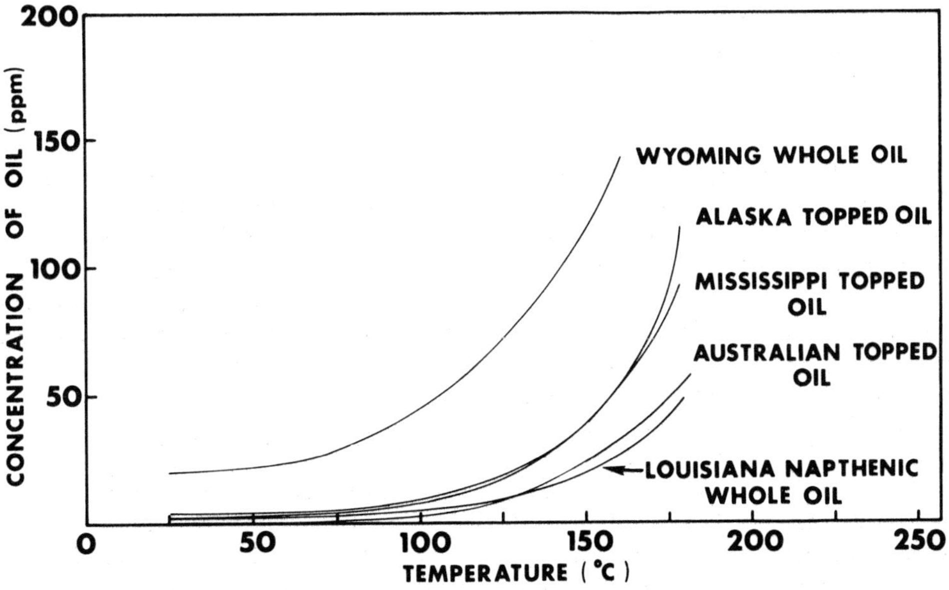

Fig. 8. From Price, 1973.

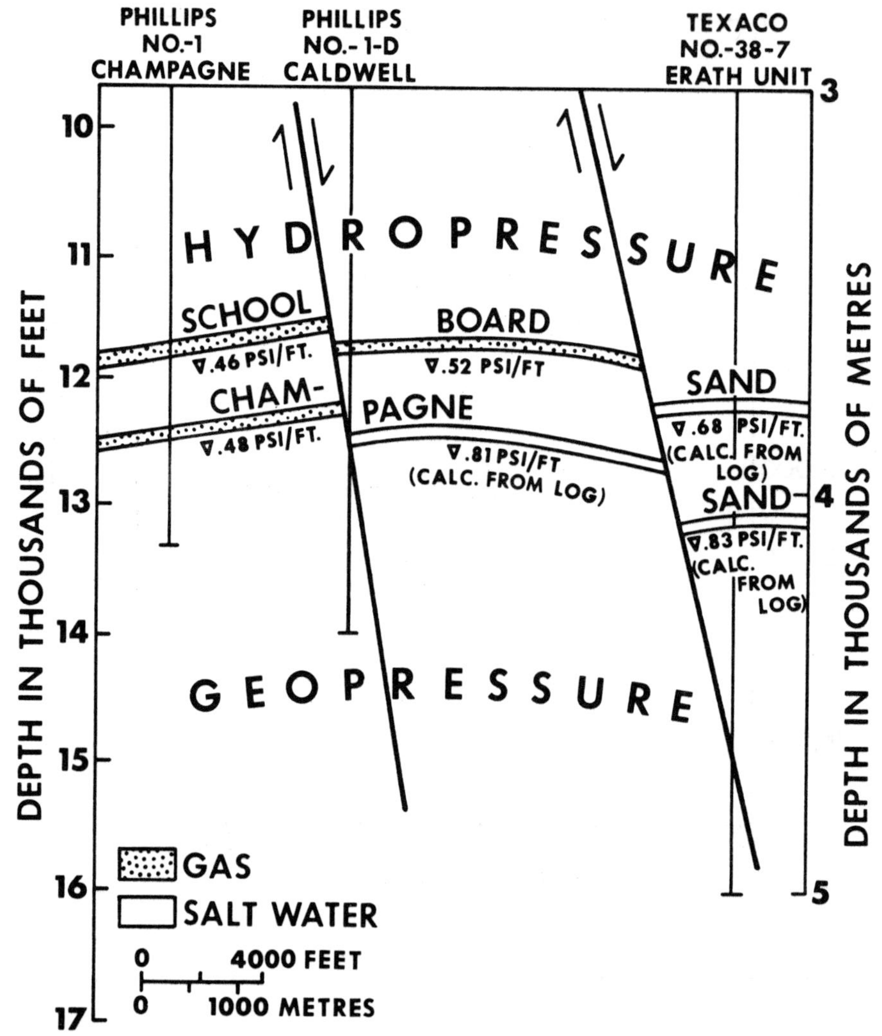

Fig. 9.

sure increase with height above the free water level—can be calculated by subtracting the oil pressure gradient from the water pressure gradient, and can range from about 0.2 to about 0.5 psi/ft.

Hydrodynamic force results from the motion of subsurface water; its effect is to change the buoyant force and thus the migration potential of a fluid hydrocarbon column of a given height. Downdip flow reduces buoyancy force, whereas updip flow increases it, for any oil-phase component in a sediment. Hubbert (1953) demonstrated the great importance of hydrodynamic forces in the migration and entrapment of petroleum, but unfortunately did not apply his concepts to the geopressure transition zone described above.

Schowalter (1979) has explained the mechanics of secondary hydrocarbon migration and entrapment in an excellent treatise; these aspects of petroleum hydrogeology need be discussed no further here.

HYDROGEOLOGY OF NATURAL GAS

Pressure-volume-temperature and solubility relations for natural gas-water mixtures were described by Dodson and Standing in 1945. Their findings regarding methane solubility as a function of water salinity are illustrated in

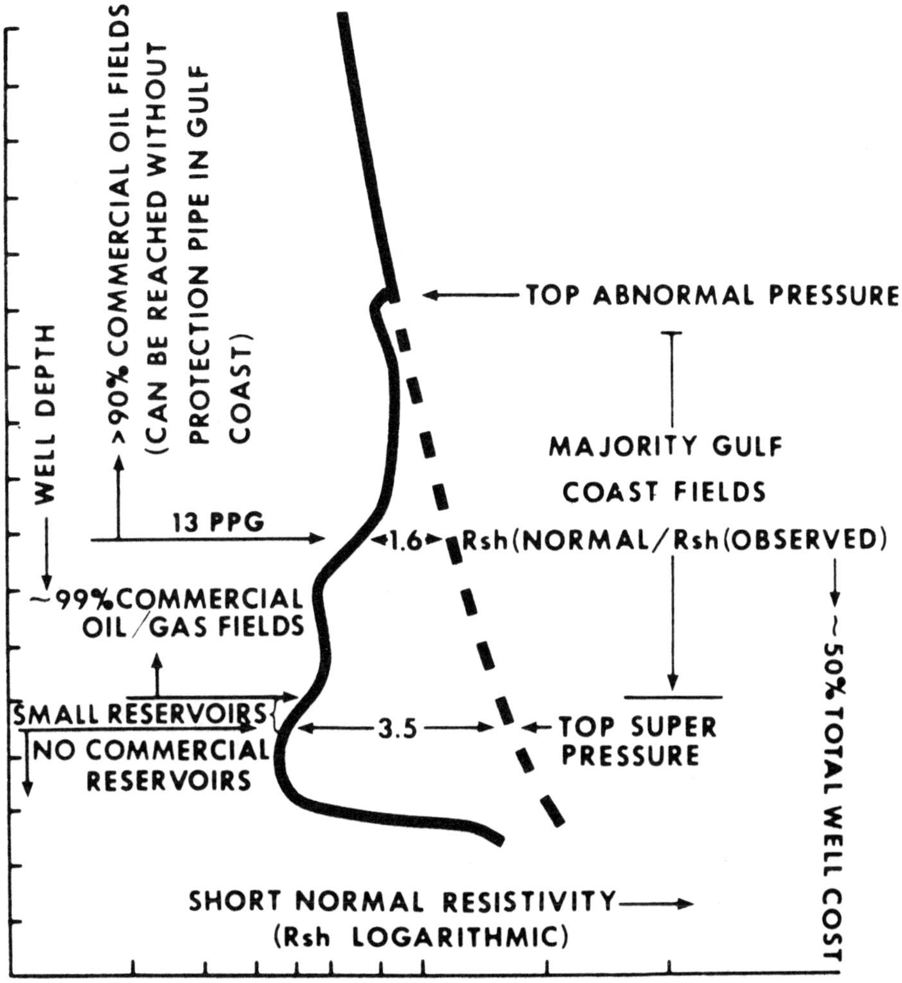

Fig. 10.

Figure 11. Culberson and McKetta reported in 1951 on the solubility of methane in fresh water at pressures to 10,000 psia and temperatures to 350°F (176°C) (Figure 12). Sultanov, Skripka, and Namiot (1972) confirmed the earlier work of Culberson and McKetta and extended solubility data to 16,000 psi and 660°F (350°C). (See Figure 13.) Using their data, Bonham (1978) described the effects of three different pressure gradients on the possible dissolved gas content of formation waters to a depth of 20,000 ft, with assumed geothermal gradients. (See Figure 14.) Price (1979) measured methane solubility in water at selected temperatures from 154° to 354°C and pressures from 100 to 28,600 psi; at temperatures above 300°C and pressures above 10,000 psi his results do not agree with the Russian data. The highest methane solubility measured by Price was 828 SCF/bbl at 28,600 psi and 354°C.

Methane is produced in enormous quantities in petroliferous basins, not only during the thermal conversion of kerogen to fluid hydrocarbons at temperatures above 150°F (66°C), but also by the thermal cracking of liquid hydrocarbons that fail to escape from shale source rocks as temperatures rise above 300°F (150°C) (Pusey, 1973, p. 195). The realm of "thermal gas," as defined by depth and geothermal gradient, is shown in Figure 15. Methane is taken

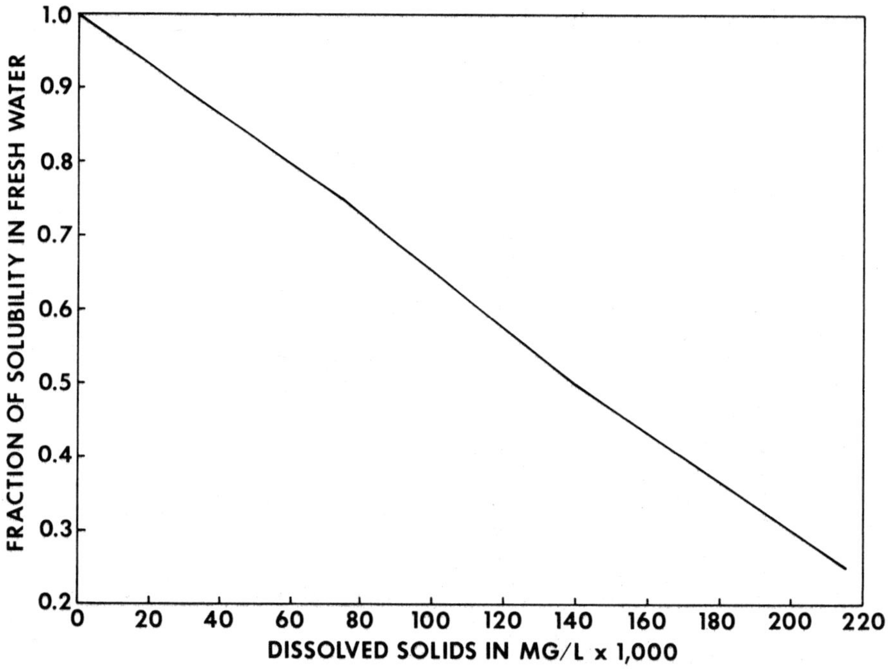

Fig. 11.

into water solution and escapes from source rocks with the water as compaction and consolidation continue with deepening burial.

Where the kerogen content of source rocks exceeds about 1% and the top of the geopressure zone is perhaps 10,000 ft to 15,000 ft below

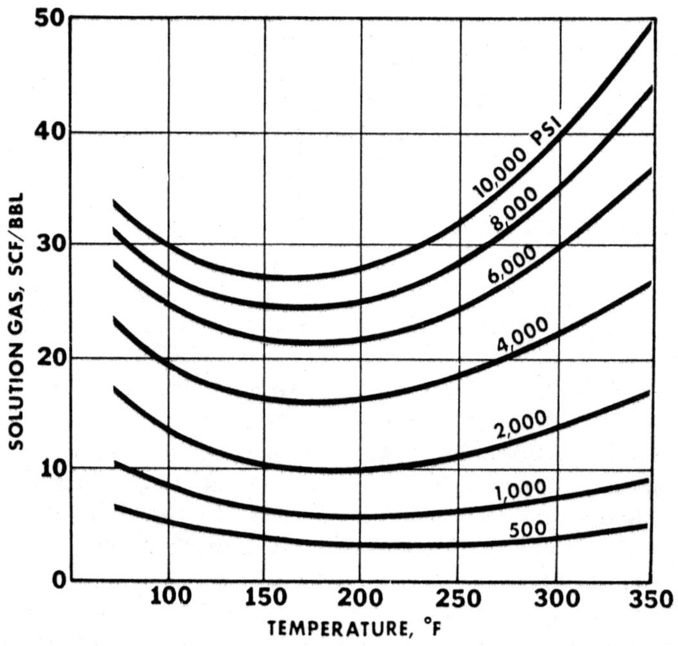

Fig. 12.

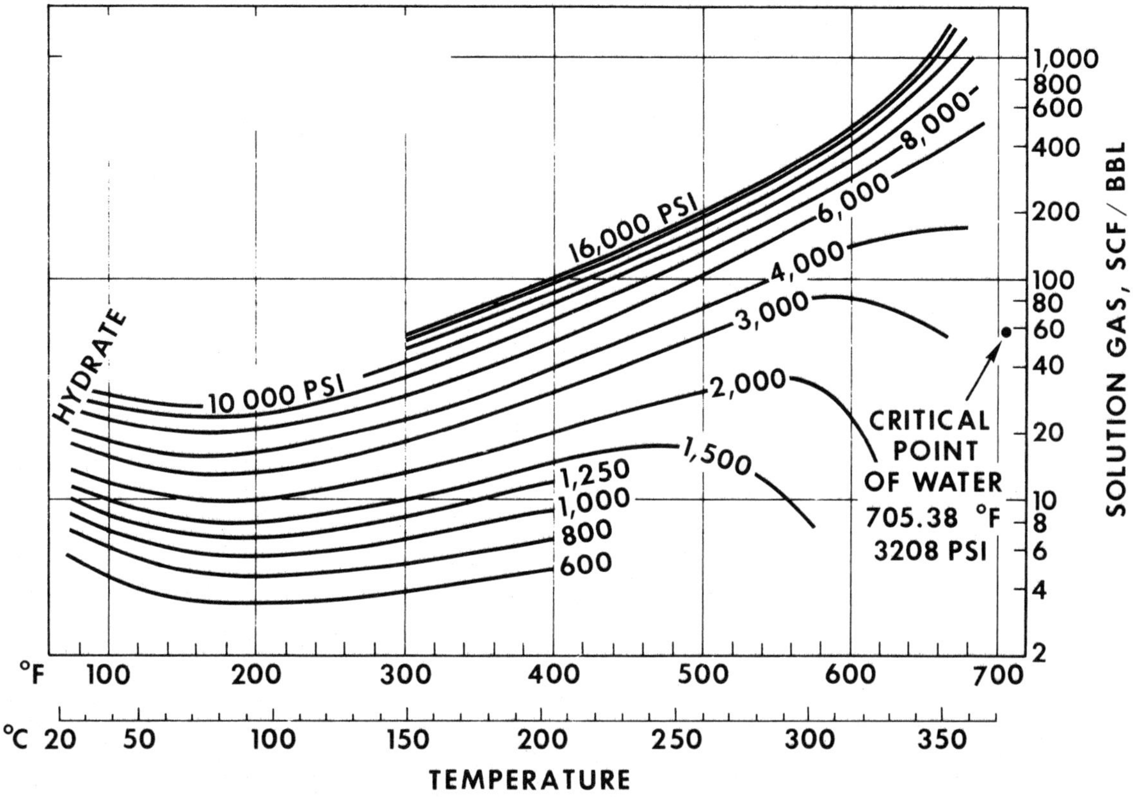

Fig. 13.

land surface, geothermal gradients are great enough to result in temperatures above 300°F (150°C) at depths less than 20,000 ft. In such settings, "thermal gas" is produced in such great quantity that all of the accompanying saline formation waters are saturated in methane; that which cannot be dissolved forms large nonassociated vapor phase gas reservoirs in the structural traps. Gas reservoirs may also be found wherever water has escaped from zones of very high fluid pressure (as shown in Figure 9 above); they are formed by gas exsolved from the water as the fluid pressure drops. Very large differences in methane solubility occur as a consequence of pressure changes, in the range 4,000 to 16,000 psi, at temperatures greater than 300°F (150°C). These are shown in Table 1, which is based on Figure 13.

TABLE I

Solubility of Methane in Water at Selected Temperatures and Pressures, in scf/bbl (values approximate).

PRESSURE PSI	TEMPERATURE °F 200	300	400	500	600	656
2,000	10	12	20	30	17	
3,000	13	17	30	52	80	
4,000	15	23	40	76	135	
6,000	20	29	52	105	230	380
8,000	24	35	64	130	285	440
10,000	28	41	77	149	340	620
12,000		47	86	168	400	800
14,000		53	95	186	440	900
16,000		58	104	200	480	1,000

SUMMARY

Whole petroleum crude and natural gas (methane) are highly soluble in low-salinity water at temperatures and pressures common at depth in most Mesozoic and Cenozoic petro-

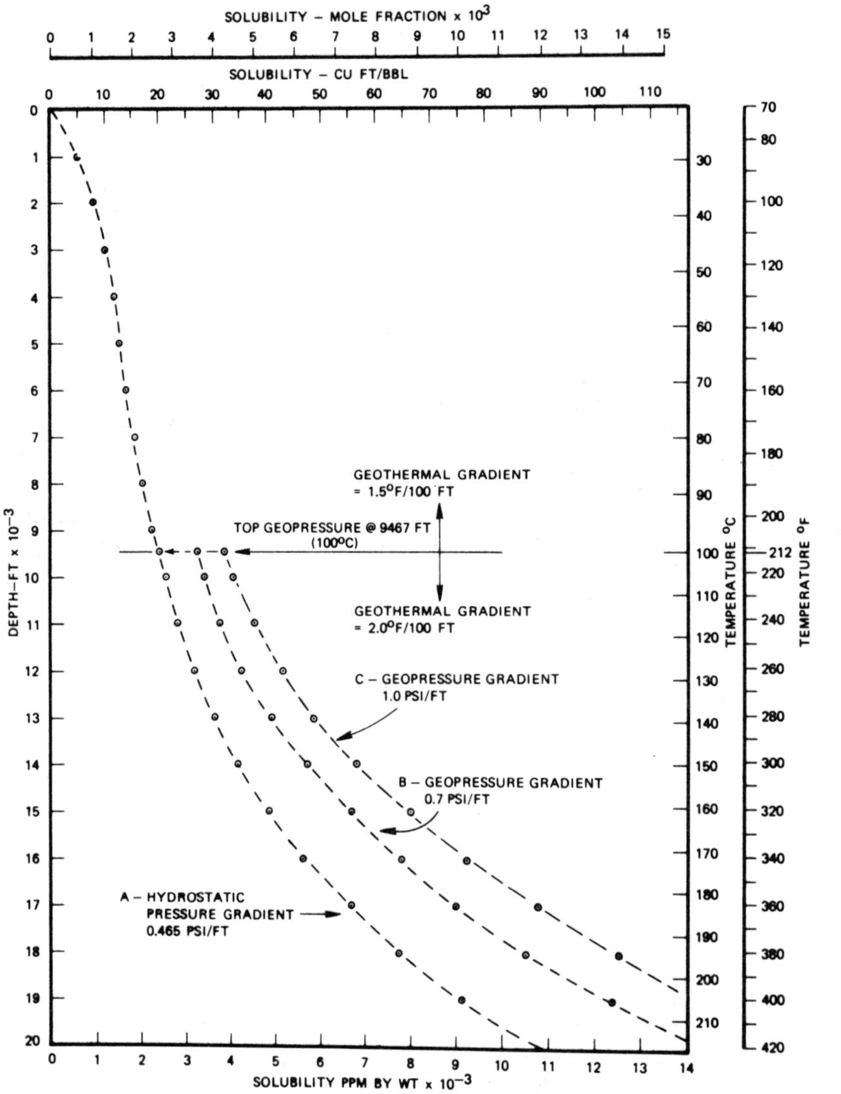

Fig. 14.

liferous basins of the world. Saturation of water with dissolved methane (or carbon dioxide) increases petroleum solubility. Kerogen conversion to fluid hydrocarbon in smectite-rich source rock and the release of bound and intracrystalline water in the smectite-illite transformation occur simultaneously, in the temperature range −80° to 120°C. Large amounts of methane and carbon dioxide are formed in categenesis; these dissolve in the newly released low-salinity pore water, which in turn dissolves and removes from reaction sites the hydrocarbon liquids formed by kerogen conversion. An increase in porosity, with a very large increase in permeability, occurs in the source rock as a consequence of the smectite-illite transformation. These changes reduce the load-bearing strength of shale source rock; the interstitial fluid pressure in the source rock rises as pore fluids share a larger part of the overburden load; and source-rock pore waters with their dissolved hydrocarbons are expelled into the adjacent, more competent sand-bed aquifers. This is the dominant mode of primary migration of oil and gas in sedimentary basins.

Within the geopressure zone in petroliferous basins, oil is present only in water solution where pressure exceeds about 0.7 psi/ft and

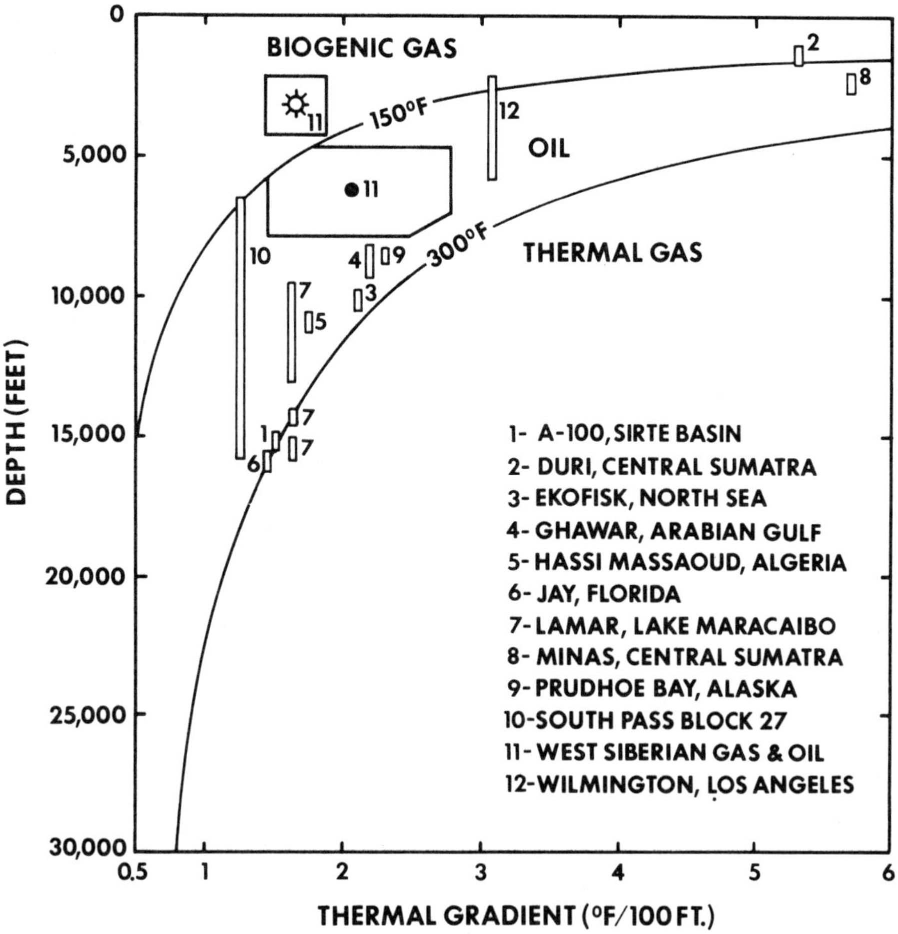

Fig. 15.

temperature is greater than 150°C. The solubility of whole petroleum in such settings can range up to about 10 weight %. Natural gas, on the other hand, can exist in vapor phase at all pressures and temperatures known to occur in sedimentary basins. In the northern Gulf of Mexico basin, about 80% of the vapor phase gas reserves occur in the geopressure zone, where the petroleum which did not escape from source beds is subjected to progressively higher temperature and pressure; natural catalytic "cracking" converts heavier to lighter hydrocarbon molecules. Methane, a stable end-product in this process, forms giant vapor phase gas accumulations in spite of its greater solubility. The largest reserves are found in gas reservoirs having initial pressures reflecting pressure gradients greater than 0.8 psi/ft and temperatures above 150°C.

Formation water saturated in oil and gas releases dissolved hydrocarbons as it escapes from the geopressure zone. Exsolution of gas from water reduces the oil solubility, and both gas and oil accumulations of commercial size are common in the zone of transition from geopressure to hydropressure. Structural or hydrodynamic traps that define these accumulations may be filled to the spill point, may cease to be traps as a result of subsequent structural deformation, or may lose the oil and gas to water solution if the reservoir trap is subsequently exposed to moving formation water at elevated temperature and pressure. Only such solution, migration, and exsolution—repeated again and again during the geologic history of structural basins of regional dimensions—could accumulate enough oil and gas to form the world's giant fields.

REFERENCES

Bonham, L. C., 1978, Solubility of methane in water at elevated temperatures and pressures: American Association of Petroleum Geologists Bulletin, v. 62, p. 2478-2488.

Bray, E. E., and Foster, W. R., 1979, Process for primary migration of petroleum in sedimentary basins (abs.): American Association of Petroleum Geologists Bulletin, v. 63, p. 697-698.

Burst, J. F., 1969, Diagenesis of Gulf Coast clayey sediments and its possible relation to petroleum migration: American Association of Petroleum Geologists Bulletin, v. 53, p. 73-93.

Culberson, O. L., and McKetta, J. J., 1951, Phase equilibrium in hydrocarbon-water systems, III: The solubility of methane in water at pressures to 10,000 psia: American Institute of Mining and Metallurgical Engineers Petroleum Transactions, v. 192, p. 223-226.

Dodson, C. R., and Standing, M. B., 1944, Pressure-volume-temperature and solubility relations for natural gas water mixtures: Drilling and Production Practices, p. 173-179.

Fertl, W. H., and Timko, D. J., 1972-1973, How Downhole temperatures, pressures affect drilling: World Oil, June 1972-March 1973.

Foster, W. R., and Custard, H. C., 1979, The smectite-illite transformation: Its role in generating and maintaining geopressure: unpublished abstract of paper presented at American Association of Petroleum Geologists Research Symposium, Mechanisms of Abnormal Pressure Generation, Sept. 1979, C. E. Hottman, Convenor.

Hubbert, M. K., 1953, Entrapment of petroleum under hydrodynamic conditions: American Association of Petroleum Geologists Bulletin, v. 37, p. 1954-2026.

Jones, P. H., 1969a, Hydrology of Neogene deposits in the northern Gulf of Mexico basin: Louisiana Water Resource Research Institute Bulletin, GT-2.

———, 1969b, Hydrodynamics of geopressure in the northern Gulf of Mexico basin: Journal of Petroleum Technology, v. 21, p. 803-810.

———, 1975, Geothermal and hydrocarbon regimes, northern Gulf of Mexico basin: Geopressured Geothermal Resources of the Gulf Basin, 1st conference, Proceedings, University of Texas, Austin, p. 15-89.

———, 1978, The role of geopressure in the fluid hydrocarbon regime, *in* Exploration and economics of the petroleum industry: New York, Matthew Bender and Co., Inc., v. 16.

LaPlante, R. E., 1974, Hydrocarbon generation in Gulf Coast tertiary sediments: American Association of Petroleum Geologists Bulletin, v. 58, no. 7, p. 1281-1289.

McAuliffe, C., 1979, Oil and gas migration—chemical and physical constraints: American Association of Petroleum Geologists Bulletin, v. 63, p. 761-781.

Meyers, J. D., 1968, Differential pressures, a trapping mechanism in Gulf Coast oil and gas fields: Gulf Coast Association of Geological Societies Transactions, v. 18, p. 66-80.

Price, L. C., 1973, The solubility of hydrocarbons and petroleum in water as applied to the primary migration of petroleum: University of California, Riverside, Ph.D. dissertation, 298 p.

———, 1976, Aqueous solubility of petroleum as applied to its origin and primary migration: American Association of Petroleum Geologists Bulletin, v. 60, p. 213-244.

———, 1978a, Crude oil and natural gas dissolved in deep, hot geothermal waters of petroleum basins—a possible significant new energy source: Geothermal-Geopressured Energy, 3rd annual conference, v. 1, p. GI 167-GI 249

———, 1978b, New evidence for hot, deep origin and migration of petroleum (abs.): American Association of Petroleum Geologists Bulletin, v. 62, p. 556.

———, 1979, Aqueous solubility of methane at elevated pressures and temperatures: American Association of Petroleum Geologists Bulletin, v. 63, p. 1527-1533.

Pusey, W. C., 1973, Paleotemperatures in the Gulf Coast using the ESR-kerogen methods: Gulf Coast Association of Geological Societies Transactions, v. 23, p. 195-202.

Schmidt, 1973, Interstitial water composition and geochemistry of deep Gulf Coast shales and sands: American Association of Petroleum Geologists Bulletin, v. 57, no. 2, p. 321-337.

Schowalter, T. T., 1979, Mechanics of secondary hydrocarbon migration and entrapment: American Association of Petroleum Geologists Bulletin, v. 63, p. 723-760.

Stuart, C. A., 1970, Geopressures: Second Symposium on Abnormal Subsurface Pressure, Louisiana State University, Baton Rouge, La., 121 p.

Sultanov, R. C., Skripka, V. C., and Namiot, A. Yu., 1972, Solubility of methane in water at high temperatures and pressures: Gazova Promyshlennost, v. 17, May, p. 6-7 (in Russian).

Weaver, C. E., 1969, Possible uses of clay minerals in search for oil: American Association of Petroleum Geologists Bulletin, v. 44, p. 1505-1518.

White, D. E., 1965, Saline waters of sedimentary rocks, *in* Fluids in subsurface environments: American Association of Petroleum Geologists Memoirs, no. 4, p. 342-366.

USE OF LANDSAT GEOMORPHIC AND TONAL ANOMALIES IN PETROLEUM PROSPECTING

Donald F. Saunders

TEXAS INSTRUMENTS, INC.

ABSTRACT

The general applications of drainage, landform, and tonal analysis of LANDSAT imagery in reconnaissance for potential petroleum traps are summarized. "Curvilinear" circular to subcircular drainage and/or landform patterns influenced by domal or anticlinal structures, and tonal patterns suggestive of surface geochemical soil-salt anomalies are studied in particular. Maps of suggested prospects prepared in 1974-1975 as part of a nonexclusive LANDSAT-based morphostructural map series in the Rocky Mountain area are compared with approximately four to five years of subsequent oil and gas developments in an attempt to estimate the prediction effectiveness of such geomorphic studies. Preliminary results indicate that about 1 in 3.6 of the suggested prospects which were tested by drilling resulted in one or more new producing wells.

The region covered by a 1958 paper showing suggested prospects in part of the Powder River Basin, Wyoming, based on detailed drainage analysis at 1:63,360 scale using aerial photos, is studied in light of subsequent exploration developments. Seven out of twelve 1958 suggested prospects tested during the past twenty years resulted in new field discoveries. The 1958 analysis was compared with recent geomorphic analysis at 1:250,000 scale and 1:500,000 scale to study the relative efficiency of working at different effective resolutions of detail. It was concluded that the most cost-effective approach would be to use drainage, landform, and tonal analysis of LANDSAT imagery at 1:250,000 scale giving priority to combined geomorphic and tonal anomalies.

Halbouty (1976) stated: "The LANDSAT (ERTS) project is the most significant mission ever flown by NASA. The use of LANDSAT imagery by the mineral and energy industries in the United States can improve the nation's domestic resource base in a shorter time and at a more reasonable cost than would have been possible otherwise. . . .

"LANDSAT imagery provides the explorationist a most rapid and inexpensive tool which could add immeasurably to his geologic knowledge."

INTRODUCTION

OBJECTIVES

The principal objective of this report is to present the results of tests of the effectiveness of drainage, landform, and tonal analysis of LANDSAT imagery in reconnaissance for potential petroleum traps. In addition, the basic characteristics of the satellites and the data products are summarized (See Appendix).

The effectiveness tests are based on follow-up study of petroleum developments in the Rocky Mountain area subsequent to the preparation of maps of suggested prospects based on geomorphic analyses of satellite imagery and aerial photography. In 1974 Texas Instruments designed a nonexclusive "morphostructural" map series based on LANDSAT imagery and intended to serve two general purposes in the area of petroleum exploration:

1. To perform as an inexpensive "first step" reconnaissance tool to indicate possible new and previously undetected prospecting areas based on newly discernible linear and curvilinear geomorphic features mapped from satellite imagery.

2. To provide a regional structural "framework" on which other exploration data, including seismic information, gravity, magnetics, lithology, and geochemistry, can be superimposed possibly to reveal new relationships significant to petroleum occurrence.

Areas mapped in 1974-75 have been studied in conjunction with December 1978 petroleum production and dry hole maps to evaluate how well the suggested prospects predicted new field developments over the ensuing four to five years.

Buttorff (1958) and Elliot (1958) presented geomorphic anomalies derived from a detailed drainage analysis (at 1:63,360 scale) of two areas in the Powder River Basin, Wyoming, based on aerial photography. These have been compared with subsequent exploration results, and with related analyses of LANDSAT imagery at 1:250,000 scale and 1:500,000 scale, to

Fig. 1a. Annular drainage pattern.

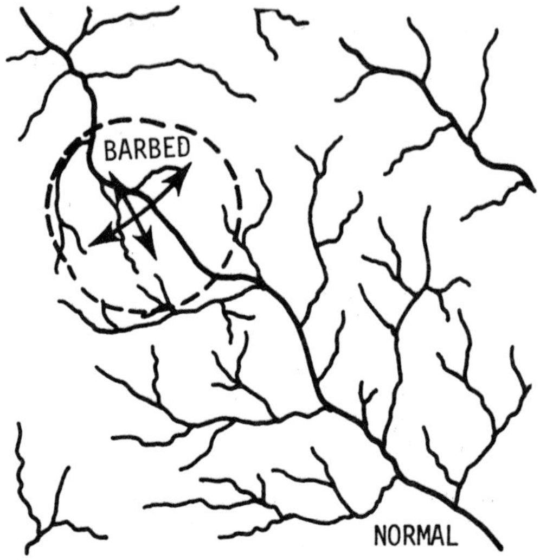

Fig. 1c. Barbed and normal dendritic drainage pattern.

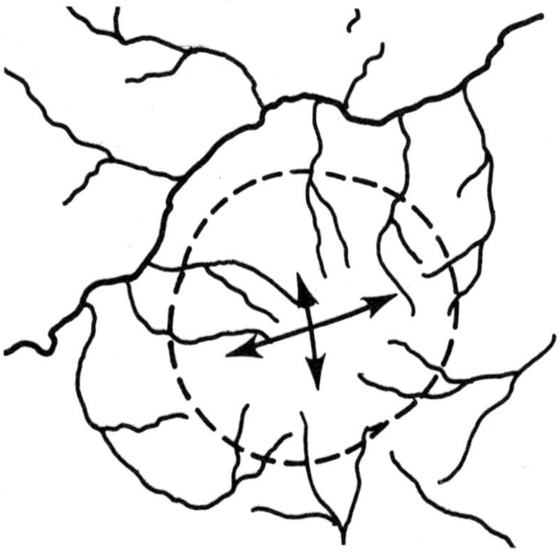

Fig. 1b. Radial drainage.

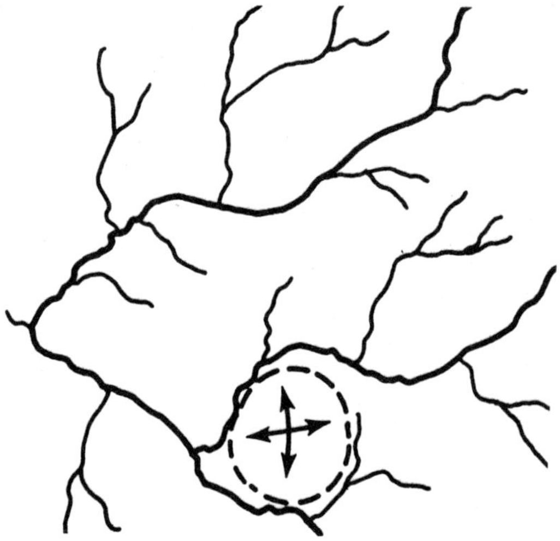

Fig. 1d. Deflected drainage pattern.

Fig. 1. Examples of drainage influenced by domal or anticlinal structures (after Doeringsfeld and Ivey, 1964).

study the relative efficiency of working at different effective resolutions of detail.

Acknowledgements

Thanks are extended to Texas Instruments for support of this work and release of the information. Additionally, the author is grateful for critical reviews of this paper and helpful suggestions by David D. Van Wart, Keith Thompson, and Jack M. Proffitt. The author thanks Mary Roberts for her assistance in preparing this manuscript.

STREAM DISSECTION, LANDFORM AND TONAL ANOMALIES

General

Geomorphic studies using aerial photographs came into popular use after the mid-1950s and utilize four basic categories of analysis which generally, in the order of their importance, are (Doeringsfeld and Ivey 1964): (1) drainage analysis; (2) landform analysis; (3) fracture pattern analysis; and (4) tonal characteristic analysis. These characteristics of the earth's surface reflect the rock materials which underlie each part of the surface and the agencies of change which have acted upon them, and geologists who understand and can recognize the surface expressions of the materials and structures associated with oil and gas traps have used these geomorphologic studies effectively in reconnaissance exploration for many years. The applications of photogeomorphology were reviewed by Tator et al. (1960), Thornbury (1969), and Reeves et al. (1975) with comprehensive bibliographies on the subject. Recently introduced analyses of stream dissection and tonal anomalies on LANDSAT imagery use similar methodology from the more regional viewpoint of the satellite to provide reconnaissance for possible petroleum traps at a lower cost (about \$0.03—\$0.05 per square mile) than detailed analyses of aerial photos (approximately \$15.00 per square mile).

Drainage and Landform Analysis

General Geological Significance

The drainage system which develops on a regional surface is controlled by the types and attitudes of the underlying rock as well as by the slope of the surface, and therefore reflects to varying degrees the lithology and structure of the region. Diagnostic drainage patterns have been described by Zernitz (1932), Smith (1943), Melton (1959), Tator et al. (1960), von Bandat (1962), Doeringsfeld and Ivey (1964), Strandberg (1967), Howard (1967), Reeves et al. (1975), and others. The influence of underlying structure on drainage patterns appears to be quite general. Melton (1959) estimated that between 25% and 75% (either by length or number of segments) of all continental drainage in unglaciated regions is tectonically influenced or controlled. He indicated even higher figures were probable for some tectonic mountain districts and in lowland "plains" which lie above deeply eroded but not too deeply buried old mountain ranges. Doeringsfeld and Ivey (1964) reported that a thorough examination of over 25,000 sq mi of east Texas showed that more than 75% of the structural features reported in the published literature can be observed at the surface because of their geomorphic effects.

Basic curvilinear drainage patterns of primary interest in petroleum exploration include those commonly formed over domes or anticlines (after Doeringsfeld and Ivey, 1964):

1. Annular (Figure 1a)
2. Radial (Figure 1b)
3. Barbed (Figure 1c)
4. Deflected (Figure 1d)

1958 Prospects Reexamined

As an introduction to the use of drainage analysis let us examine some twenty-year-old data from the Powder River Basin, Wyoming. Buttorff (1958) presented a map showing geomorphic anomalies in the Dead Horse Creek area derived from 1:63,360 scale photo index sheets, and Elliot (1958) provided a drainage analysis in the Donkey Creek area based on 1:20,000 scale aerial photography (see Figure 2). Petroleum development in those areas was minimal at the time the data were published. Figure 3 shows a comparison of the Donkey Creek area in 1958 and 1978 with the 1958-annotated geomorphic anomalies superimposed. Figure 4 provides a similar comparison for the Dead Horse Creek area.

Anomalies 1 and 3 in Figure 3 were developed into producers over the past twenty years. The dashed unnumbered anomaly was not annotated by Elliot, but is clearly visible in his data as a drainage anomaly (Figure 5), and is included as a "find." In summary, all six anomalies have been drilled and three resulted in production. It should be noted that the Donkey Creek field (northeast corner of area) is a stratigraphic trap producing from erosional remnants of Minnelusa sandstone between shale-filled channels. The geomorphic anomaly may be related to differential compaction giving drainage control similar to a positive structure.

In Figure 4, anomalies 1, 4, 7, and 10 resulted in new production; anomalies 2 and 8 are productive just inside their margins and show promise, but they are counted as not tested along with anomalies 2, 6, and 12. Anomalies 5 and 11 are counted as tested dry. In summary, four out of the six anomalies tested resulted in production, and when added to the Donkey Creek results, the grand total for both areas is

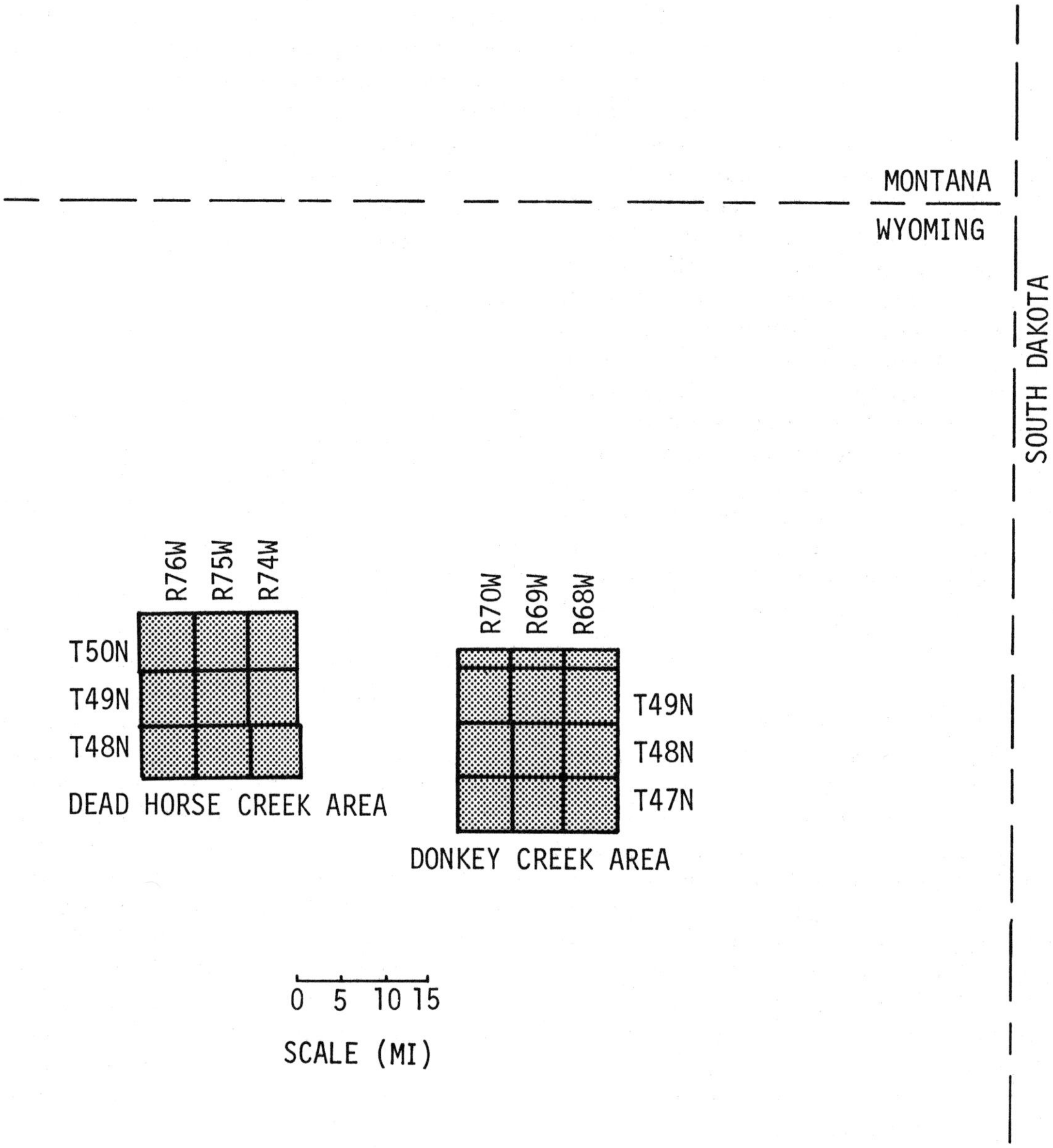

Fig. 2. Index map, Dead Horse Creek and Donkey Creek areas, Powder River Basin, Wyoming.

seven finds out of twelve prospects or about one success for 1.7 attempts. Dead Horse Creek field (anomaly 9) is a stratigraphic trap formed by pinchout of a local sand in the Parkman Member of the Mesaverde Formation along positive updip structures (Buttorff, 1958). Anomaly 5 apparently was caused by one of those updip structures. In light of this, testing is suggested along the southwest flank of anomalies in this region.

It may be concluded that, although many of the oil accumulations in this region do not show easily recognized drainage anomalies, a very significant proportion of the stronger anomalies are apparently associated with productive traps. The use of drainage analysis for petroleum reconnaissance in this region should increase the chances of success by allowing follow-up efforts to be concentrated in the most promising regions.

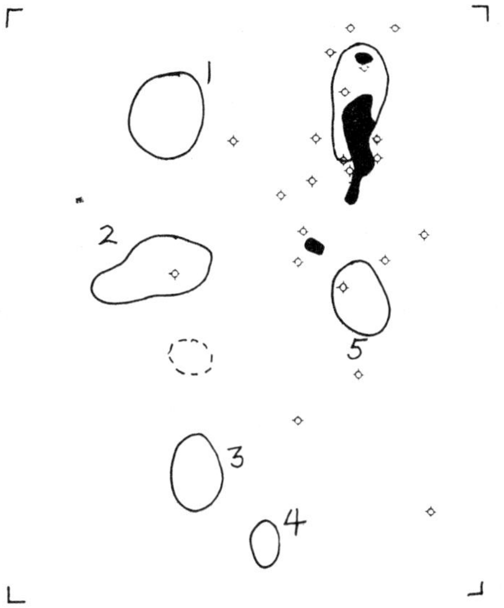

1958 (from Elliot, 1958).

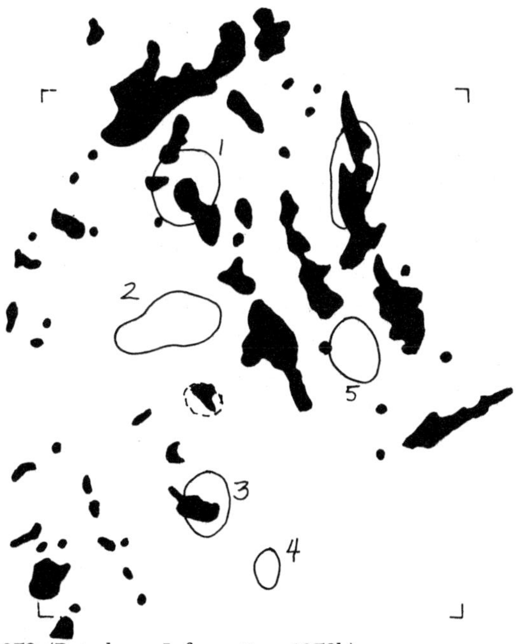

1978 (Petroleum Information, 1978b).

Fig. 3. Donkey Creek area oil development (scale: 1 in = 8 mi).

Advantages of Satellite Viewpoint

Henderson and Swann (1976) expressed a general conclusion of nearly all LANDSAT investigators when they pointed out that the greatest advantage of space-derived images for petroleum and mineral exploration is the greater synoptic perspective over that from the surface or even the aerial photographic viewpoint (see also Saunders and Hicks, 1979; Fischer, 1976; and others). Just as the aerial camera expanded the geologist's perspective for analyzing the character of the earth's crust, space imagery further enlarges this perspective.

Many investigators sponsored by NASA and other organizations have studied LANDSAT imagery and found that the distant satellite view permits the effective perception of many previously unseen linear* and circular† to subcircular arrangements of drainage and/or tonal features. Short (1974, 1974a) indicated that many of the linear features prove to be faults, joints, or fracture zones upon field examination and the closed (circular to subcircular) patterns represent such structures as ring dikes, calderas, intrusive plutons and stocks, salt domes, and astroblemes. Saunders et al. (1973) pointed out that in many cases the relationship of the linears to one another implied major geologic structural significance and suggested that many of the regional lineaments represented near-vertical basement faults of probable great age and lengths ranging to hundreds or even thousands of miles (see also Saunders and Hicks, 1979). Comparison of the locations of mineral deposits and petroleum accumulations with the linear elements, lineaments and curvilinear elements implied genetic relationships that can be used to predict new possible prospects (NASA, 1973; Short, 1974; 1974a; Short et al. 1976; Fischer and Lathram, 1973; Saunders et al., 1973; Saunders and Hicks, 1979; Collins et al., 1974)

These new linear elements, lineaments, and curvilinear elements can be interpreted using relatively standard photogeologic or geomorphic methods much as has been done for years in interpreting data from aerial photos or

*In this report, "linear elements" are those linear or gently curved alignments of topographic features or tones identified on LANDSAT imagery, and "lineaments" are those linear elements or groups of aligned linear elements which are interpreted to have geologic structural significance (see O'Leary et al., 1976 for a discussion of these terms).

†Circular or subcircular curved features are termed "curvilinear elements" or "curvilinear geomorphic anomalies."

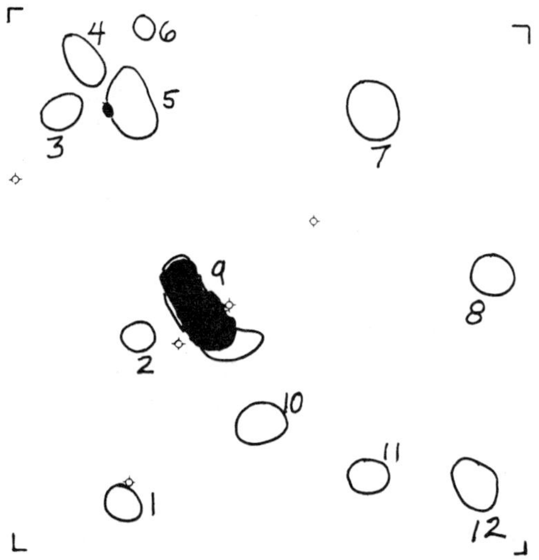

1958 (from Buttorff, 1958).

1978 (Petroleum Information, 1978b).

Fig. 4. Dead Horse Creek area oil development (scale: 1 in = 8 mi).

mosaics (Tator et al., 1960; Reeves et al., 1975). However, until the advent of LANDSAT, it was not economically feasible to gather wide imagery coverage in the form of aerial photos in several wavelength bands during different seasons of the year, which is desirable for mapping the large linear features with optimum detectability. An additional disadvantage in using aerial photos has been the very large number of photographs necessary for continental-scale coverage; the work is very tedious, and the large number of edges between photos can be misleading or can obscure the surface lineament indications needed for mapping. One LANDSAT image is equivalent in coverage to 448 1:48,000 scale aerial photos.

LANDSAT images already acquired provide the capability and opportunity of producing relatively good base maps for essentially 90% of the earth's land surface. This is especially important in some of the less-developed areas of the world, and the use of LANDSAT imagery is becoming widespread in cartographic and photogeologic mapping applications at scales of 1:250,000 or smaller.

General Interpretive Approach

Halbouty (1968) outlined for the four basic steps in petroleum exploration: (1) regional geologic reconnaissance and cursory appraisal of geomorphic features; (2) detailed geologic surface and subsurface mapping with aerial geophysical surveys; (3) ground geophysical exploration; and (4) drilling.

The interpretation of satellite imagery is most applicable in the regional reconnaissance stage of investigation, and can be helpful also in the detailed mapping stage as an interpretive aid in aerial geophysical surveys.

LANDSAT imagery can be used empirically by noting the characteristics of known oil or gas fields and searching for similar characteristics to indicate new prospective areas. This approach can be improved, however, by applying proved models to relate the observed data to geologic processes involved in the formation of traps or deposits.

Linear Elements and Lineaments: Geological Significance

For many years, geologists have interpreted linear features or lineaments detected visually or in geophysical data to represent the surface expression or effects of buried faults or fracture zones. Hodgson (1976) reviewed the early history of the development of lineament tectonics which had its start in Britain in the early 1800s (see also Saunders and Hicks, 1979). Hobbs (1911), in his description of the Atlantic border region, was among the earlier writers to use the term "lineament"; however, the study of linear

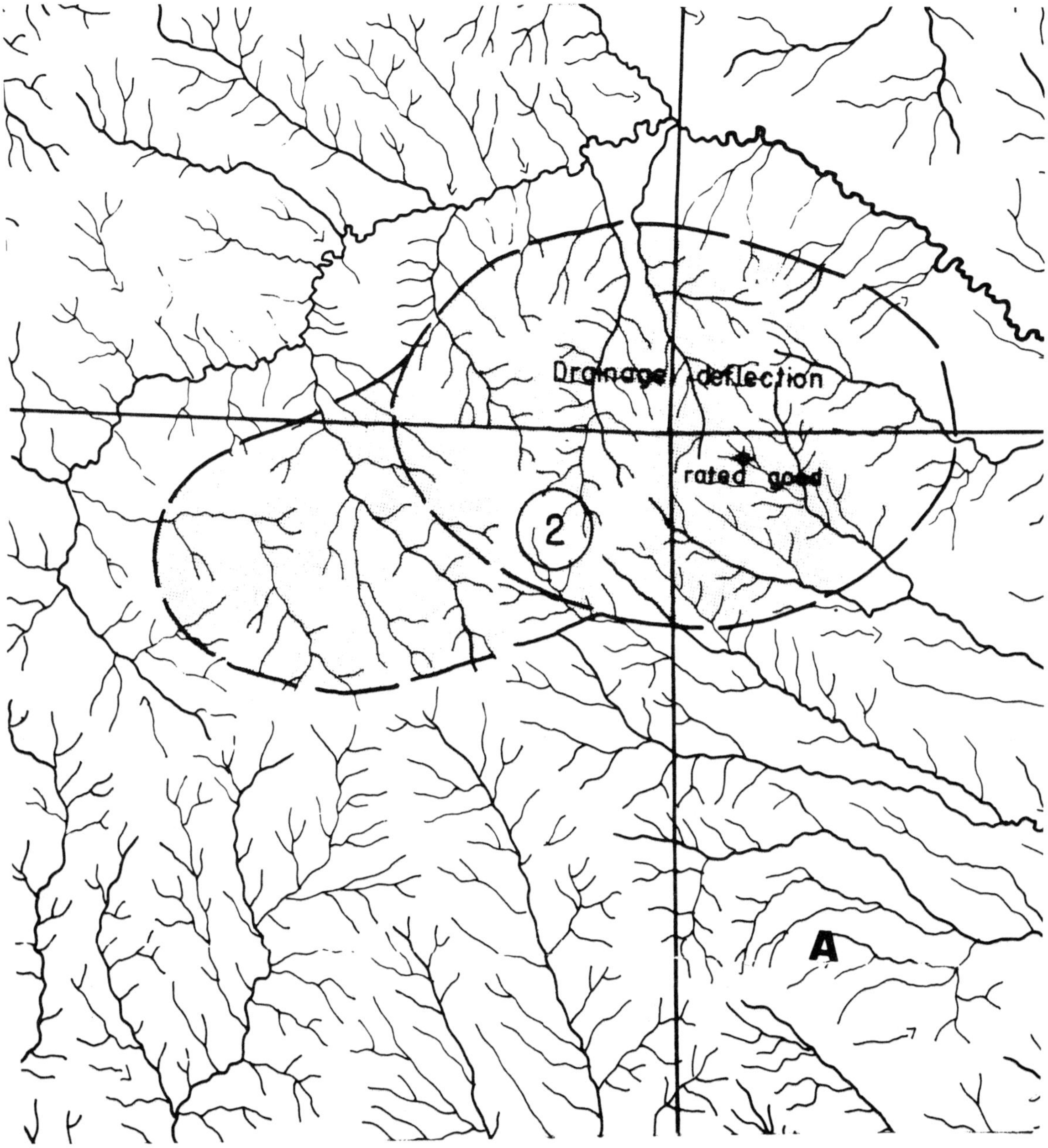

Fig. 5. Portion of Elliot's 1958 drainage map showing unannotated anomaly—centered on "A" (scale: 1 in = 1 mi).

features and their geologic significance was pursued by John Phillips as early as 1828 (Umgrove, 1947). Lattman (1958) described the methodology of mapping these features using aerial photography, and Tator et al. (1960) summarized their use in geologic photointerpretation (see also Reeves et al., 1975). O'Leary et al. (1976) discussed the evolution of the terms *lineament*, *linear*, and *lineation* and their use.

Since 1972, hundreds of investigators have studied LANDSAT imagery to determine its utility in many fields, and a large fraction of these have been concerned with geological or mineral exploration problems. Their findings have been summarized in many documents

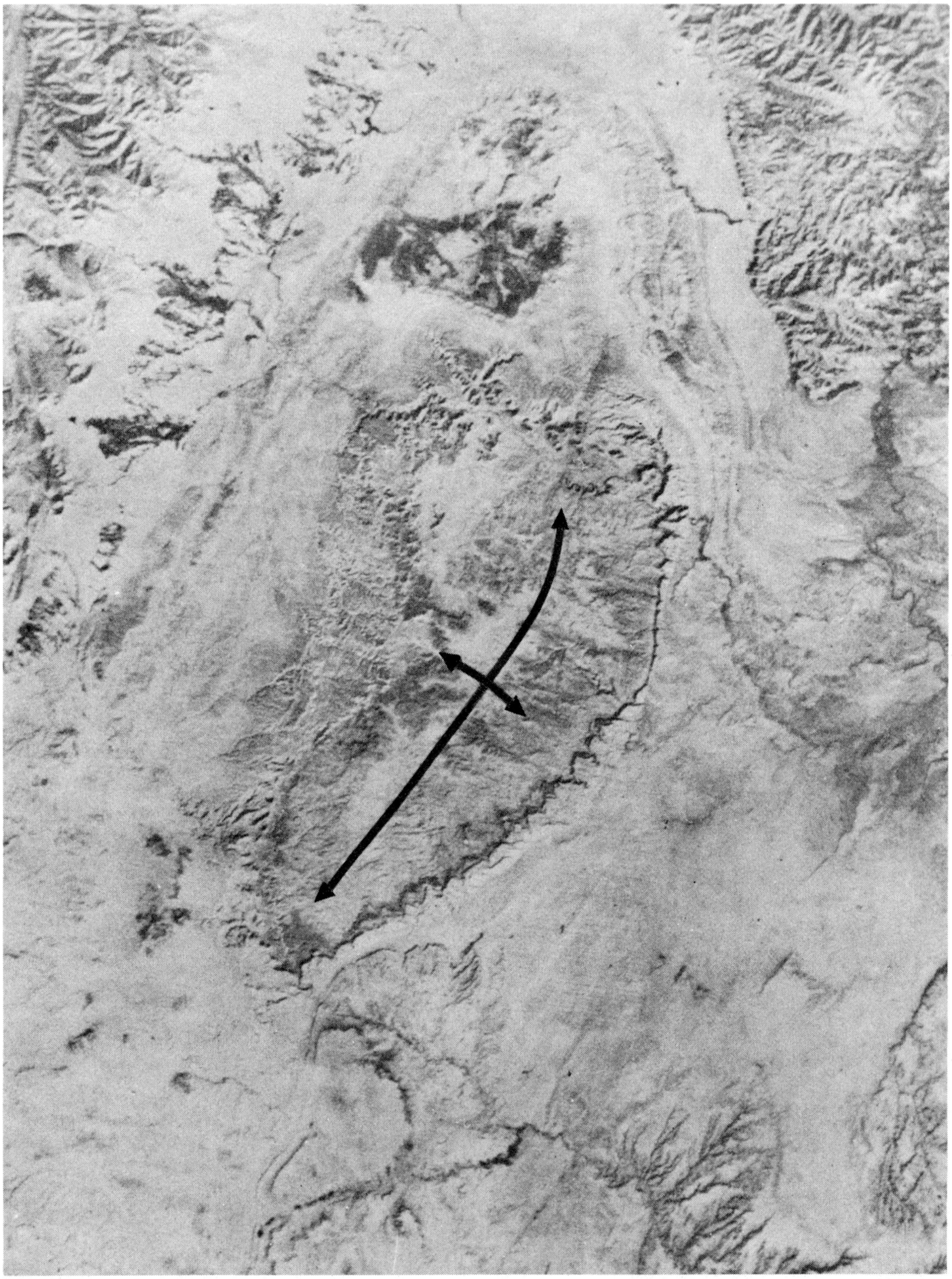

Fig. 6. San Rafael Swell, Utah. MSS band 5 (scale: 1 in = 10 mi).

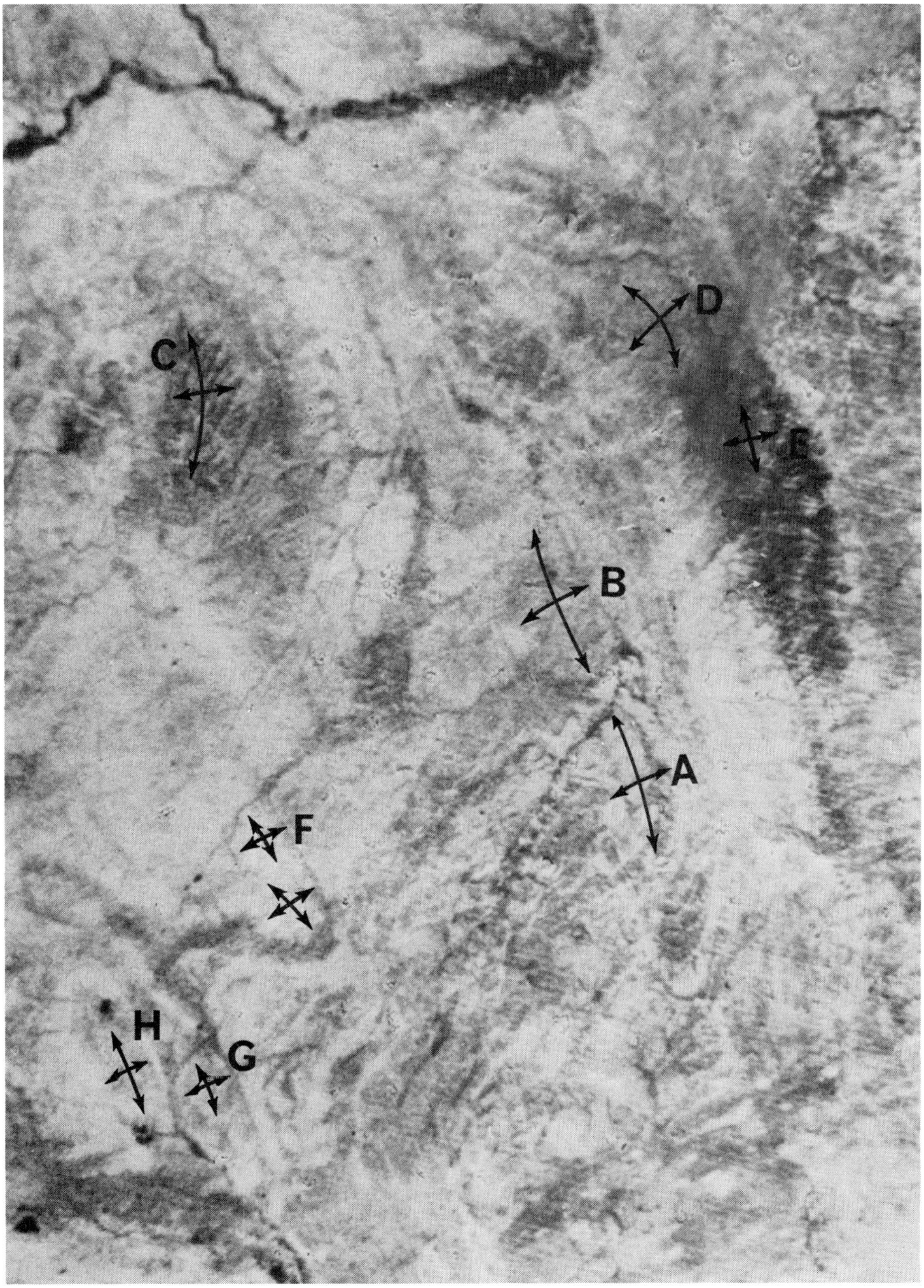

Fig. 7. Some typical anticlines along the Casper Arch. MSS band 5 (scale: 1 in = 6 mi).

(see Halbouty, 1976). The results of NASA-sponsored LANDSAT studies concerned with geology or minerals were presented by NASA (1973), Short (1974, 1974a), Short et al. (1976), and in individual reports generally available from NTIS (National Technical Information Service, Department of Commerce, Springfield, Virginia 22151). Williams and Carter (1976) summarized findings of U.S. Geological Survey studies of LANDSAT applications.

Reports specifically concerned with petroleum prospecting using linear element interpretations of LANDSAT imagery include: Saunders et al. (1973), Everett and Petzel (1974), Collins et al. (1974), Halbouty (1976), Lathram (1976), and Saunders and Hicks (1978).

Several studies sponsored by NASA and recent morphostructural mapping programs by Texas Instruments have revealed that the synoptic viewpoint of LANDSAT (ERTS) imagery can be used to map a system of regional lineaments of statewide or continental extent (Sanders and Hicks, 1979). This system is interpreted to represent the basement tectonic framework of North America. The lineaments are believed to be the surface expression of very old (i.e., Precambrian), deep-seated planes of weakness which divide the continental crustal plate into blocks. They are visible on the satellite imagery as linear stream drainage segments and/or tonal or topographic features which imply the presence of possible faulting in the subsurface.

It was concluded that these lineaments have been of primary importance in controlling tectonics, sedimentation, and mineralization. The preparation of regional lineament maps and their careful study should be of great assistance in (1) unraveling the complex plate tectonic history of Earth, and (2) providing valuable reconnaissance guidance in the search for energy and other mineral sources.

Saunders and Hicks (1979) summarized several possible relationships of lineaments to the origin of petroleum traps:

1. Lateral adjustments along basement faults can cause drag folding in overlying sediments.
2. Vertical adjustments can provide boundaries of basins and uplifts (likely places for both structural and stratigraphic traps), and can control sedimentation.
3. Vertical movements can stimulate massive reef growth along the basement line of weakness.
4. Increased rock fracturing in the sedimentary rock above the basement weakness zone (because of lateral reactivation of the basement feature in response to orogenic stress) can have a pronounced effect on rock porosity and permeability and on the ultimate recovery of hydrocarbons.

Curvilinear Anomalies on LANDSAT Imagery

Typical examples of the appearance of domal or anticlinal landforms on LANDSAT imagery are shown in Figures 6 and 7. The San Rafael Swell (Figure 6) is a good example of a breached anticline, and Figure 7 shows the following petroleum field structures along the Casper Arch:

A — Teapot Dome
B — Salt Creek
C — Tisdale
D — Meadow Creek
E — Sherwood
F — Castle Creek
G — Horse Ranch
H — N. Casper Creek

In the preparation of the Morphostructural Map Series, the curvilinear drainage and landform anomalies were annotated first, using 1:500,000 scale MSS band 7 imagery mosaics, which show the best contrast for dissection anomalies, and then the data were supplemented by further annotation on MSS band 5 mosaics, which show the best tonal contrasts associated with landforms, etc.

Geologic features which may show this type of anomaly include:

- Volcanic calderas
- Intrusives
- Domes
- Anticlines
- Synclines
- Playas

Tonal Anomalies

Geological Significance

Everett and Petzel (1974), in their investigation of the suitability of LANDSAT imagery for petroleum exploration, found that 59 out of 76 total geomorphic tonal and "hazy" anomalies in the Anadarko Basin correlated with producing oil and gas fields, while 11 correlated with known but nonproducing structures; the remaining six could not be correlated with known features. Of 35 "hazy" anomalies (characterized as areas of "smudged" or "partly erased" imagery detail which are not artifacts of the imagery processing), 33 correlated with producing or drilled structures. Examples presented by those authors on a color composite image appear to be generally lighter in tone than do their immediate surroundings. Examination of these "hazy" anomalies on black and white LANDSAT mosaics show them to be light tonal anomalies on both MSS bands 5 and 7. This information, coupled with the author's experience as outlined in the following paragraphs, provided the basis for analysis of tonal anomalies.

Saunders (1969) suggested that new developments in the remote detection of surface and near-surface materials might provide a possible new approach to "direct" oil-finding. Evidence that supports this contention includes (1) much better than average drilling success ratio experienced by several companies using near-surface geochemical techniques, and (2) certain aerial photographic tonal anomalies which resemble geochemical and fracture-density anomalies and correlate closely with oil- and gas-bearing structures. Davidson (1967) reported that new field wildcats drilled on prospects where geochemical prospecting played a vital role yielded from two to seven times as many producers as the national average. Tator et al. (1960) has summarized descriptions of certain types of photographic tonal anomalies which are associated with oil structures and cut across or at least do not coincide with soil or rock formational boundaries or outcrops. These include dark-toned anomalies over the Nottingham, Huntoon, and steelman oil fields in Saskatchewan, and light ones around the Buttonwillow gas field in the southern San Joaquin Valley of California, as well as others. No simple explanation for their origin has been generally accepted, although several authors have suggested that there may be a connection with escaping hydrocarbon gases.

It is suggested that more pathways for hydrocarbon gases to migrate to the surface could be created by increased fracturing around the edges of petroleum structures caused by structural stresses (or draping of sediments over reefs, etc.) rather than at a distance from the accumulation or even directly over it. The extra gas leakage could provide soil-gas anomalies and cause enhanced evaporation of near-surface groundwaters, producing caliche-like evaporite concentrations of soluble soil salts. These anomalies have been observed to appear as "edge-leakage halos" partly or completely surrounding the petroleum accumulation. Fracture-density studies based on photo linears often show similar halo-shaped anomalies. During a research study concerned with geochemical prospecting for petroleum (Saunders and Dunlap, 1962), the author found concentrations of soil salts associated with all six major producing areas studied in semiarid regions. Saunders (1969) summarized earlier observations of aerial photographic tonal anomalies of both light and dark tones and their occurrence around the edges or over producing areas. Donovan et al. (1974) described calcitization and bleaching of sandstones and other surface formations attributed to seeping hydrocarbon gases over the Cement and Davenport oil fields and suggested models to explain the observed effects (See also Donovan, this volume).

Inspection of the LANDSAT data for several map areas shows an apparent correlation between regions with a relative high density of oil fields and generally lighter tones in both the MSS bands 5 and 7 imagery. Some of the detailed anomalies show an apparent rough "edge-leakage halo" shape. Some apparently are related to land-use patterns, and inasmuch as land use reflects soil composition, at least to some extent, then perhaps these light-toned anomalies may be related to geochemical soil-salt patterns and consequently may provide a rough guide to the presence or absence of petroleum at depth.

On the basis of the previous studies and the observed LANDSAT data relationships to

TABLE 1

SUMMARY OF GEOMORPHIC ANOMALIES AND PETROLEUM DEVELOPMENTS

Description	3 and 4	14	26	25	37	Totals or Averages
I. Summary of all developments to 1979:						
Total Number Anomalies	591	699	292	303	279	2164
Number of Anomalies						
Tested to 1979	309	470	174	101	123	1177
Percent	52%	67%	60%	33%	44%	54%
Number of Anomalies with Production						
in 1979	108	306	110	50	63	637
Percent of Tests	35%	65%	63%	50%	51%	54%
Ratio to Tests	1/2.9	1/1.5	1/1.6	1/2.0	1/2.0	1/1.8
Percent of Fields with Anomalies	52%	75%	35%	68%	46%	56%
Number of Untested Anomalies	282	229	118	202	156	987
II. 1974-1975 Developments						
Number of Suggested Prospects (1974)	492	416	202	267	222	1599
Numbers Prospects Tested (1974-1979)	106					
Number of New Producers Found						
(1974-1979)	30	23	20	14	6	93
Percent of Tests	28%					
Ratio to Tests	1/3.5					(1/2.4)
Number Prospects Tested (1976-1979)	41	32	16	19	12	120
Number of New Producers Found						
(1976-1979)	9	8	8	6	2	33
Percent of Tests	22%	25%	50%	32%	17%	28%
Ratio to Tests	1/4.6	1/4.0	1/2.0	1/3.2	1/6.0	1/3.6

known production, geomorphic dissection anomalies have been interpreted to indicate potential oil-bearing structures; the light tonal anomalies are considered a surface "caliche-like" indication of possible hydrocarbon leakage. Therefore, combined anomalies should offer the most promise.

The author is well aware of the highly controversial nature of the whole subject of geochemical prospecting for petroleum and suggests that the reader examine the LANDSAT data for himself and form his own conclusions as to the significance of the apparent relationships of the light-toned anomalies. If future experience in exploration substantiates that these LANDSAT data light tonal anomalies are genetically related to the presence of subsurface hydrocarbons, the use of their new approach could be a major factor in discovering new stratigraphic oil and gas accumulations in the arid or semiarid onshore areas.

Light tonal anomalies may be annotated using the following procedure: The mosaic for MSS band 7 is superimposed in register on the band 5 mosaic on a bright light table, and the areas which are light-tones on both are roughly outlined.

AN EVALUATION OF LANDSAT GEOMORPHIC ANOMALIES

1974-1979 Exploration Results

Inspection of mapped LANDSAT data and known producing areas has revealed that many known fields are apparently associated with curvilinear dissection and/or tonal anomalies. A summary comparison of drilling results with LANDSAT geomorpic anomalies was prepared for a large portion of the petroleum-producing region of the Rocky Mountains, including the western portion of the Williston Basin. The six map sheets indicated in Figure 8 by the solid

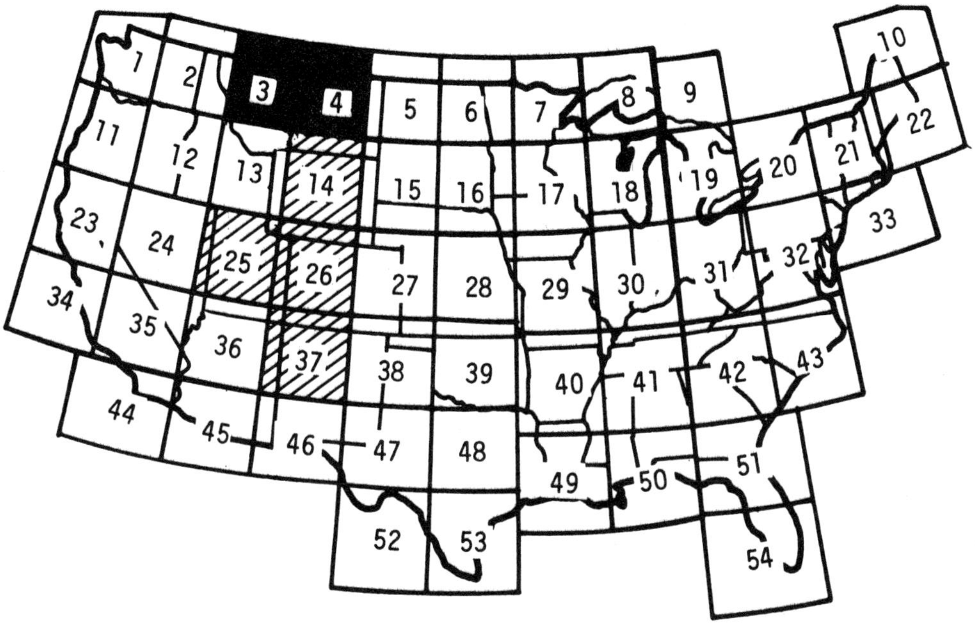

Fig. 8. Index map of area investigated.

black and shaded patterns were compared with new discovery and dry hole maps (Petroleum Information 1978a,b) to examine relationships between recent petroleum developments and geomorphic anomalies annotated in 1974. Table 1 summarizes the observed relationships.

A total of 2,164 geomorphic anomalies were annotated in 1974 on LANDSAT imagery over the area of the six map sheets. Of these, about half have been tested by drilling during the exploration history of the region, and over half of those tested have known petroleum production associated with them. From another viewpoint, over half of the total number of known fields present at the time of mapping (1974) had mappable geomorphic anomalies associated with them. One may conclude that these types of geomorphic anomalies which generally reflect the effects of uplift or doming could have provided a reconnaissance guide to more than half of all the fields discovered to date, and used as such would have provided one new field for every two anomalies investigated. It must be noted, however, that the geomorphic anomalies were annotated with knowledge of 1973 known field locations; and in spite of a major effort to be objective in the annotation, this could have influenced the data correlations to some extent. Also, the good correlations are not too surprising in light of the fact that much of the Rocky Mountain production is structure-controlled, and many fields were discovered on the basis of geologic and seismic studies which indicated anticlinal or domal structure.

The critical question is: How does it work under present exploration circumstances when the locations of newly discovered production were unknown to the annotator? Maps showing the dry hole situation at the time of the morphostructural mapping could be found only for the Williston Basin area (Map Sheets 3 and 4, shown in black on Figure 8). Comparison with the December, 1978, dry hole and production map of Montana allowed the identification of dry holes and new discoveries drilled in the 1974-1979 period on the previously mapped suggested prospects on sheets 3 and 4. A total of 106 prospects was tested and 30 new fields were found on those prospects for a "success ratio" of about 1 in 3.6 (Table 1). This is only slightly greater than the ratio of 1 anomaly with known production in 2.7 anomalies tested over the entire exploration history of the area. Assuming that the decrease in ratio from 1 in 2.7 to 1 in 3.6 with time for maps 3 and 4 is proportionately applicable to the entire Rocky Mountain area, then an estimated average ratio for 1974 to 1979 would be about 1 in 2.4 (shown in parentheses in Table 1).

Reference to Petroleum Information (1975 and 1976) new discovery maps showed that during 1974 and 1975 about 165 new discoveries were found in the test area including all six sheets, and 60 of these (i.e., about 1 in 3) were located on geomorphic anomalies listed as suggested prospects.

Developments during the 1976-1979 period are shown in the bottom section of Table 1 for all six map sheets based on a comparison of dry hole maps for April, 1976, and December, 1978. Out of 120 prospects tested during that period, 33 fields were found for a success rate of about 28% or 1 in 3.6.

Finally, it is concluded that the present use of LANDSAT geomorphic anomalies should provide reconnaissance guidance to about 1 out of 3 fields yet to be discovered (probably reflecting an increasing percentage of stratigraphic discoveries), and that between 1 in 3 and 1 in 4 geomorphic anomalies investigated can be expected to lead to new production. The overall industry new-field wildcat experience in 1974 was approximately 1 producer in 7 wells drilled (Wagner, 1975) and in 1977 it was about 1 producer in 6 wells drilled (Johnston, 1978). Although these ratios are not computed on exactly the same base, they may be compared roughly with our evaluation results to indicate that the widespread use of these geomorphic anomalies as a reconnaissance guide for more detailed geological and geophysical studies should substantially improve the chances of drilling success—perhaps by as much as a factor of two.

Comparison of Analyses at Different Scales

General

Finally, the 1958 detailed drainage analyses of the Donkey Creek and Dead Horse Creek areas at 1 inch = 1 mile scale are compared with the 1974 1 inch = 8 mile scale reconnaissance LANDSAT-based geomorphic analysis and a 1978 intermediate 1:250,000 scale (1 inch = 4 miles) analysis based on drainage annotated from the National Topographic Map Series (NTMS) Gillette sheet and from LANDSAT imagery at the same scale. Only drainage and landform anomalies were annotated in 1974 at 1:500,000 scale, and light tonal anomalies were also considered in the 1978 analysis of LANDSAT at 1:250,000 scale. Each of the later analyses is compared with the 1958 detailed analysis and success ratios are calculated for each individually and for combinations of reconnaissance and detailed anomalies.

1974 LANDSAT Analysis

Figure 9 shows the LANDSAT geomorphic anomalies (solid lines) annotated at 1:500,000 scale from MSS bands 5 and 7, black and white mosaics prepared by the U.S. Department of Agriculture, Soil Conservation Service (see Appendix). These are compared with the 1958 drainage anomalies (dashed lines). Those considered to be in agreement are shaded. Table 2 compares the success ratios based on the 1978 production shown in solid black.

TABLE 2
Comparison of Success Ratios
1958 and 1974 Analyses

	1958	1974	1958 and 1974
A. Anomalies on Production	7	12	5
B. Total Anomalies	12	20	6
A/B X 100	58%	60%	83%

1978 LANDSAT Analysis

Figure 10 shows the light tonal anomalies annotated for the Donkey Creek area (Figure

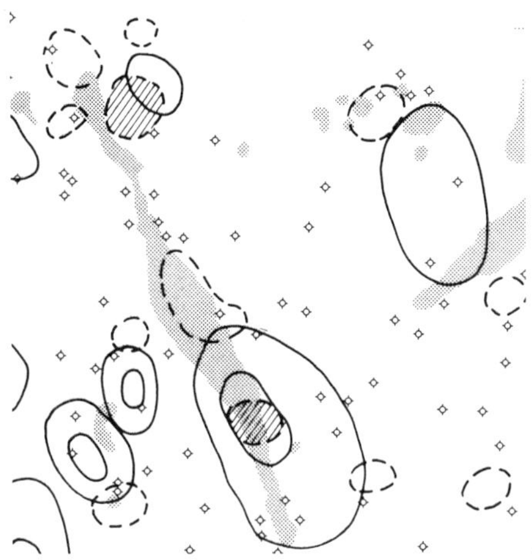

A. Dead Horse Creek area.

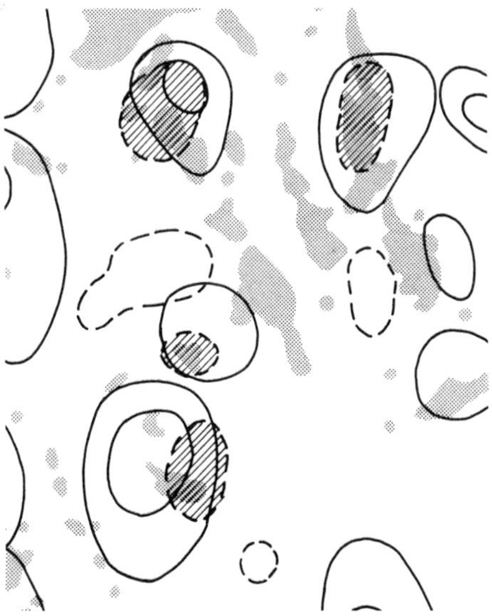

B. Donkey Creek area.

Fig. 9. 1974 LANDSAT data compared with present production (scale: 1 in = 8 mi).

10A) and the Dead Horse Creek Area (Figure 10B) by superimposing MSS Bands 5 and 7 imagery backlighted in register on a light table and outlining those areas appearing light on both. The anomalies shown were drawn assuming the light-toned areas represented edge-leakage patterns of soil salts. Ten out of the 22 anomalies appear to be associated with oil production as shown in Table 3, Case 1.

A. Donkey Creek area.

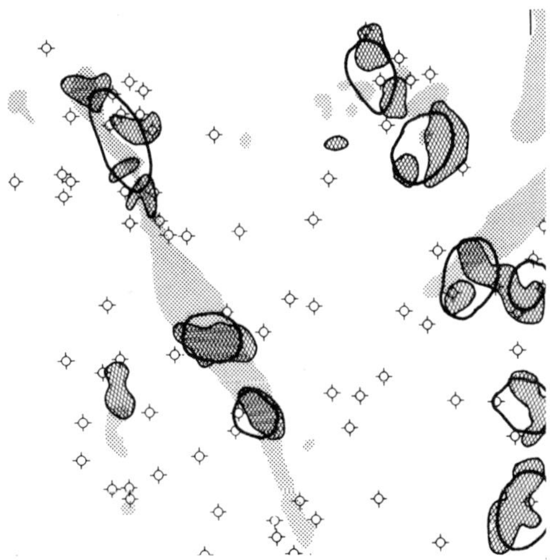

B. Dead Horse Creek area.

Fig. 10. 1978 LANDSAT data—tonal anomalies (scale: 1 in = 8 mi).

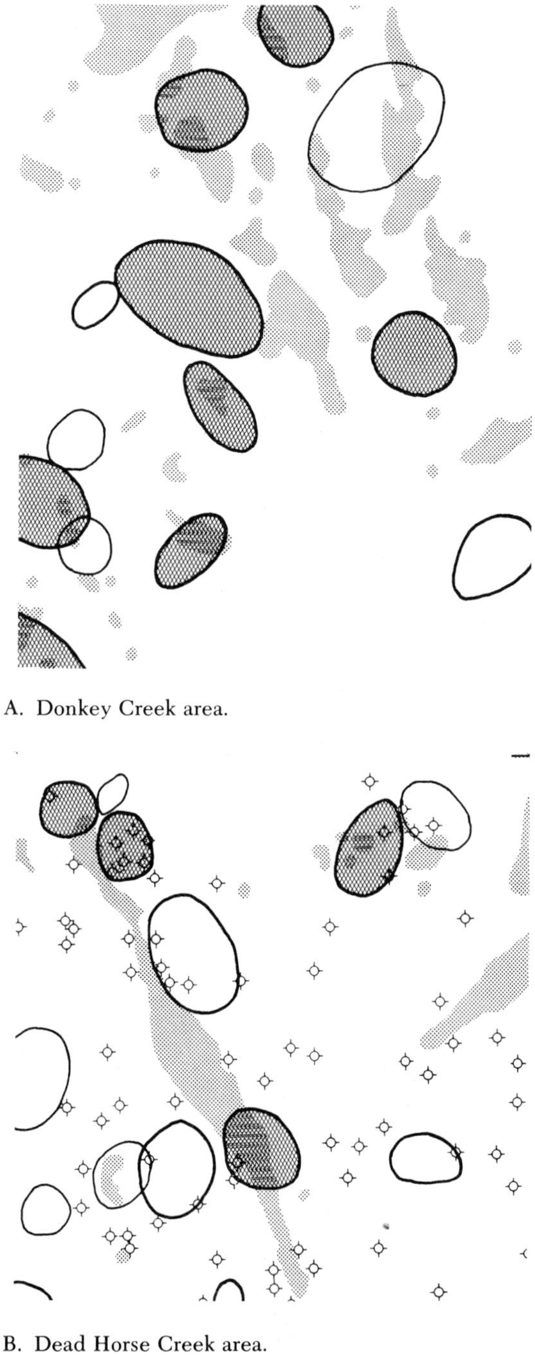

A. Donkey Creek area.

B. Dead Horse Creek area.

Fig. 11. 1978 LANDSAT data—geomorphic anomalies (scale: 1 in = 8 mi).

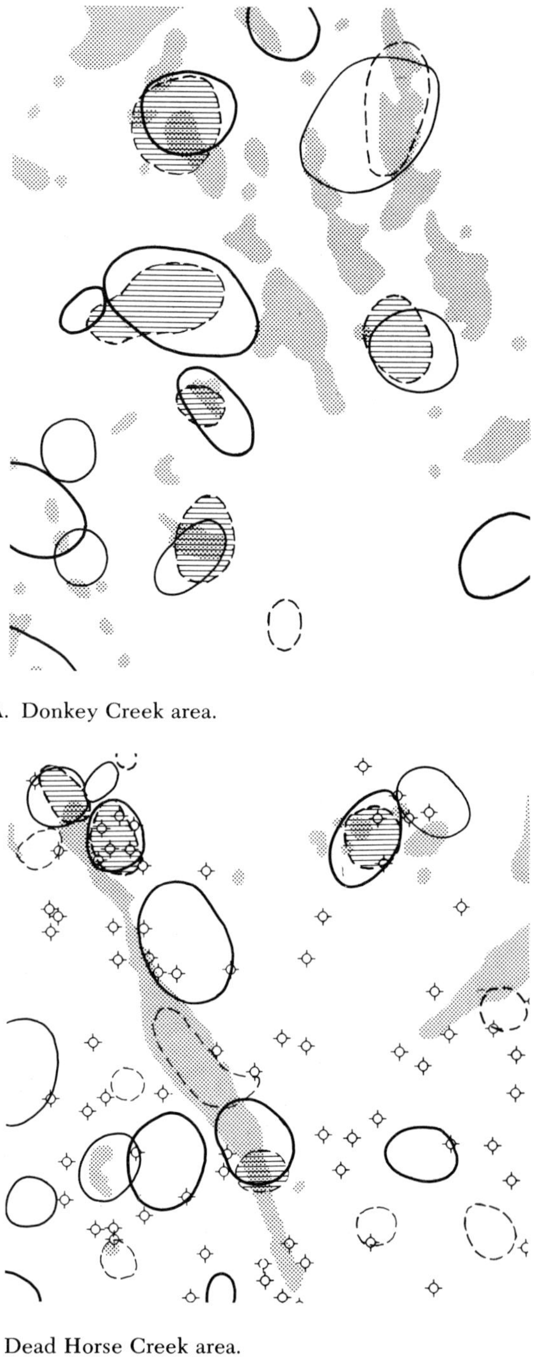

A. Donkey Creek area.

B. Dead Horse Creek area.

Fig. 12. 1978 LANDSAT data compared with 1958 drainage analysis (scale: 1 in = 8 mi).

Figure 11 shows the drainage anomalies annotated from the streams shown on the 1:250,000 scale NTMS topographic maps (heavy solid lines) and landform anomalies based on LANDSAT imagery (light lines). These were compared with the light tonal anomalies in Figure 10, and those drainage or landform anomalies also showing light tonal

TABLE 3

COMPARISON OF 1978 1:250,000 SCALE ANOMALIES WITH 1958 DRAINAGE ANALYSIS

Data	Cases 1	2	3	4	5	6	7	8
A	10	8	5	13	9	7	7	6
B	22	16	11	27	12	12	10	9
C	45%	50%	45%	48%	75%	58%	70%	67%

A = Anomalies Over Production
B = Total Anomalies
C = A/B × 100 (Success Ratio)

Case 1. Tonal Anomalies only
Case 2. Drainage Anomalies from Topographic Map at 1:250,000 scale
Case 3. Added Landform Anomalies from 1:250,000 scale LANDSAT
Case 4. Total of 2 plus 3
Case 5. Combined 1 and 4: Drainage and/or Landform Anomalies with Tonal Anomalies
Case 6. 1958 Drainage Anomalies from Aerial Photography
Case 7. Combined 4 and 6: LANDSAT Drainage and/or Landform Anomalies confirmed by Detailed Drainage Anomalies on Photos
Case 8. Combined 5 and 6: Drainage and/or Landform with Tonal Anomalies confirmed by Detailed Drainage Anomalies on Photos

anomalies are shaded. The results are summarized in Table 3 under cases 2 through 5.

The anomalies annotated at 1:250,000 scale in 1978 are compared in Figure 12 with those reported on aerial photos at 1:20,000 or 1:63,360 scale in 1958, and the results are summarized in Table 3 under cases 6, 7, and 8.

CONCLUSIONS

It is concluded that LANDSAT drainage *or* landform *or* tonal anomalies (from 1:250,000 scale images) *or* detailed drainage anomalies (from air photos) are all about 40-50% effective in the Powder River Basin when used alone. Anomalies confirmed by more than one approach (LANDSAT drainage and/or landform anomalies confirmed by LANDSAT tonal anomalies) or those which agree when annotated at widely different scales (LANDSAT anomalies at 1:500,000 scale confirmed by detailed air photo analysis) are significantly more effective than when each is used alone. Confirming the combined LANDSAT 1:250,000 scale geomorphic *and* tonal anomalies with detailed air photo analysis did not significantly change the success ratio for better or for worse.

Drainage and landform anomalies annotated from early (1971-1972) 1:500,000 scale LANDSAT mosaics with relatively low resolution of surface detail developed production on about one out of 3.6 anomalies tested in the 1976-1979 time period in the Rocky Mountains. This suggests that, even under these less favorable conditions, use of geomorphic analysis may significantly increase the effectiveness of petroleum exploration in that region.

It is suggested that the most cost-effective reconnaissance of this variety probably is the use of drainage, landform, and tonal analysis of LANDSAT bands 5 and 7, and false color composites at 1:250,000 scale with priority given to the combined geomorphic and tonal anomalies.

APPENDIX

LANDSAT CHARACTERISTICS AND PRODUCTS

Details on the LANDSAT series characteristics, operations, and available data products are provided in the "LANDSAT Data Users' Handbook" published by NASA and current issues of the "LANDSAT Data Users' Notes." Address inquires to:

User Services
EROS Data Center
U. S. Geological Survey
Sioux Falls, SD 57198

Information is available also in NASA (1976) and Reeves et al. (1975).

Three LANDSATS have been launched, and a fourth is in the planning stage. The first satellite in this program was launched in July, 1972, and designated ERTS-1. It was followed by a second (ERTS-2) in January, 1975, and the two were renamed LANDSAT 1 and 2 to emphasize their prime area of interest, the resources of land masses, and to differentiate them from a new planned series, SEASAT, to be focused on oceanographic problems. LANDSAT 3 was launched March 5, 1978, and LANDSAT 1 was retired from operation January 6, 1978, after five years in orbit, producing some 300,000 images of earth's surface. The next satellite in this series is LANDSAT D, which will become LANDSAT 4 after its launch tentatively planned for late 1981.

All LANDSATS are placed in slightly inclined polar orbits covering from about 85°N to 85°S latitude at an altitude of about 500 miles.

Imagery sensors on the LANDSATS include both return beam vidicon (RBV) television type cameras in three wavelength bands and multispectral scanners (MSS). The morphostructural analyses were based on multispectral scanner data.

The LANDSAT 1 and 2 multispectral scanner is a line-scanning device that uses an oscillating mirror to scan continuously perpendicular to the spacecraft orbit path covering a swath about 185 km wide. Radiation is sensed in each of four spectral bands from 0.5 to 1.1 micrometers (NASA 1976):

Wavelength (micrometers)	Band No.	Color	Remarks
0.5-0.6	4	green	
0.6-0.7	5	red	good for tonal anomalies
0.7-0.8	6	near infrared	
0.8-1.1	7	near infrared	good for dissection patterns

Experience in interpreting black and white images in these bands for geomorphic interpretations indicates that cultural items (roads, cities, etc.) are suppressed in bands 6 and 7, liquid water is emphasized (dark black tones), and the images are more "crisp" than in bands 4 and 5. Tonal variations due to vegetation or soil differences are emphasized in band 5 over bands 6 and 7. Band 4 has a "washed out" appearance and appears to be the least useful in this application. It is concluded generally that a combination of bands 5 and 7 is most cost-effective in interpreting geomorphic anomalies. False color composites of bands 4, 5, and 7 are helpful in that vegetation, which often is a clue to concealed fracture, appears red and is therefore emphasized over its appearance in the black and white images. A thermal infrared band was present in the MSS of LANDSAT 3 until its failure in July, 1978, and its value to these interpretations remains to be evaluated on future missions.

On LANDSAT 1 and 2 the return beam vidicon camera systems operate by shuttering three independent cameras simultaneously, each sensing a different spectral band in the range of 0.48 to 0.83 micrometers (NASA 1976):

Wavelength (micrometers)	Band No.	Color
0.47-0.59	1	Blue-green
0.57-0.68	2	Orange-red
0.66-0.82	3	Red-near infrared

The three cameras are aligned to view the same 185 km (100 nautical miles) square ground scene. On LANDSAT 3, two identical RBV cameras are employed to provide increased ground resolution for mapping. Each has a broad band spectral response (yellow to near infrared) of 505 to 750 nanometers. The two provide side-by-side pictures, each approximately 98 km (53 nmi) on side covering a total width of about 183 km (99 nmi).

Henderson and Swann (1976) provided suggestions for a satellite sensor system specifically dedicated to geological remote sensing. Improvements over the current LANDSAT series would include: stereo coverage, higher resolution sensor spectral bands "tuned" to geological sensing, more complete high quality color photography, etc. It is hoped that these suggestions will be incorporated in future LANDSAT missions.

LANDSAT data in worldwide coverage are currently available (late 1978) to the public through the EROS Data Center,* in the following formats:

1. *Computer-compatible tapes,* geometrically corrected or uncorrected, in band-sequential (SQ) or band interleaved by line (BIL) formats for MSS data, and subscene sequential (SSQ) format for LANDSAT 3 RBV data.
2. *Black and white positive or negative film transparencies,* at scales ranging from 1:369,000 to 1:250,000 for MSS data, 1:684,000 to 1:250,000 for LANDSAT 3 RBV data.
3. *Black and white and false color composite paper prints of MMS or RBV data.*

Black and white mosaics of the United States, including Alaska, are available for MMS bands 5 and 7 in several scales from 1:500,000 to 1:5,000,000 from:

Cartographic Division
Soil Conservation Service
Federal Center Building
Hayattsville, MD 20782

Mosaics of selected areas, including some color and false color composites, are available from:

U.S. Geological Survey
Branch of Distribution
1200 Eads Street
Arlington, VA 22202

and from:

General Electric Company
Photographic Engineering Laboratory
5030 Herzel Place
Beltsville, MD 20705

*Address inquiries to User Services, EROS Data Center, Sioux Falls, South Dakota 57198 (Phone: 605-594-6511, Ext. 151).

REFERENCES

Buttorff, C. L., 1958, Geomorphic anomalies, Dead Horse Creek area, Wyoming: Wyoming Geological Association Annual Field Conference, 13th, Casper, Wyoming, Guidebook, p. 215.

Collings, R. J., McCown, F. P., Stonis, L. P., Petzel, G. P., and Everett, J. R., 1974, An evaluation of the suitability of ERTS data for the purposes of petroleum exploration, Final Report, NASA Contract No. NAS 5-21735, Eason Oil Company, Oklahoma City, Oklahoma, Goddard Space Flight Center, Greenbelt, Maryland, 149 p.

Davidson, M. J., 1967, Petroleum geochemical exploration from space—a radical new technology: NASA Manned Spacecraft Center, Houston, 78 p.

Doeringsfeld, W. W., Jr., and Ivey, J. B., 1964, Use of Photogeology and geomorphic criteria to locate subsurface structure: Mountain Geologist, v. 1, no. 4, p. 183-195.

Donovan, T. J., Friedman, I., and Gleason, J. D., 1974, Recognition of petroleum-bearing traps by unusual isotopic compositions of carbonate-cemented surface rocks: Geology, July, p. 351-354.

Elliot, D. H., 1958, Drainage analysis, Donkey Creek area; Powder River Basin, Wyoming: Wyoming Geological Association Annual Field Conference, 13th, Casper, Wyoming, Guidebook, p. 214.

Everett, J. R., and Petzel, G., 1974, An evaluation of the suitability of ERTS data for the purposes of petroleum exploration, *in* Freeden, S. C., et al., eds., Third earth resources technology satellite symposium: NASA SP-356, Goddard Space Flight Center, Greenbelt, Maryland, v. 2, p. 50-61.

Fischer, W. A., 1976, Applications to geology and geophysics, *in* Williams, R. S., Jr., and Carter, W. D., eds., ERTS-1, A new window on our planet: U. S. Geological Survey Professional Paper 929, p. 48-49.

———, and Lathram, E. H., 1973, Concealed structures in Arctic Alaska identified on ERTS-1 imagery: Oil and Gas Journal, May 28, p. 97-102.

Halbouty, M. T., 1968, Giant oil and gas fields in the United States: American Association of Petroleum Geologists Bulletin, v. 52, p. 1115.

———, 1976, Application of LANDSAT imagery to petroleum and mineral exploration: American Association of Petroleum Geologists Bulletin, v. 60, no. 5, p. 745-793.

Henderson, F. B., and Swann, G. A., 1976, Geological remote sensing from space: The Geosat Committee, Inc., 690 Market Street, Suite 1400, San Francisco, California.

Hobbs, W. H., 1911, Repeating patterns in the relief and in the structure of the land: Geological Society of America Bulletin, v. 22, p. 123-176.

Hodgson, R. A., 1976, Review of significant early studies in lineament tectonics, *in* Hodgson, R. A., Gray, S. P., Jr., and Benjamins, J. Y., eds., First international conference on the new basement tectonics: Utah Geological Association, Salt Lake City, Proceedings, no. 5, p. 1-10.

Howard, A. D., 1967, Drainage analysis in geologic interpretation, a summation: American Association of Petroleum Geologists Bulletin, v. 51, no. 11, p. 2246-2259.

Johnson, R. R., 1978, North American drilling activity in 1977: American Association of Petroleum Geologists Bulletin, v. 62, no. 8, p. 1275.

Lathram, E. H., 1976, Clues to geologic structure possibly indicating oil and gas sources, *in* Williams, R. S., Jr., and Carter, W. D., eds., ERTS-1, A new window on our planet: U. S. Geological Society Professional Paper 929, p. 99-101.

Lattman, L. H., 1958, Technique of mapping geologic fracture traces and lineaments on aerial photographs: Photogrammetric Engineering, v. 24, no. 4, p. 568-576.

———, and Matzke, R. H., 1961, Geological significance of fracture traces: Photogrammetric Engineering, v. 27, no. 3, p. 435-438.

Melton, F., 1959, Aerial photographs and structural geomorphology: Journal of Geology, v. 67, no. 4, p. 351-370.

NASA, 1973, Symposium on significant results obtained from the Earth Resources Technology Satellite-1: NASA, Goddard Space Flight Center, Greenbelt, Maryland, Report SP-327, v. 1, 2, 3.

———, 1976, LANDSAT Data Users' Handbook, Doc. No. 76SDS4258, National Aeronautics and Space Administration, Goddard Space Flight Center, Greenbelt, Md.

O'Leary, D. W., Friedman, J. D., and Pohn, H. A., 1976, Lineament, linear, lineation: some proposed new standards for old terms: Geological Society of America Bulletin, v. 87, p. 1463-1469.

Petroleum Information, 1975, 1974 oil and gas discoveries map of the Rocky Mountain region: Petroleum Information Corp., Denver, scale: 1 in. = 26 mi.

———, 1976, 1975 oil and gas discoveries map of the Rocky Mountain region: Petroleum Information Corp., Denver, scale: 1 in. = 26 mi.

———, 1978a, State dry hole maps for Montana, Colorado, Utah, and New Mexico: Petroleum Information Corp., Denver (posted to 4 December, 1978), scale 1:500,000.

———, 1978b, State dry hole map for Wyoming: Petroleum Information Corp., Denver (posted to 4 December, 1978), scale 1:500,000.

Reeves, R. G., Anson, A., and Landen, D., 1975, Manual of remote sensing, *in* Janza, F. J., Blue, H. M., and Johnston, J. E., eds., Theory, Instruments and Techniques, v. 1; and Bowden, L. W., and Pruitt, E. L., eds., Interpretation and applications, v. 2: American Society of Photogrammetry, Falls Church, Virginia, 2144 p.

Saunders, D. F., and Dunlap, R. C., Jr., 1962, Geochemical reconnaissance in petroleum exploration: Texas Instruments Inc., Dallas, 30 p. (out of print).

Saunders, D. F., 1969, Airborne sensing as an oil reconnaissance tool, *in* Heroy, W. B., ed., Unconventional methods in exploration for petroleum and natural gas: Institute for the Study of Earth and Man, Southern Methodist University, Dallas, p. 105-125.

———, Thomas, G. E., Kinsman, F. E., and Beatty, D. F., 1973, ERTS imagery use in reconnaissance prospecting: NASA Contract NAS5-21796, Texas Instruments Inc., Dallas, Final Report.

———, and Hicks, D. E., 1976, Regional lineament map of the United States: Texas Instruments Inc., Dallas, scale: 1:2,500,000.

———, and Hicks, D. E., 1979, Regional Geomorphic lineaments on satellite imagery—their origin and application, *in* International Conference on the New Basement Tectonics, 2d, Basement Tectonics Committee, Inc., Denver, Proceedings, 80201, p. 326-352.

Short, N. M., 1974, ERTS-1 investigator status review—geology, *in* Smith, O. G., and Granger, H., eds., Earth resources program results and projected applications: ERTS-1 Applications Investigations, NASA, Lyndon B. Johnson Space Flight Center, Houston, p. 4-2 to p. 4-14.

———, 1974a, Mineral resources, geological structure and landform surveys, *in* Freden, S. C., and Mercanti, E. P., eds., Third earth resources technology satellite symposium: Discipline Summary Reports, NASA SP-357, Goddard Space Flight Center, Greenbelt, Maryland, v. 3, p. 33-51.

———, Smith, A., and Baker, R., 1976, Excerpts from selected LANDSAT-1 final reports in geology: Goddard Space Flight Center, Greenbelt, Maryland, Document X-923-76-74, 72 p.

Smith, H. T. U., 1943, Aerial photographs and their applications: New York, D. Appleton-Century Co., 372 p.

Strandberg, C. H., 1967, Aerial discovery manual: New York, John Wiley & Sons, p. 75-156.

Tator, B. A., et al., 1960, Photointerpretation in geology; Manual of photographic interpretation: American Society of Photogrammetry, Washington, D.C., p. 169-342.

Thornbury, W. D., 1969, Principles of geomorphology: New York, John Wiley & Sons, 594 p.

Umgrove, J. H. F., 1947, The pulse of the earth: The Hague, Martinus Nijhoff, p. 358.

Von Bandat, H. F., 1962, Aerogeology: Houston, Gulf Publishing Company, p. 39-55.

Wagner, F. J., 1975, North American drilling activity in 1974: American Association of Petroleum Geologists Bulletin, v. 59, no. 8, p. 1275.

Williams, R. S., Jr., and Carter, W. D., 1976, ERTS-1, A new window on our planet: U. S. Geological Survey Professional Paper 929, 362 p.

Zernitz, E. R., 1932, Drainage patterns and their significance: Journal of Geology, v. 40, no. 6, p. 498-521.

HYDROCARBON GEOCHEMICAL PROSPECTING AFTER FORTY YEARS

Leo Horvitz

HORVITZ RESEARCH LABORATORIES, INC.

ABSTRACT

A significant level of acceptance has finally been attained in the United States by the geochemical exploration method that is based on the determination of light, saturated hydrocarbons in the near-surface. Increased attention was earned after the method was applied to offshore areas where surveys could be evaluated soon after completion. When many hydrocarbon anomalies were confirmed by the drill, explorationists accepted more readily the basic requirement of the method: that hydrocarbons migrate vertically from petroleum accumulations and produce recognizable near-surface distribution patterns.

Added evidence of the validity of geochemical exploration came from carbon isotope data. Ratios of C^{13} to C^{12} of near-surface methane, associated with hydrocarbon geochemical anomalies, were found to be almost the same as those of methane from reservoir gas.

The same evidence that substantiates the geochemical method which measures near-surface hydrocarbons directly also tends to validate other prospecting methods. Among these are a method that is based on fluorescence measurements, remote sensing technques, and those geochemical methods that measure inorganic substances and alteration products which are present in the near-surface as a result of hydrocarbon migration.

INTRODUCTION

In 1933, Laubmeyer reported the results of an experiment he had conducted in Germany in 1929 that turned out to be the beginning of what we now know as Geochemical Prospecting for Petroleum. Laubmeyer found that near-surface soil air, collected over gas fields, contains more methane than does the soil air over barren areas. Russian workers, under the direction of Sokolov (1935), improved Laubmeyer's sampling technique and found heavier hydrocarbons, as well as methane, in near-surface soil air.

The first statement that research in this field had been undertaken in the United States was made by Rosaire in 1938. I was privileged to have been involved in this early work. From the outset, we concentrated on using the soil itself instead of the soil air as the sampling medium, because sampling of the latter is limited to arid and sandy areas and is not applicable in areas of clay soil or in water covered areas. Furthermore, soils and sediments were expected to contain larger quantities of hydrocarbons in the adsorbed state than were reported for soil air. Now, the Russians too use the adsorbed hydrocarbon technique. And we in turn have come to measure the interstitial, or free hydrocarbons, where applicable, as well as the adsorbed components, and have found that the two sets of data often complement each other. In applying hydrocarbon geochemical prospecting, it is essential that all hydrocarbons, whether they are in the free state, loosely bound to sand particles, or tightly adsorbed on clay particles, are measured and used in evaluating areas for their petroleum-bearing potential.

Fifty years have elapsed since Laubmeyer's original experiments were conducted, and more than forty years of work in the field of geochemical exploration have been completed in our own country. The method gained some acceptance in the United States during its first thirty years, but in recent years its status has greatly improved. The purpose of this paper is to submit data accumulated during the last ten years which is responsible, to a large degree, for this change. The main contribution of these data is that they validate the basic requirement of all geochemical prospecting methods: that hydrocarbons move from subsurface petroleum deposits to the near-surface.

CARBON ISOTOPE DATA

Some of the most important evidence that substantiates vertical migration of hydrocarbons from subsurface reservoirs has come from carbon isotope studies. Our first experience with this type of data demonstrates how it can be used to validate a near-surface hydrocarbon anomaly.

In 1970-1971, a survey was conducted in Jackson County, Texas, which includes the collection of hydrocarbon data over the Francitas

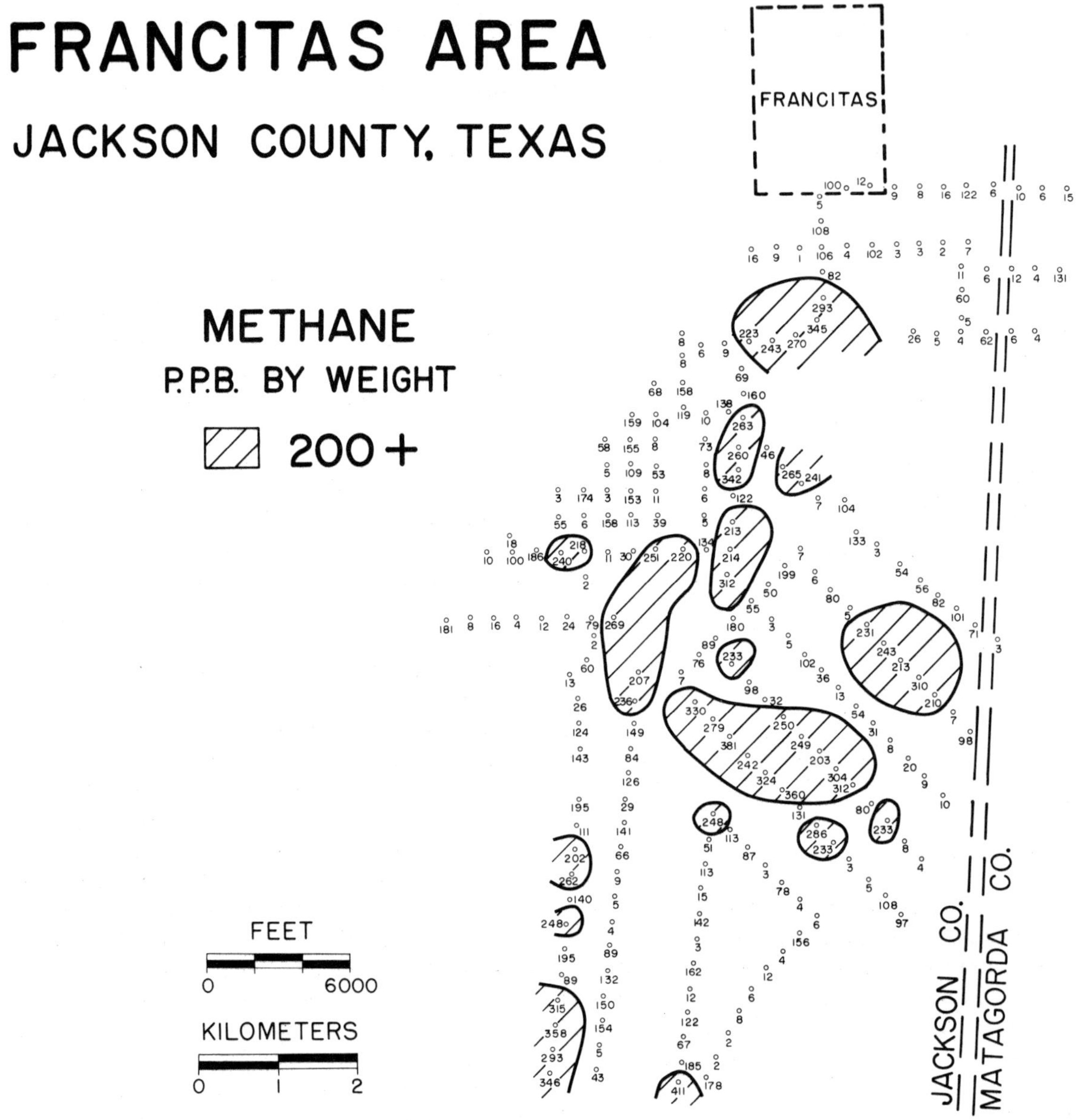

Fig. 1. Hydrocarbon survey conducted in Francitas area, Jackson County, Texas. Adsorbed methane values, obtained from 12-ft samples, are expressed in parts per billion by weight on dry sample basis and appear on map near station locations. Shaded areas contain anomalous values of 200 and above.

Oil Field. Figure 1 shows the methane data for samples of soil that were collected in this area at depths of 12 ft from 229 locations and analyzed for the light, saturated hydrocarbons ranging from methane through pentane. The details of the extraction and analytical techniques used have previously been presented (Horvitz, 1969, 1972), but, in brief, the hydrocarbons were removed by heating the fine-grained portions of the samples in a partial vacuum with phosphoric acid at 100°C. The extracted gases were injected into a hydrogen flame chromatograph, which separated them into their individual components. The methane values which range from 1 to 411 parts per billion by weight were calculated on the dry sample basis. The contoured areas include those samples containing 200 or more parts per bil-

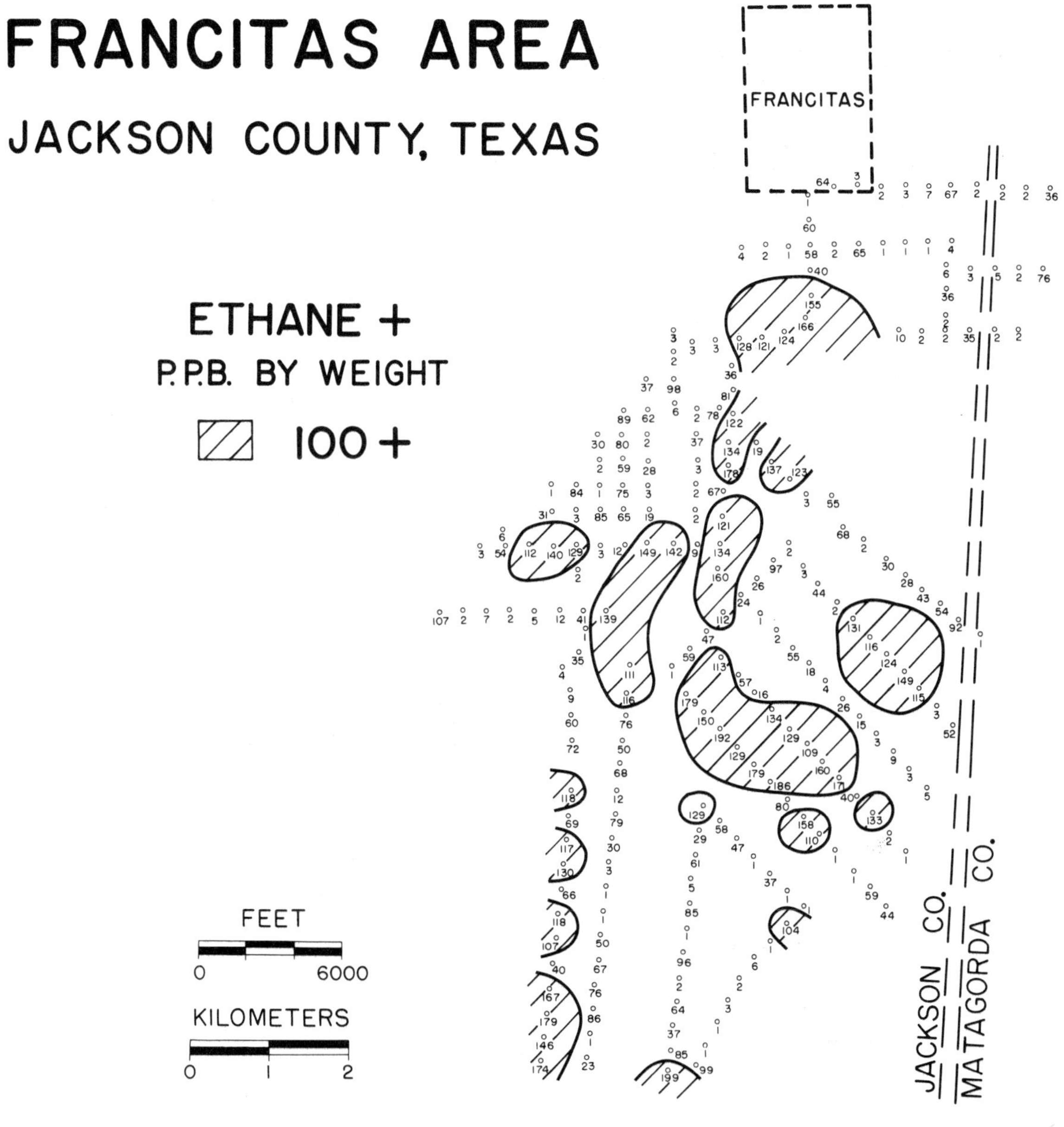

Fig. 2. Adsorbed ethane and heavier (C_2-C_5) saturated hydrocarbons produced a near-surface pattern that is very similar to that of the methane. Same source is suggested for both hydrocarbon fractions.

lion by weight of methane. The ratio of the anomalous values to the background is 4.8, indicating an anomaly of high contrast. An area of low concentration, almost entirely surrounded by high values, suggests an anomaly of the halo type.

Of the light hydrocarbons in natural gas, methane is present in largest quantities. In the near-surface, however, methane can be present as a result of bacterial action on organic matter. Therefore, when conducting a hydrocarbon survey, ethane and heavier hydrocarbons must also be measured. Data for this fraction (C_2 - C_5) are shown in Figure 2. The values range from one to 199 parts per billion by weight and the contoured areas include concentrations of 100 parts per billion and higher. The ratio of the anomalous to background values is 4.9, and

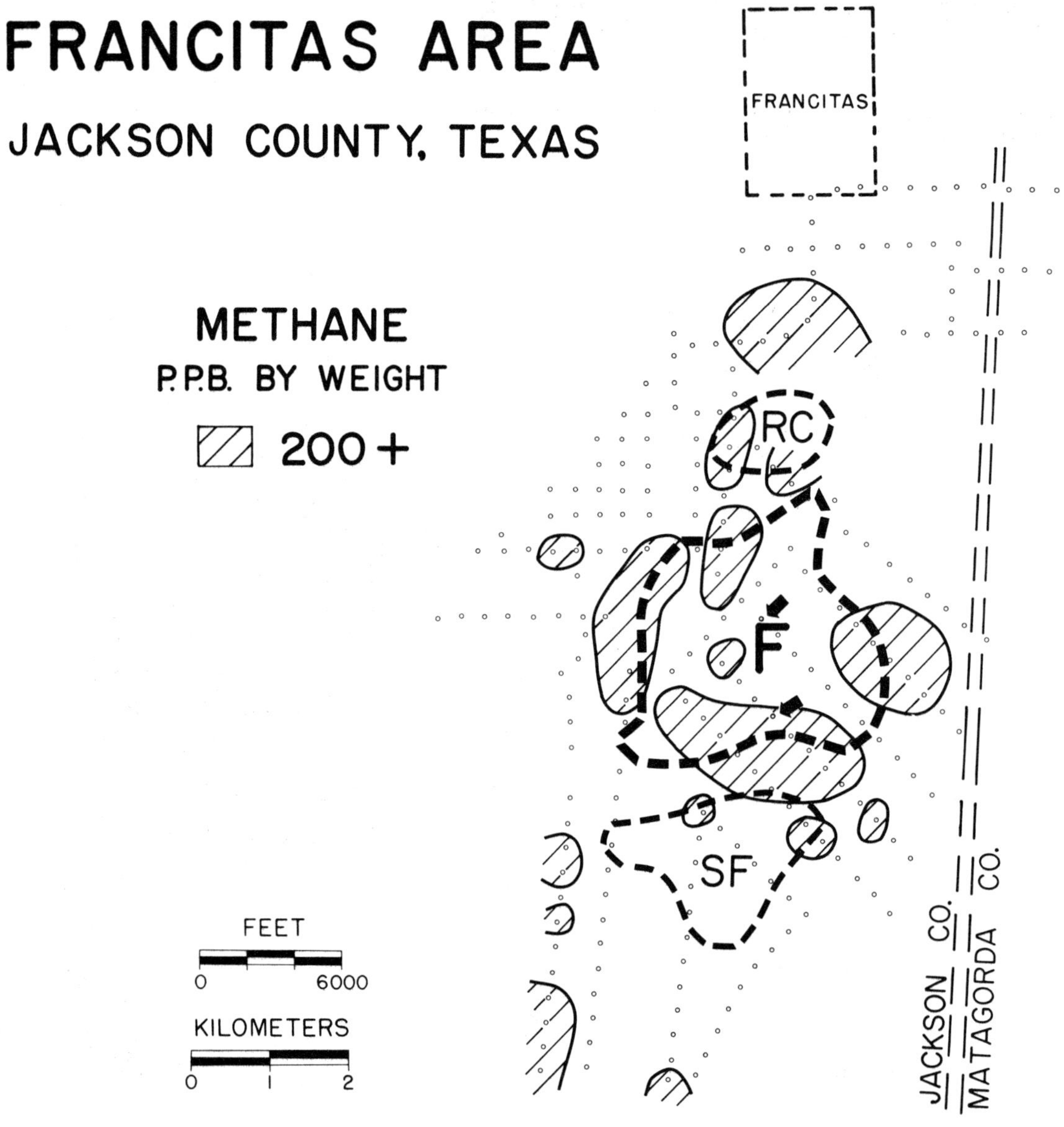

Fig. 3. Adsorbed methane anomaly, of halo type, bears close relationship to outline of Francitas (F) Oil Field while unimportant South Francitas (SF) gas field produced few near-surface anomalous values. Small Reed Creek (RC) gas field, discovered after survey was completed, was indicated by data. Arrows point to locations of samples, used in carbon isotope measurements, which showed C^{13} to C^{12} ratios that are nearly the same as those of methane in reservoir gas of Francitas Field.

a high contrast anomaly, which resembles closely that of the methane, is apparent.

Figure 3 shows the methane map (Figure 1) to which outlines of producing areas have been added. Obviously, a very close relationship exists between the near-surface methane distribution pattern and the outline of the Francitas Oil Field (F). The high near-surface methane concentrations appear to overlie the oil-water contact of the field which has produced more than 8,000,000 barrels of oil and about 100 billion cu ft of gas. The unimportant South Francitas (SF) gas field produced only a few anomalous values. Reed Creek (RC), a small gas field, to the north of the Francitas field, was found in 1975 after the survey was completed.

Although a mass of empirical geochemical data, collected from many areas, suggests that near-surface hydrocarbons originate in subsurface deposits, corroboration of the phenomenon of vertical migration by additional evidence has been needed to provide more confidence in geochemical exploration techniques. Carbon isotope studies have provided some of this evidence. It is now possible to determine from isotope data whether different hydrocarbon samples come from the same source or from different sources, if their ratios of C^{13} to C^{12} are known. Briefly, the technique involves the removal, from a hydrocarbon whose carbon ratios are desired, of any carbon dioxide initially present, and then the conversion of the resulting purified hydrocarbon to carbon dioxide. Carbon isotope ratios of this carbon dioxide are determined in a sensitive mass spectrometer and the results are expressed as d-values, or deviations, in parts per thousand, of the C^{13}/C^{12} ratio of a sample from that of a standard. A standard which is often used is belemnite, a carbonate from the Peedee formation of South Carolina, and it is known as the PDB standard.

Figure 3 includes the locations (indicated by arrows) of two soil samples of the Francitas survey which used for C^{13}/C^{12} determinations. The sample located near the south side of the Francitas field (F) yielded a $d^{13}C°/oo$ value of -44.0 for the adsorbed methane. The other sample was low in adsorbed methane but high in interstitial methane. The $d^{13}C°/oo$ value for this interstitial, or free, methane was found to be -40.8. Samples of the reservoir gas, taken from the Francitas field, yielded $d^{13}C°/oo$ values for methane which ranged from -41.0 to -43.8, practically the same as those of the near-surface methane. On the other hand, $d^{13}C°/oo$ values for methane from biologic sources range from -50 to -80. Thus, the source of the anomalous near-surface methane, as well as the ethane and heavier hydrocarbons, is undoubtedly reservoir gas.

We are indebted to Amoco Production Company for the carbon isotope ratios of the near-surface and reservoir methane samples.

OFFSHORE GEOCHEMICAL EXPLORATION

Carbon isotope data strengthened the case for hydrocarbon geochemical exploration, but the method was also enhanced considerably by its application of offshore areas around the world. Among others, the areas included the Gulf of Mexico, the East Coast of the United States, the North Sea, and the Sulu Sea. During the last decade, most of the work was done in the Gulf of Mexico, where many hydrocarbon anomalies were encountered prior to drilling. The resulting confirmation of more than 50% of these anomalies provided evidence of the usefulness of the technique in distinguishing potentially productive structures from those that are dry. The value of geochemical data in the lease bidding process also became apparent.

Briefly, offshore samples are taken from the upper six feet of sediment with a piston-type coring device equipped with a six-foot core barrel, three inches in diameter, into which a plastic liner is inserted. The bottom part of the sample is removed from the liner on board the ship and packaged in plastic bags in such a manner as to allow a minimum of air to come in contact with the sample. Stored in this manner, no significant losses of adsorbed hydrocarbons have been noted even after many months.

Offshore samples are prepared for adsorbed hydrocarbon analysis by the same procedure that is now used for onshore samples; that is, coarse-grained material larger than 63 microns is removed prior to analysis (Horvitz, 1978). The same analytical techniques (Horvitz, 1969, 1972) are also used.

Experiments, conducted with samples taken from the same locations at several intervals down to about twelve feet, indicated that the depth of sampling is not as critical in water-covered areas as it is onshore. Offshore samples show fairly uniform concentrations of hydrocarbons from the water-sediment interface to twelve feet. Often, the uppermost foot of sediment contains somewhat higher concentrations than does the deeper part of the hole. On land, the first few feet of soil, especially along the Gulf Coast, contain very low concentrations of hydrocarbons, and significant values do not usually appear above six feet; therefore, samples are taken well below this depth, generally at ten to twelve feet. Offshore the water serves as a blanket which prevents the rapid escape of hydrocarbons from the sediment. Nevertheless, offshore samples are taken at six feet, whenever attainable, instead of at the immediate surface, because surface sediments may be moved around by water currents or storms. At six feet the sediment is more likely to have been in place for many years. Because of the water blanket, background values in offshore petroliferous areas are usually much higher than those in onshore areas. In the former, background ethane and heavier hydrocarbon values are, usually, at least thirty-five to fifty parts per billion by weight, while onshore they are often below ten. On the other hand, in nonpetroliferous areas, offshore values for the ethane through pentane fraction are also very low; they are frequently well below ten parts per billion.

In areas where the sea bottom is composed of clay or sandy clay, sampling poses no problem, but in areas of hard bottom, samples are often missed, or only small quantities of rock or reef material are retrieved. In the case of very sandy bottoms, the penetration depth is often less than six feet; frequently only two to three feet. There are usually available, however, sufficient amounts of such samples from which adequate quantities of fine-grained material can be extracted. Furthermore, in such areas an additional analysis is conducted for hydrocarbons, which may be loosely held by the sands and which are driven from them during the sand-removal process. These loosely bound hydrocarbons are often very helpful in evaluating hydrocarbon surveys. Analyses of bottom-water samples, collected with the sediment, often yield clues which are also useful in evaluating areas with sand bottoms.

An Offshore Louisiana Survey

During the fall of 1973 and winter of 1974, a hydrocarbon survey was conducted offshore Louisiana to evaluate some areas prior to the March, 1974, Offshore Federal Lease Sale. One of the areas that was included is in the Mobile South No. 2 Area, about 100 miles southeast of New Orleans.

Figure 4 shows the locations of 256 stations from which sediment samples were collected at six feet, under water ranging in depth from 318 ft in the northwest corner of the survey to 1,860 ft in the southeast corner. The stations were spaced at one-half mile intervals along north-south profiles one mile apart, resulting in a density for the area of one sample to 320 acres. The ethane and heavier hydrocarbon values, shown on the map at their respective station locations, range from 39 to 537 parts per billion by weight, and average 68. Values from 75-89 are in the lightly shaded areas, and values of 90 and above are in the more heavily shaded areas.

Figure 5 shows the interpretation that was made on the basis of the ethane and heavier hydrocarbon data. A well-defined, halo-type anomaly developed, which was indicated by the dashed outline and graded A. The central part of the anomaly lies under about 1,000 ft of water.

The methane data is presented on Figure 6, and a pattern resembling that of the ethane and heavier hydrocarbons is apparent. Values of 250-299 parts per billion by weight are in the lightly shaded areas, and values of 300 and above are in the more heavily shaded areas. The methane values range from 123 to 1,011 and average 220.

The pentane values, which range from 0 to 101 parts per billion and average 9.9, are shown on Figure 7. The near-surface pattern is very similar to those produced by the methane and by the ethane and heavier hydrocarbons. The pentane data are useful in determining if the reservoir contains liquid hydrocarbons. If pentane, in anomalous amounts, appears in the near-surface sediment and produces an anomaly, liquid hydrocarbons may be expected to be present in the potential reservoir. The absence of pentane, however, does not exclude the presence of liquids in the reservoir, because gas associated with crude oil is often relatively dry and contains very little pentane.

In July, 1975, Shell announced the results of an exploratory well drilled about one mile west and a few hundred feet south of the northeast corner of block N660-E62, shown on Figure 8. Several gas and oil zones were encountered which indicated that a potentially important discovery had been made. Since then, eleven

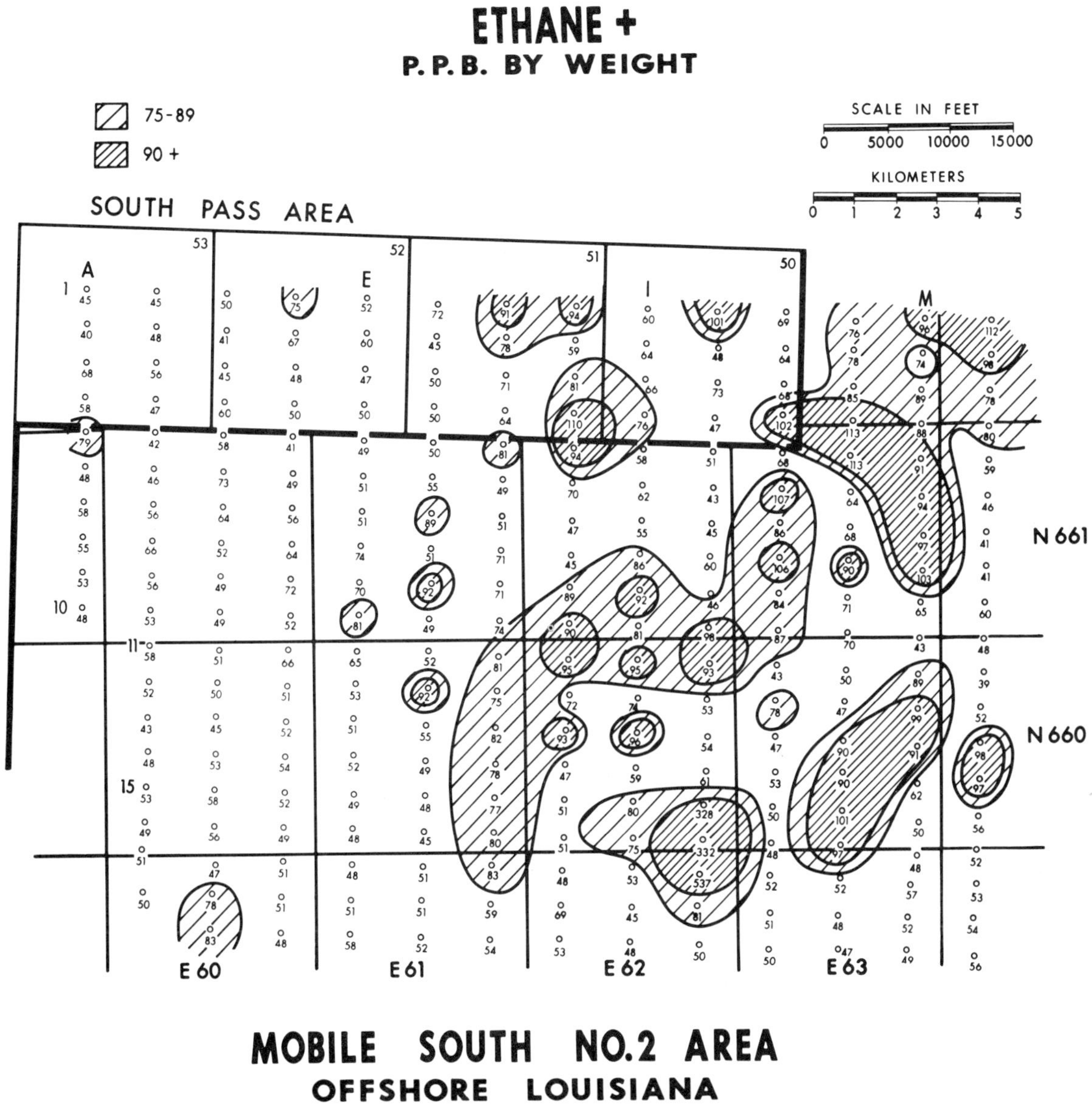

Fig. 4. Halo-type concentration pattern produced by ethane and heavier hydrocarbons from sediment samples taken at 6 ft in Mobile South No. 2 area, Gulf of Mexico. Water depth ranges from 318 ft at northwest corner to 1,860 ft at southeast corner of survey. From Horvitz, 1978, reprinted by permission of AAPG.

additional exploratory wells were drilled to depths ranging from 9,000 ft to 13,930 ft within the area enclosed by the dotted outline. How many of these wells are potentially productive is not yet known, but the field, named Cognac, is believed to be a major reserve and will be on stream soon.

The area containing the exploratory wells bears a very close relationship to the hydrocarbon anomaly. If the anomaly were shifted about one mile to the northwest, an almost perfect fit with the estimated productive limits would result. Such displacements are frequently noted. Perhaps hydrodynamic effects are involved in the process of moving hydrocarbons from depth to the near-surface. If this could be resolved, it would improve the accuracy with which potentially productive areas could be outlined by near-surface hydrocarbon data. Even without such a correction, it is significant that 75% of the

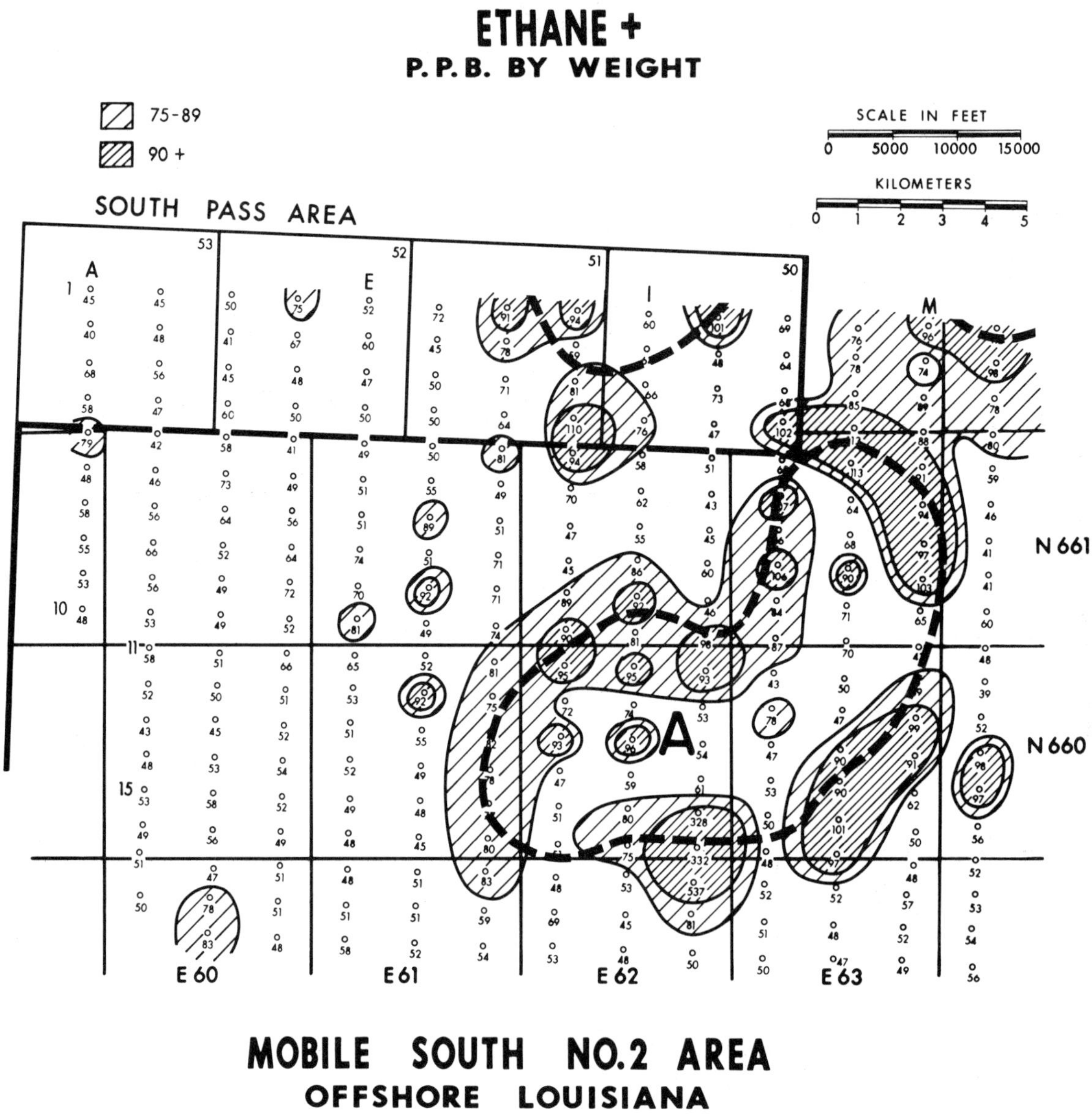

Fig. 5. Interpretation of ethane and heavier hydrocarbon data of Mobile South No. 2 area made prior to March, 1974, Gulf of Mexico lease sale. Area considered most prospective is outlined by heavy dashed line and graded A. From Horvitz, 1978, reprinted by permission of AAPG.

potentially productive area is included within the hydrocarbon anomaly. It is also of interest to note that, by volume, the methane represents 89% of the total light hydrocarbon fraction (C_1-C_5) in the near-surface of this area and is within the concentration range normally expected in reservoir gas. The ratios of the anomalous values to the background values are 1.8 for methane and for ethane and heavier hydrocarbons, and 2.0 for pentane. While these are significant ratios for offshore anomalies, they are much lower than those of most land anomalies.

Several miles to the west of the Cognac field, five wells were drilled (within the small area shown on Figure 8 by a dotted outline), but no official results have been released. According to the interpretation presented, no anomaly de-

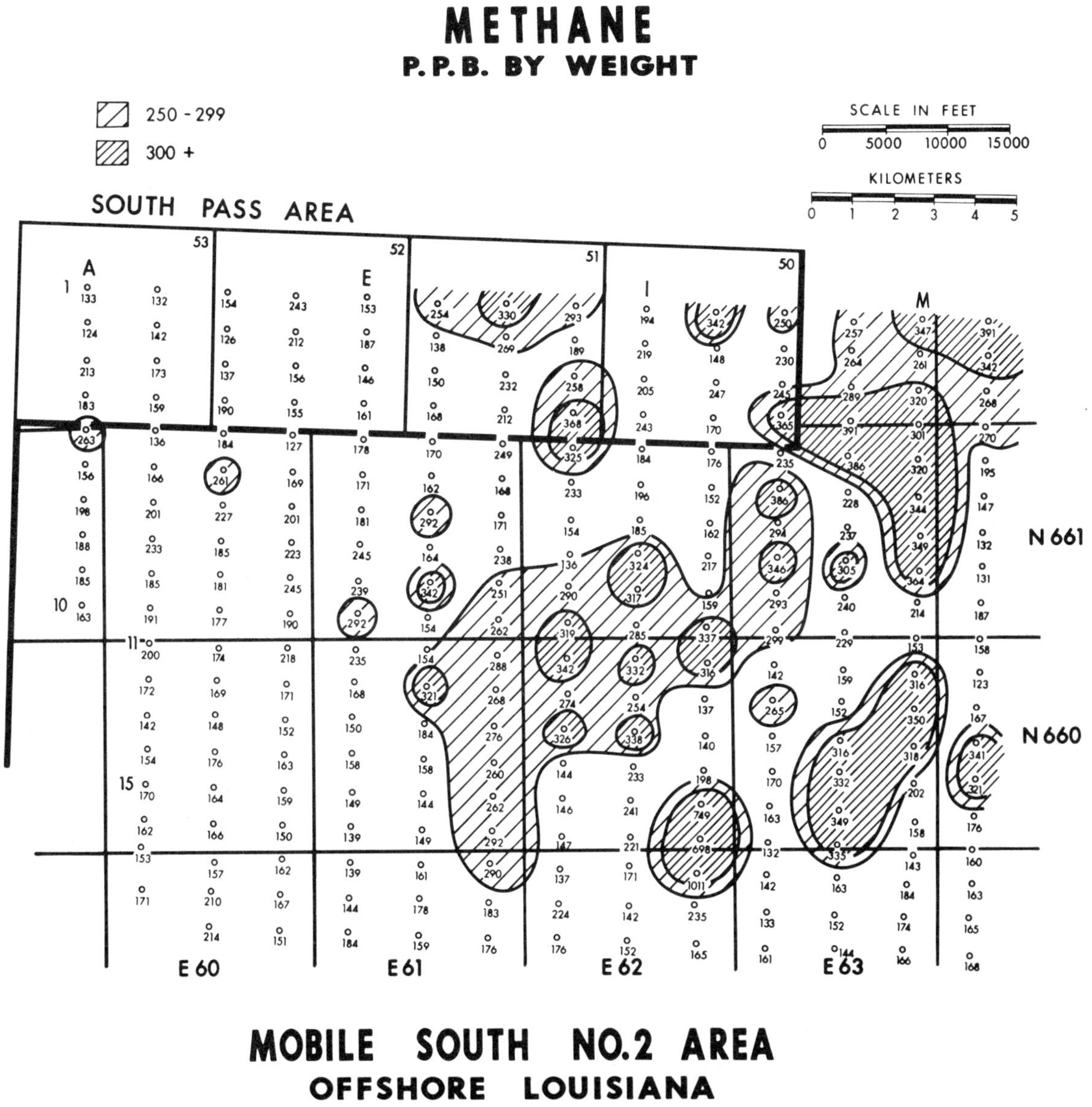

Fig. 6. Methane data in Mobile South No. 2 area produced a halo-type anomaly in eastern part of survey, which resembles closely the one produced by the ethane and heavier hydrocarbons. From Horvitz, 1978, reprinted by permission of AAPG.

veloped and, on that basis, any accumulation that exists there should be unimportant, at least when compared to Cognac.

Carbon Isotope Ratios–Cognac Anomaly

Carbon isotope ratios of methane extracted from three samples, two from the northeast edge (M-3 and M-4, Figure 6) of the Cognac anomaly and one near the southeast edge (M-14), were determined. Samples M-3 and M-4 yielded d^{13}C°/00 values of -38.1 and -39.2, respectively, on the PDB scale and sample M-14, near the southeast edge, yielded a d^{13}C°/oo value of -37.3. Just as in the case of Francitas, the d-values fall in the range of those of methane from petroleum gases. The carbon isotope ratios of the samples from the Cognac survey were obtained through the courtesy of Tenneco Oil Company.

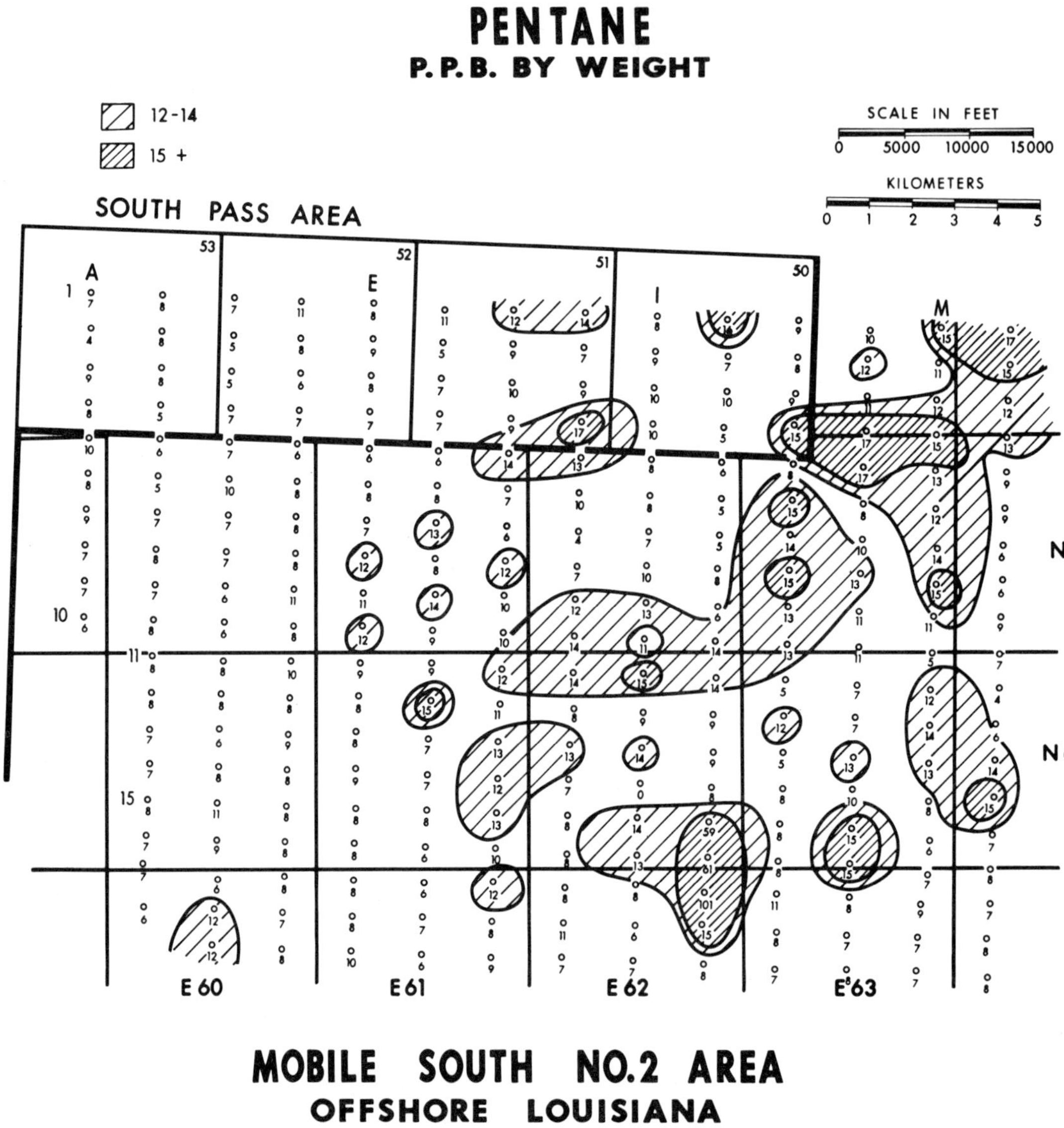

Fig. 7. Pentane data also produced a halo-type anomaly in Mobile South No. 2 area which is similar to that of ethane and heavier hydrocarbons (Fig. 4). From Horvitz, 1978, reprinted by permission of AAPG.

FLUORESCENCE MEASUREMENTS

An interesting development in the field of geochemical prospecting, introduced in the last decade, involves the determination in soils and sediments of substances that fluoresce when excited by ultraviolet light. The same types of fluorescence spectra are produced by soils and sediments in the near-surface over oil fields as are produced by crude oils. When crude oils are excited with ultraviolet light having a wavelength of 265 nanometers, emissions are produced at 320 or 365 nanometers, respectively, or at both. Using a sensitive fluorescence spectrophotometer, the emission intensities can be measured and the resulting spectra can be recorded.

The upper curve on Figure 9 was produced when a very small amount of crude oil was dissolved in spectroscopically clean hexane and exposed to ultraviolet light having a wavelength of 265 nanometers. Emission peaks

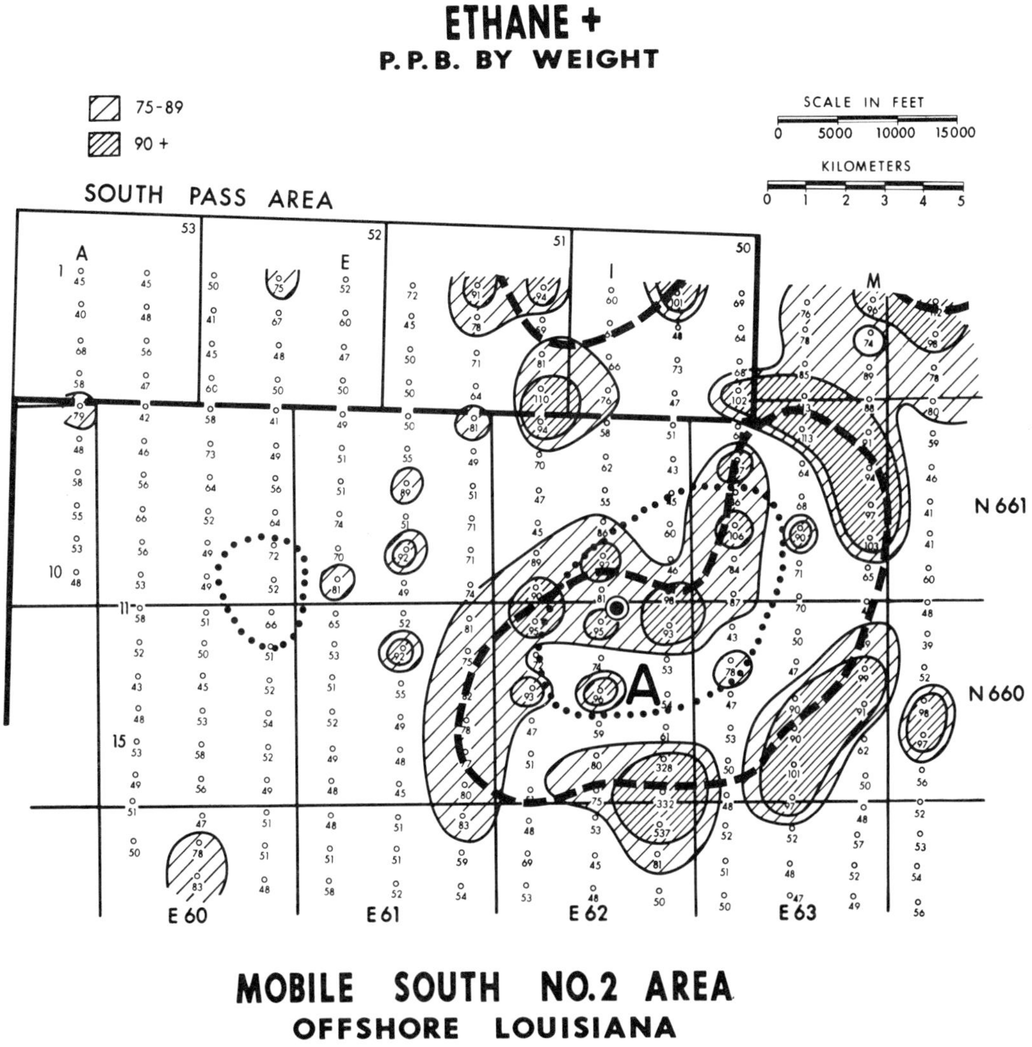

Fig. 8. Ethane and heavier hydrocarbon map (Fig. 5) to which the discovery well ((●)) of the new Cognac field has been added. Within the dotted outline, 11 additional exploratory wells have been drilled. The small area (dotted outline) to the west of the A anomaly contains 5 exploratory wells, some of which may have penetrated a gas deposit. No significant hydrocarbon anomaly is apparent in this part of the survey. From Horvitz, 1978, reprinted by permission of AAPG.

at wavelengths of 320 and 365 nanometers were produced. The middle curve represents the spectrum of a hexane extract of a soil sample taken from 12 ft near an oil field. Only a minor peak is seen at 320 nanometers, but a significant one appears at 365 nanometers. The bottom curve is that of the pure solvent, hexane, which contains a negligible amount of the component that emits at 320 nanometers and none of the one that emits at 365.

The hydrocarbons of natural gas and distillates do not fluoresce, nor do the heavy straight chain saturated hydrocarbons. Experiments conducted on a series of ten hydrocarbon fractions that were distilled form a crude oil indicated that the material which fluoresces occurs

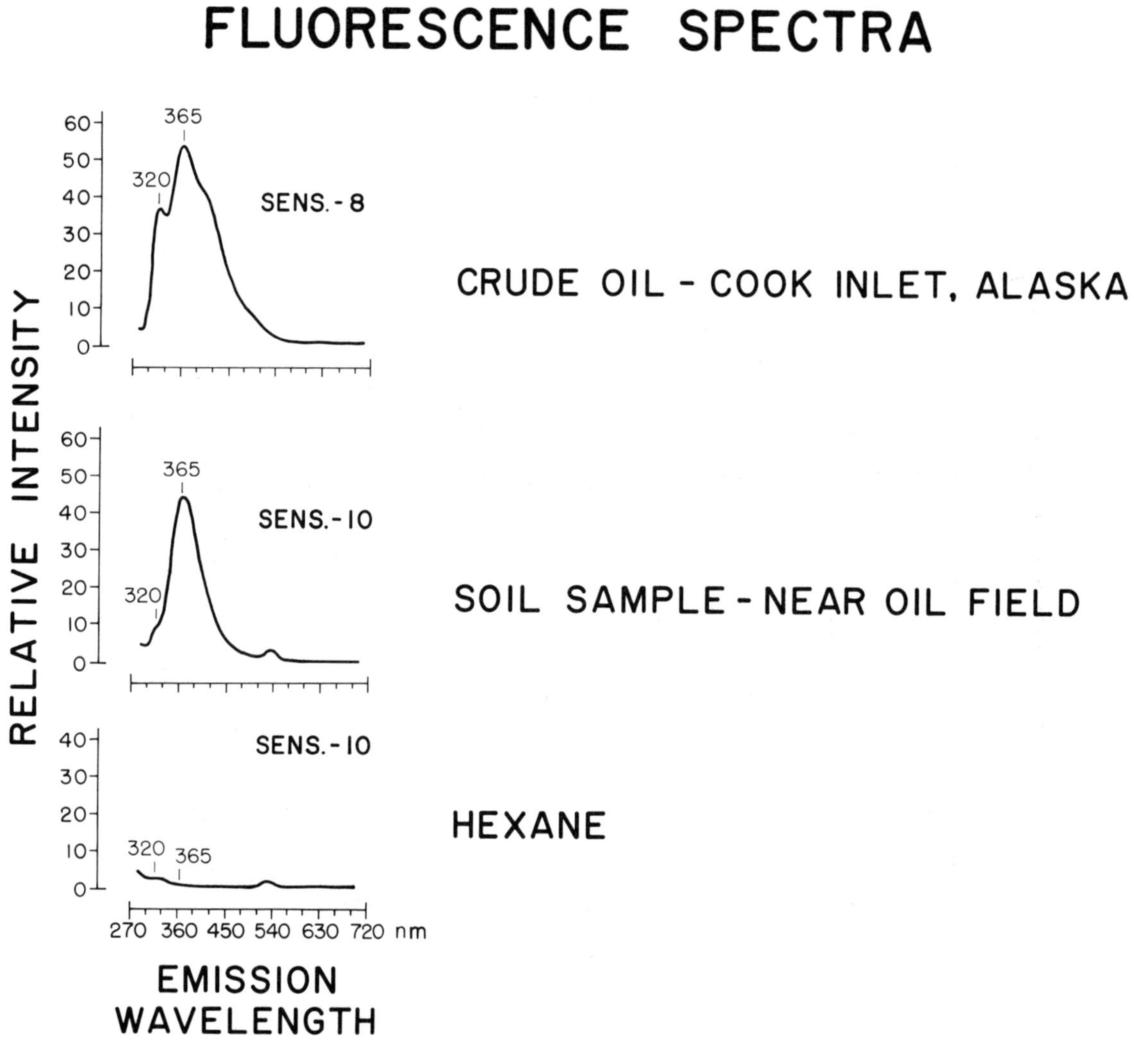

Fig. 9. Fluorescence spectrum of hexane extract of crude oil compared with that of a 12-ft soil sample taken near an oil field. High molecular weight aromatic hydrocarbons may be source of fluorescence.

in the two heaviest fractions, those that showed an amber or darker color. The exact compositions of the substances that fluoresce are not yet known, but they are thought to be members of the high molecular weight, aromatic hydrocarbon family.

Unfortunately, the refining process does not eliminate the materials that fluoresce; therefore, it is impossible to distinguish crude oil from refined oils. Because of this fact, fluorescence data should not be used alone in exploration. When used in conjunction with a light hydrocarbon survey, however, fluorescence data can be very useful. A fluorescence anomaly which conforms to that of the light hydrocarbons suggests a subsurface oil reservoir, while a light hydrocarbon anomaly, in the absence of a fluorescence anomaly, would suggest a gas or gas-distillate accumulation. Purvis (1978) was apparently the first to apply this interesting geochemical technique to oil exploration.

NEAR-SURFACE HYDROCARBON STUDIES BY GULF

Geochemical exploration received encouraging support in a recent paper by Jones and

Drozed (1979). The paper revealed near-surface hydrocarbon data, acquired by Gulf Research and Development Company, which showed close relationships to subsurface petroleum accumulations. Furthermore, the authors reported that by using percent compositions of near-surface soil gases and ratios of methane to ethane and propane, one can predict if a reservoir contains oil or gas.

CONCLUSIONS

Geochemical exploration has made much progress during the past decade, especially in substantiating its basic premise: that near-surface hydrocarbons originate in subsurface gas and oil deposits. This is important not only to hydrocarbon geochemical prospecting, but also to other geochemical techniques, including those that measure inorganic substances such as carbonates, sulfates, ferrous iron, etc., in the near-surface, and those that rely on hydrocarbon migration as the cause of near-surface concentration changes. Remote sensing techniques that purport to measure hydrocarbons at the very surface of the earth also gain in credibility through proof of hydrocarbon migration from subsurface oil and gas reservoirs.

It is not my intention to create the impression that geochemical methods represent the panacea of oil exploration. In the difficult search for oil and gas, all valid methods should be used, whether geological, geophysical, or geochemical; and they should all be integrated intelligently and efficiently into a complete and viable exploration system. In the event of one circumstance, however, geochemistry may have to stand alone: that is, when a strong, well-defined, near-surface hydrocarbon anomaly is encountered which is not supported by structurally oriented methods, serious consideration must be given to the possibility that it is reflecting a very important stratigraphic trap.

REFERENCES

Horvitz, L., 1969, Hydrocarbon geochemical prospecting after thirty years, *in* Unconventional methods in exploration for petroleum and natural gas: Dallas, Texas, Southern Methodist University, p. 205-218.

———, 1972, Vegetation and geochemical prospecting for petroleum: American Association of Petroleum Geologists Bulletin, v. 56, p. 925-940.

———, 1978, Near-surface evidence of hydrocarbon movement from depth. Read before the Annual Meeting, American Association of Petroleum Geologists, Oklahoma City, Oklahoma: To be published by AAPG in Special Volume on Hydrocarbon Migration. I am indebted to AAPG for permission to use much material from this paper.

Jones, V. T. III, and Drozd, R. J., 1979, Predictions of oil or gas potential by near-surface geochemistry: American Association of Petroleum Geologists, Annual Meeting, Houston, Texas.

Laubmeyer, G., 1933, A new geophysical prospecting method, especially for deposits of hydrocarbons: Petroleum, v. 29, no. 18, p. 1-4.

Purvis, D. B., 1978, Personal Communication. Began research in 1964.

Rosaire, E. E., 1938, Shallow stratigraphic variations over Gulf Coast structures: Geophysics, v. 3, p. 96-115.

Sokolov, V. A., 1935, Summary of the experimental work of the gas survey: Neftyanoye Khozyaystvo, v. 27, no. 5, p. 28-34.

GEOCHEMICAL PROSPECTING FOR OIL AND GAS FROM ORBITAL AND SUBORBITAL ALTITUDES

Terrence J. Donovan

U.S. GEOLOGICAL SURVEY, FLAGSTAFF, ARIZONA

ABSTRACT

A growing body of diverse evidence indicates that in surface rocks and soils of many oil fields there can be found, as a result of leakage, a variety of subtle to obvious chemical anomalies. Hydrocarbons appear to escape their reservoirs by one of, or in some combination of, three mechanisms: effusion, solution, and gas diffusion through an aqueous phase. Given slow but relatively long-continued leakage, chemical processes and reactions in the near-surface among hydrocarbons, pore waters, solute, and rocks result in a variety of diagnostic diagenetic products. An exciting prospect is the possibility of extracting this chemical information by inexpensive, rapidly gathered, remote means. Such a concept leads to intriguing possibilities for exploration both in inaccessible, undeveloped provinces and in well-developed ones.

A series of remote-sensing and geochemical experiments was conducted over four widely separated oil fields, each trapping oil by a different control. The Recluse field, Wyoming, is a large stratigraphic-trap accumulation. Calcite cement in Tertiary clastic rocks cropping out over the field displays a variety of stable isotopic and trace-metal anomalies; fine particulates, collected by an airborne sampler, appear to reflect at least partially the diagnostic elemental composition of plant leaves along the track of the aircraft. Exposed Permian rocks over the Cement field, a large multizone anticlinal accumulation in Oklahoma, are substantially altered. An important seepage-related diagenetic phenomenon there, detectable by aeromagnetic surveying, is the reduction of hematite to magnetite in the shallowing overburden. Airborne-magnetic and gamma radiation data sets collected at Cement have been converted into digital images for computer enhancement, facilitating correlation with surface and near-surface diagenetic phenomena. Triassic red beds over the combination-trap Garza field of West Texas are bleached because seeping hydrocarbons caused reduction, dissolution, and mobilization of iron. The characteristic discoloration aureole is mappable from computer-enhanced LANDSAT images, because of the scanner's sensitivity to the amount and oxidation state of iron. Preliminary results of analysis of airborne particulates wafted from Tertiary outcrops and their overlying soils over the Red Wing Creek structure, a probable astrobleme in South Dakota, are tenuous, although distinct surface chemical and helium soil-gas anomalies occur there.

INTRODUCTION

Geochemical prospecting for oil and gas is predicated on the axiom that in nature there is no such thing as a perfect seal; thus most oil and gas traps should leak. Preliminary evaluation of empirical data suggests that indeed there are three fundamental ways in which petroleum or natural-gas compounds can escape their reservoirs (Donovan and Dalziel, 1977). Effusion involves leakage, generally of Macro amounts of hydrocarbons, through poorly compacted roofrocks and along fractures and faults. Solution escape occurs when hydrocarbons are dissolved in associated waters whose salinity is diminished through mixing with meteoric or clay-interlayer water. This water containing dissolved hydrocarbon species is transported by a variety of driving forces: chemical potential differences on opposite sides of semipermeable clay beds, hydrodynamic flow, and compaction pressure. Gas diffusion through an aqueous phase has been shown to be independent of the pore size of the containing rock (Pandey, Tek, and Katz, 1974.)

In porous rock at or near the surface, microbes interact with seeping hydrocarbons, and ultimately a variety of diagnostic diagenetic phenomena result. These phenomena include isotopically and chemically distinctive porefilling carbonate cement and reduction, dissolution, and partial removal of iron from iron-bearing minerals in superincumbent rocks (Donovan, Friedman, and Gleason, 1974; Henry and Donovan, 1978; Donovan and Dalziel, 1977). These kinds of alteration phenomena can be considered fossil evidence of petroleum microseepage. Present-day seepage may also be considered fossil evidence of petroleum microseepage. Present-day seepage may also be detected directly (Horvitz, this volume) or indirectly: for example, in an environment that is under the influence of seeping hydrocarbons, the direct uptake by plants of certain micronutrients from soils may be facilitated (Dalziel and Donovan, 1980).

Fig. 1. Structure contour map of the Garza oil field, Texas. Structure contours on top of San Andres Limestone (Permian). Contour interval: 20 ft. Dashes where control is poor. From Bailey, 1953.

In this paper, the results of several experiments expressly designed to extract remotely this kind of chemical information are reviewed. The oil-field test sites studied are: Recluse, Wyoming; Cement, Oklahoma; Garza, Texas; and Red Wing Creek, North Dakota.

LANDSAT EXPERIMENT AT GARZA OIL FIELD, TEXAS

(Donovan, Termain, and Henry, 1979)

The Garza field, discovered in 1935, is along the northeastern margin of the Permian basin in

Fig. 2. Bleached Triassic, fine-grained sandstone caprock and bleaching in proximal 0.3 m of underlying shale, Garza oil field, Texas. From Donovan, Termain, and Henry, 1979.

southwest-central Garza County, Texas. Oil is produced from irregular porous lenses in a Permian (Leonardian) anhydrite-bearing dolomite that is anticlinally folded (Figure 1). The average depth of production is about 810 m, although some deeper oil production was developed in 1955 and 1956. The productive area is about 46 km^2. Overlying rocks include red sandstone, shale, and interbedded anhydrite and salt. This sequence is overlain by outcropping Triassic red beds that in turn are partially covered by buff-colored, Tertiary nonmarine deposits of the High Plains escarpment. As of January, 1978, about 66,000,000 barrels of oil had been produced from the Garza field.

The topography is flat to rolling in the Garza area. The western part of the area is dominated by the High Plains escarpment, and isolated erosion-resistant buttes are distributed

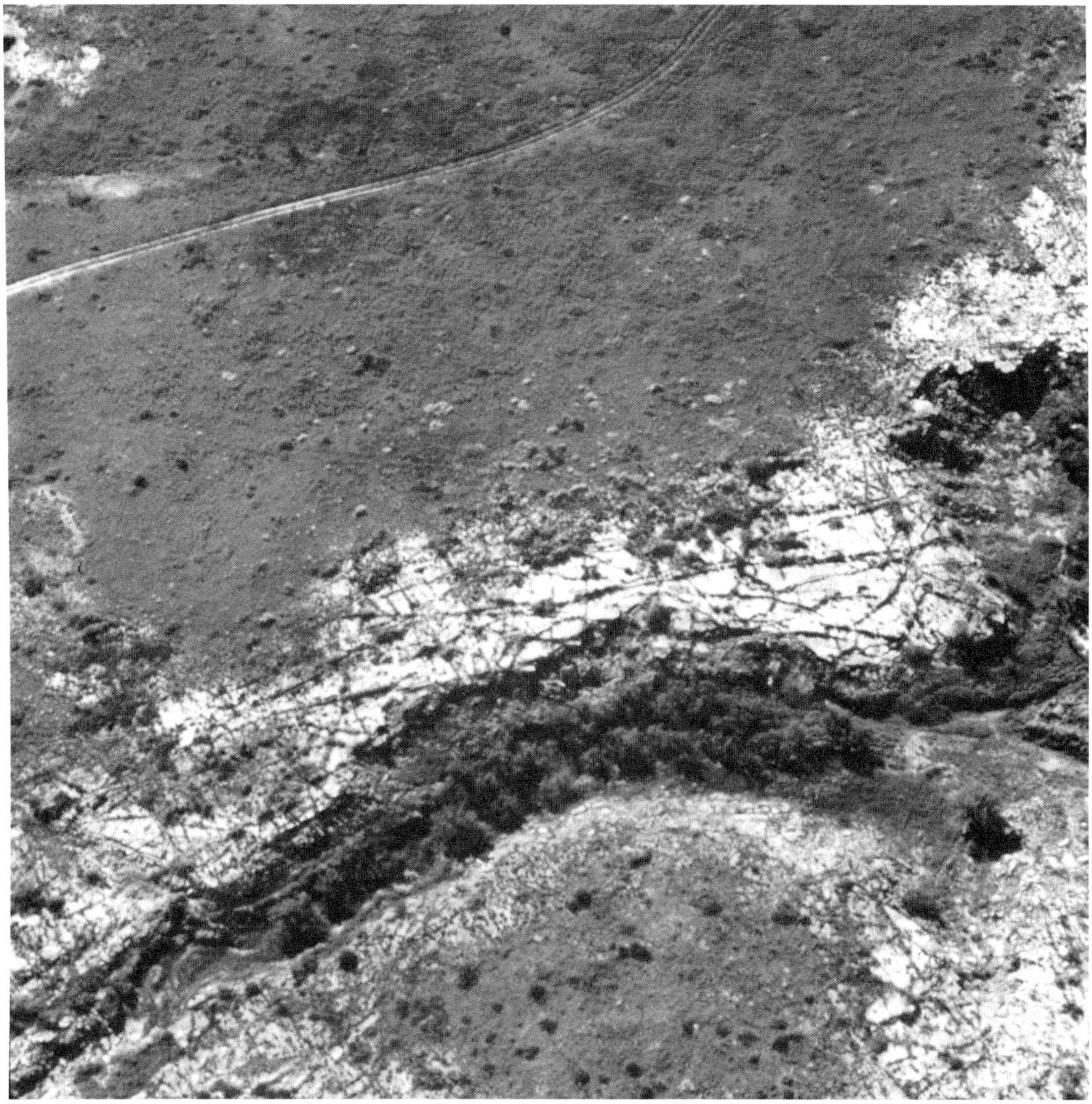

Fig. 3. Near-vertical aerial photograph of bleached Triassic sandstone and joint-controlled vegetation pattern, Garza oil field, Texas. From Donovan, Termain, and Henry, 1979.

throughout the Central Rolling Plains to the east. Drainage channels, emanating from the High Plains escarpment, transect the Central Rolling Plains. Cultivation is complete along the High Plains margin, whereas the Central Rolling Plains are dominated by a mesquite-buffalo grass association. Seven major soil associations have been mapped in the immediate Garza area (Richardson, Grice, and Putnam, 1975).

Triassic red beds overlying the Garza field are in places discolored and impregnated with isotopically anomalous carbonate cement (Donovan and Dalziel, 1977). The discolorations, the result of iron removal, are narrowly stratabound in thin sandstone units, and commonly a thin zone of red shale immediately underlying the bleached sandstones is also discolored (Figure 2). Donovan and Dalziel (1977) and Donovan, Termain, and Henry (1979)

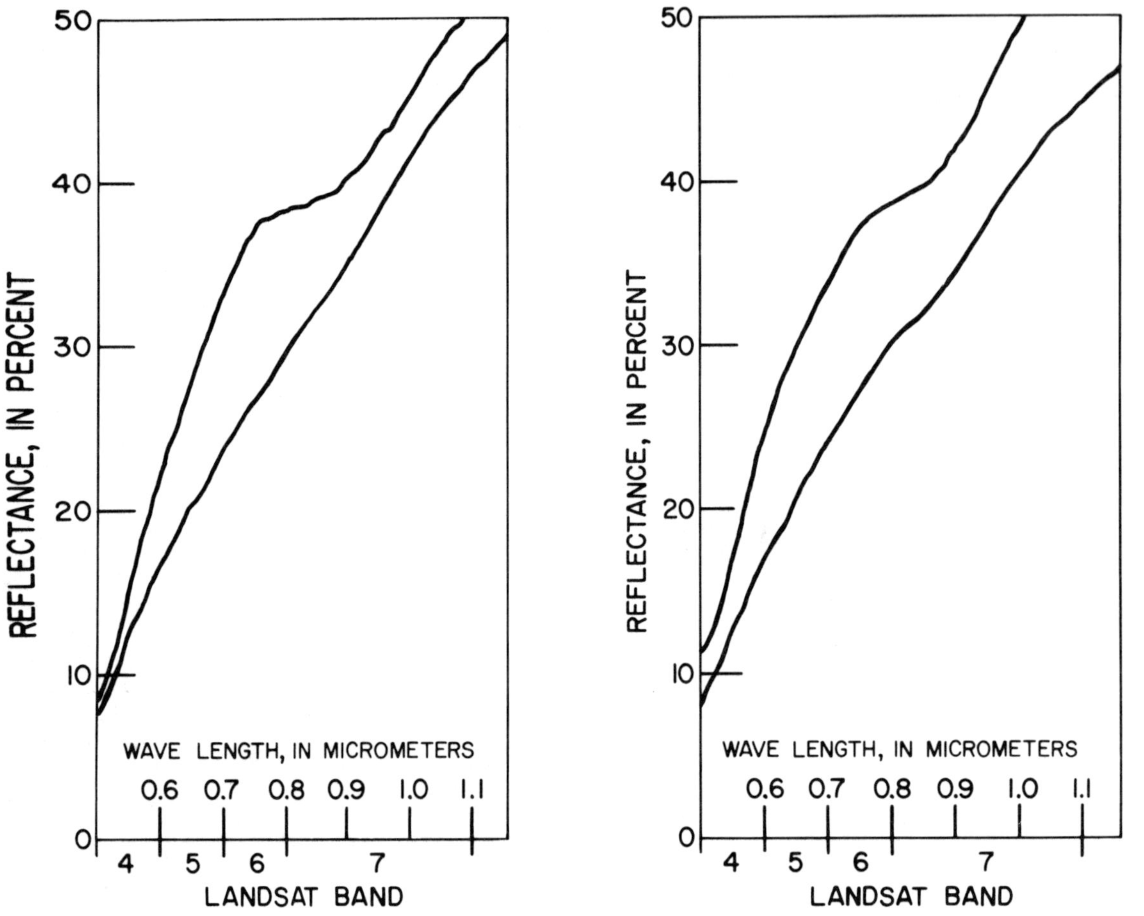

Fig. 4. Examples of reflectance spectra for variously altered surface rocks (varying iron content). From Donovan, Termain, and Henry, 1979.

suggested that seeping fluids probably have effused vertically along joints and fractures (Figure 3) and then spread laterally along permeable sandstones. Presumably parts of the shales are bleached because of their proximity to reducing agents carried in the overlying permeability channels.

Because visible and near-infrared reflectance spectra for altered and unaltered rocks differ and because there are marked differences in the shapes of spectral curves for differently altered rocks (Figure 4), an experiment was constructed to differentiate between altered and unaltered areas, using enhanced LANDSAT digital images. Chavez (1975) showed that if an effective haze removal has been accomplished, band ratios will subdue first-order topographic features, thus enabling a better display of albedos which might reflect variations in chemistry, color, or surficial materials. The band ratio 5 to 4 is best suited for discrimination within and among rock and soil types. The band ratio 6 to 5 best differentiates major material types such as rocks, soil, vegetation, and water; and the band ratio 6 to 7 best discriminates among vegetation types. However, band-ratio images did not clearly define the alteration zone at Garza, probably because the small areal extent of altered and unaltered bedrock exposures are beyond the spatial resolution capabilities of the LANDSAT I and II system (and the ratio technique suppresses any spectral averaging inherent in the unratioed images).

Study of individual-band data bases (computer processed for noise and stripping removal and geometric fidelity) did show hints of the discoloration aureole in a band-5 scene whose dynamic range (contrast) has been enhanced by a linear contrast stretch. (For detailed discussions of various image-processing techniques see Rowan et al., 1975; Eliason, Chavez, and Soderblom, 1974; Taranik, 1978; Condit and Chavez, 1979.) Each digital picture element was expanded into a 5×5 grid followed by a low-pass 5×5 boxcar filter. The same magnification and smoothing techniques were applied to band 4 and 7 scenes, thus enabling three computer-generated film data bases to be photographically composited into a color image. A light-tonal area, which corresponds to the alteration aureole and which approximates the know-oil-field boundary at Garza, can be mapped on this composite image (Figure 5)*. Because the LANDSAT scanner is particularly sensitive to the amount and oxidation state of the transition elements, especially iron (Goetz, 1975), LANDSAT imagery can be a useful tool in detecting alteration aureoles over oil deposits in carefully chosen areas.

AEROMAGNETIC AND AIRBORNE GAMMA RADIATION STUDIES, CEMENT, OKLAHOMA

(Donovan, Forgey, and Roberts, 1979; Donovan, Barringer, Foote, and Watson, 1975)

The Cement oil field is along the northeastern flank of the Anadarko basin in Caddo and Grady Counties, Oklahoma. The structure is a north-west-trending, doubly plunging, asymmetrical anticline, bounded on the north by a large thrust fault. Two pronounced domes (East and West Cement) separated by a crestal saddle distinguish the structure. Permian sandstone and Pennsylvanian carbonate and clastic rocks ranging in depth from 600 to 2,230 m yield oil and gas. By 1968, this giant multizone field had produced more than 120 million barrels of oil; more than 14.5 million barrels of recoverable oil were estimated to remain (Halbouty, 1968). An area of 75 km² is productive.

The topography at Cement is rolling, and the surface is cut by deep drainage channels typically lined with riparian hardwood trees. The anticline is topographically expressed; the crest is accentuated by the erosion-resistant Keechi Hills. Upland vegetation is a mix of cultivated crops, scrub oak, and natural grasses. Six major soil associations have been mapped in the Cement area (Moffatt, 1973).

Permian sandstone typically red brown and friable throughout the region, is discolored pink, yellow, and white along the flanks of the Cement structure, and rendered gray and hard by impregnation with carbonate cement along the crest. An overlying gypsum bed was altered to calcite along the crest of the fold in the Keechi Hills. The tinted rocks are relatively depleted in iron as the result of reduction, dissolution, and removal of iron; the calcitized gypsum and the carbonate cements of the sandstone are exceptionally deficient in C^{13}. The pattern of alteration closely coincides with the area of oil and gas production (Donovan, 1974).

AEROMAGNETIC EXPERIMENT

Low-altitude (120 m) airborne measurements of the total magnetic field over the Cement area indicate an anomalous shallow source of magnetic material (Figure 6). Separation and analysis of ferrimagnetic material in the upper 300 m of four boreholes drilled along the crest during oil-field development indicate increased amounts of magnetite near the surface (Figure 7). The obvious suggestion is that continued seepage of hydrocarbons and/or associated compounds such as H_2S has produced a near-surface chemical reducing environment, resulting in the reduction of hematite to magnetite. A postulated sequence of reduction is:

hydrated ferric oxides ⟶

hematite ⟶ **magnetite**

$$Fe_2O_3 \longrightarrow Fe^{+2}Fe_2^{+3}O_4$$

The diagenetic ferrimagnetic material produced under these conditions is a fossil chemical indicator of petroleum microseepage, a phenomenon that is readily detected by low-altitude aeromagnetic profiling.

*See atlas for Fig. 5.

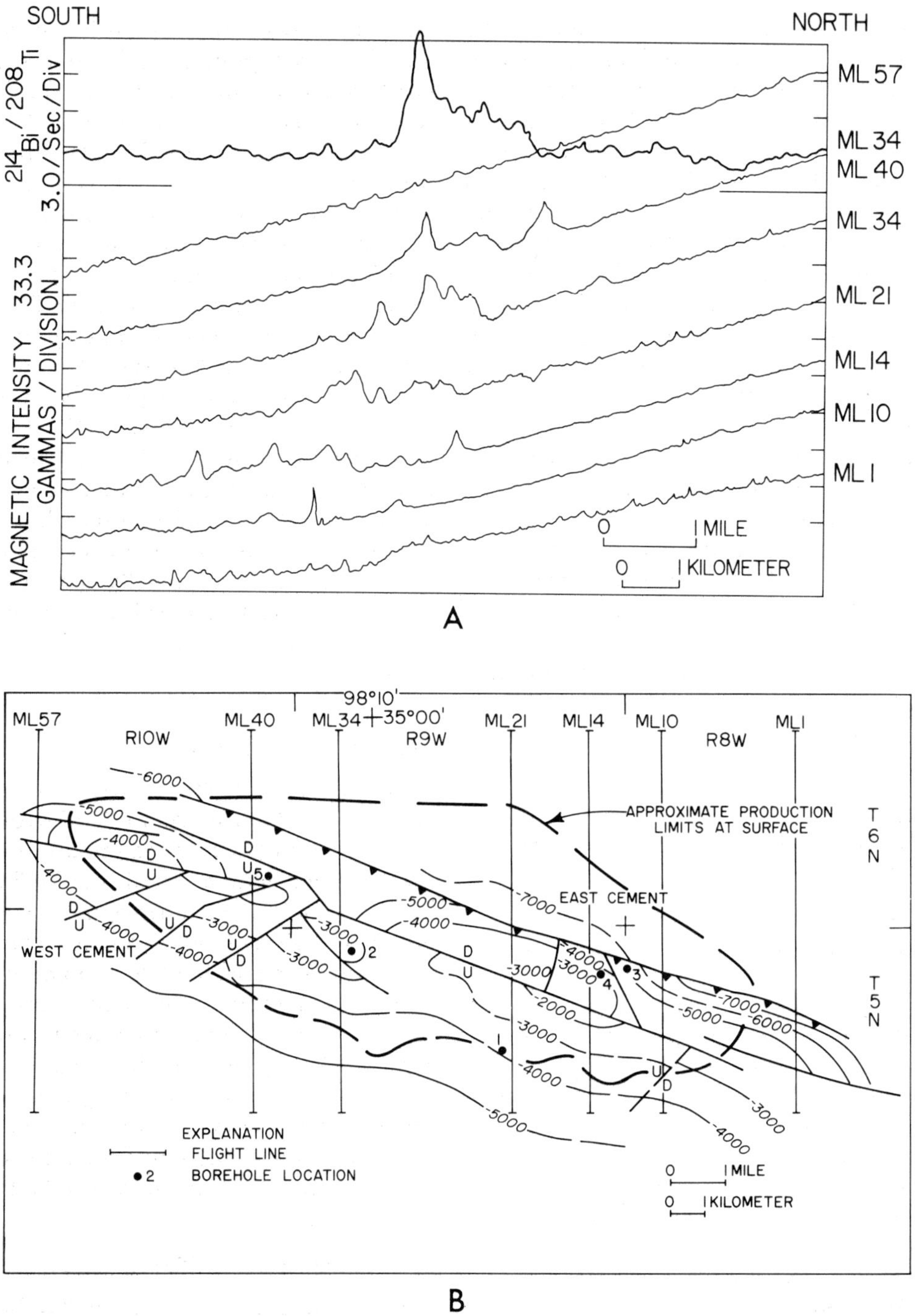

Fig. 6. A: Selected segments of magnetic profiles from Cement, Oklahoma, field aeromagnetic survey. Profiles probably reflect the rise of magnetic basement to the north. High-wave-number peaks are readily observable in profiles over the crest of the structure, presumably where maximum formation of diagenetic magnetite occurred. Anomalously high $^{214}Bi/^{208}Tl$ ratios along the crest owe their origin to late diagenetic uranium mineralization associated with petroleum microseepage.

B: Cement oil-field structure contours on top of Hoxbar Formation (Pennsylvanian) modified from Donovan, 1974, flight-line segments for magnetic profiles, and boreholes sampled for magnetite. From Donovan, Forgey, and Roberts, 1979.

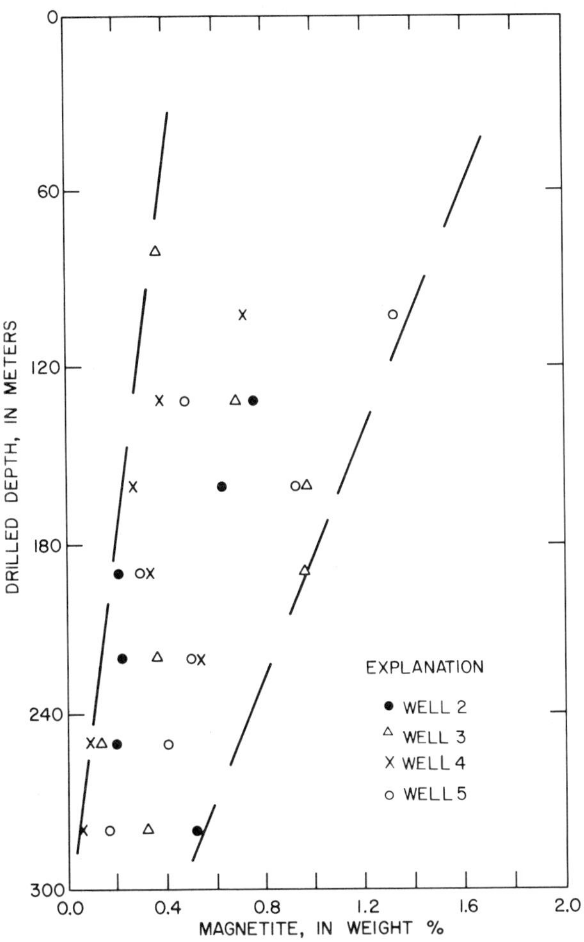

Fig. 7. Probable range of magnetite with depth in near-surface rocks at Cement, Oklahoma, oil field along and near the anticlinal crest. From Donovan, Forgey, and Roberts, 1979.

Interpretation of these kinds of data may be facilitated by transforming the data from vector space into a two-dimensional image having both fixed array dimension and resolution (Figure 8). In this format, spatial array processing techniques can be used to correct geometrically for proper registration and to interpolate areally, and also for data smoothing and display (Termain, Donovan, and Chavez, 1980). The variation in dark-to-light gray levels diagonally across the image from southwest to northeast represents the regional magnetic gradient in the Anadarko basin (compare with the maps of Jones and Lyons, 1964). High wave-number anomalies along the crestal reaches are shown as light gray-level areas.

Gamma-ray Experiment

Uranium mined in minor commerical amounts occurs along the crest at East Cement (McKay and Hyden, 1956), and other surface and subsurface anomalous radioactivity has been reported at East and West Cement (Al-Shaieb et al., 1977). The uranium occurs as carnotite and tyuyamunite. It has been proposed that the uranium was brought into the ore zone by ground waters moving upward along the flanks of the structure, and that it was sub-

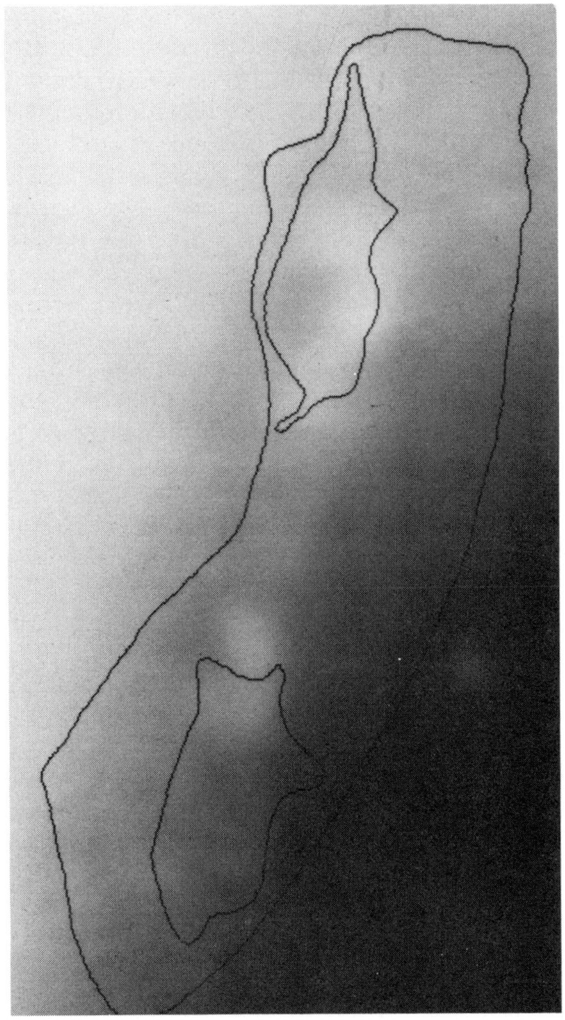

Fig. 8. Computer-generated black and white image of Cement, Oklahoma, oil-field aeromagnetic data. North is at top. Dark to light shading from southwest to northeast reflects magnetic basement. Light-gray levels along the crest reflect high-amplitude anomalies. Areas of discoloration and calcite dementation in surface rocks at Cement are stenciled on the image for reference.

sequently precipitated by reduction by seeping hydrocarbons (Olmsted and Al-Shaieb, 1975). Such postulated water flow up the structure at Cement, however, is generally contrary to ground-water-flow theory (Toth, 1962; 1963; 1972), and were it not, precipitation of uranium from dilute ground water would require funneling of enormous volumes of water through a tiny area (Gableman, 1977). Furthermore, cross-structural flow of ground water would have tended to offset in the direction of flow the hydrocarbon-induced alterations occuring at the crest. More probably, the uranium was extracted from the subsurface sedimentary column and mobilized by vertically moving brines, to be precipitated following reduction by cotransported hydrocarbons (Gableman, 1977). This process is in keeping with enrichment of rock and soil along the crest with mineral salts and, importantly, with anomalous amounts of extractable potassium (Donovan, unpublished data), all derived from hydrocarbon-bearing brines.

An image of a ^{208}Tl (thorium series) high gamma-radiation intensity along the crest detected with a sodium iodide crystal system (8380 Cm^3) flown at 120 m is shown in Figure 9*. This image was constructed in the same way as the aeromagnetic image described above. The abrupt northern boundary of relatively high gamma-radiation intensity in the image is probably controlled by the major thrust fault bounding the Cement structure, and the especially high radiation intensities shown along the crest correspond well to the reported anomalously high surface and subsurface radioactivity determined by field and gamma ray–neutron log studies (Olmsted and Al-Shaieb, 1975; Al-Shaieb et al., 1977).

ANALYSIS OF AIRBORNE PARTICULATES, RECLUSE, WYOMING AND RED WING CREEK, NORTH DAKOTA

The Recluse oil field is along the northeastern flank of the Powder River basin in Campbell County, Wyoming. It is a classical stratigraphic trap; the Cretaceous Muddy Sandstone is interpreted as originating as a northwest-southeast-trending barrier island (Woncik, 1969). Maximum drilled thickness of the Muddy at Recluse is 14 m, and the productive part of the bar is about 13 km long and ranges from 1 to 5 km wide (Figure 10). Regional dip is southwest; oil and minor gas are trapped along the updip edge of the sandstone bar. As of December 1978, 21,480,000 barrels of oil and 83,721,000 MCF of gas have been produced from Recluse.

The topography of the region is characterized by flat-topped buttes, long, narrow, and flat stream divides, and uniformly crested ridges. Pine and sage are the dominant natural vegetation, although cottonwoods grow along some of the larger streams which dissect the area.

An anomalous pattern of δC^{13} and δO^{18} values in carbonate cements in outcropping Tertiary Fort Union Formation sandstones is apparent at Recluse (Figures 11 and 12). The primary carbonate cement is calcite, but there is also ferroan dolomite and minor amounts of hematite. The δO^{18} values are more variable than the δC^{13} values, probably because of admixing of seeping and meteoric waters (Dalziel and Donovan, 1980). The general distribution of C^{13}– and O^{18}– depleted carbonate cements over and along the trend of the subsurface oil accumulation strongly suggests micropore filtration of vertically migrating formation waters carrying dissolved hydrocarbons (Donovan and Dalziel, 1977). The near-surface reducing environment resulting from influx of seeping hydrocarbons causes reduction, dissolution, and mobilization of iron and manganese. The manganese and iron are commonly coprecipitated with calcite in the presence of calcium and carbonate ions; mobilized manganese and iron may be transported together or separately depending upon the redox potential and the pH of the chemical environment and the rapidity of change (Krauskopf, 1957). Manganese distribution shows greater variability than iron in both rock cement and soil at Recluse. An especially good correlation is suggested between manganese/iron ratios in the carbonate cements and the subsurface distribution of oil. In contrast to the high manganese/iron ratios in the cement, the manganese/iron ratio data dis-

*See atlas for Fig. 9.

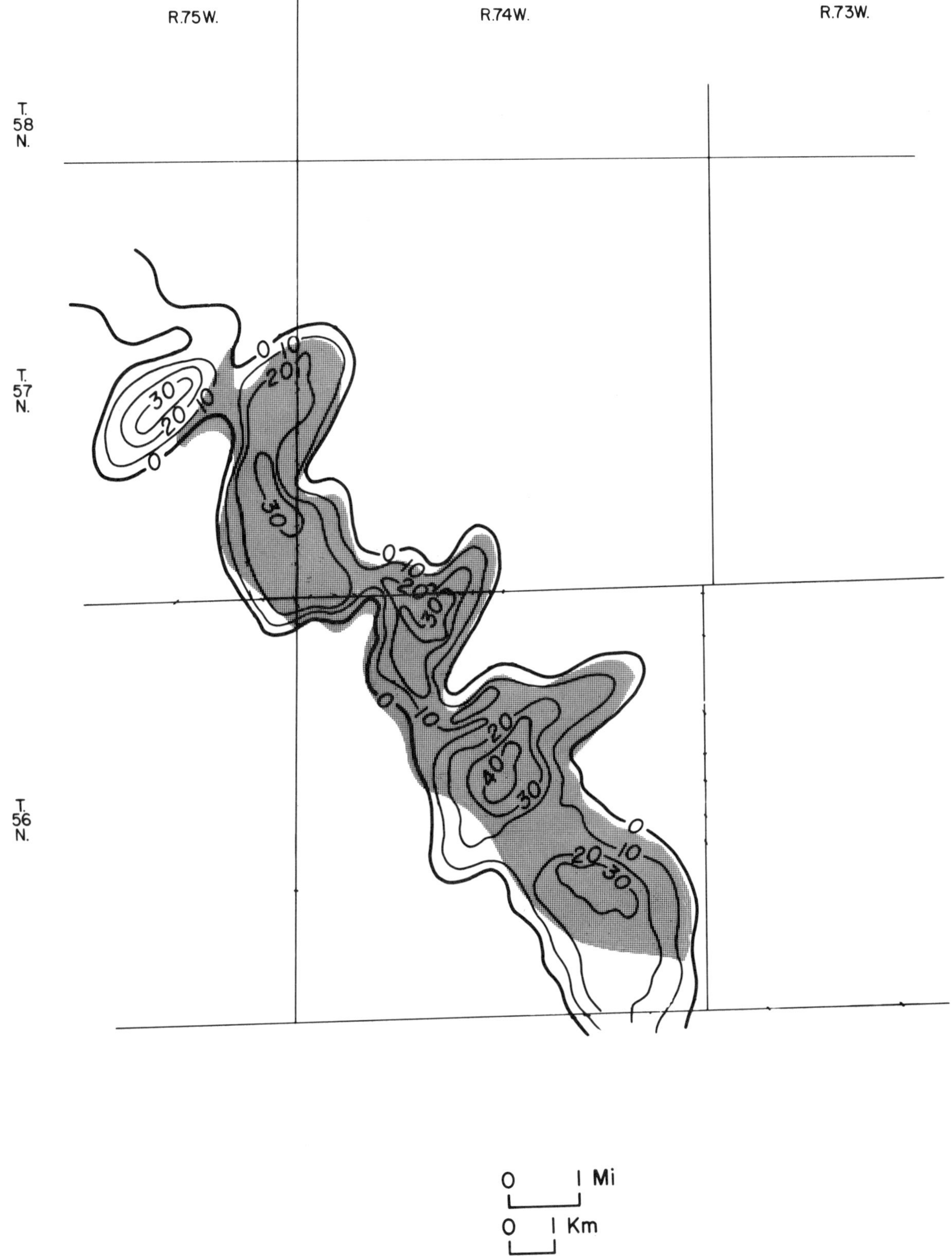

Fig. 10. Isopach map showing thickness of Muddy Sandstone (Cretaceous) reservoir, Recluse oil field, Wyoming. Isopach interval: 10 ft. Modified from Woncik, 1969. Shaded area shows approximate productive limits.

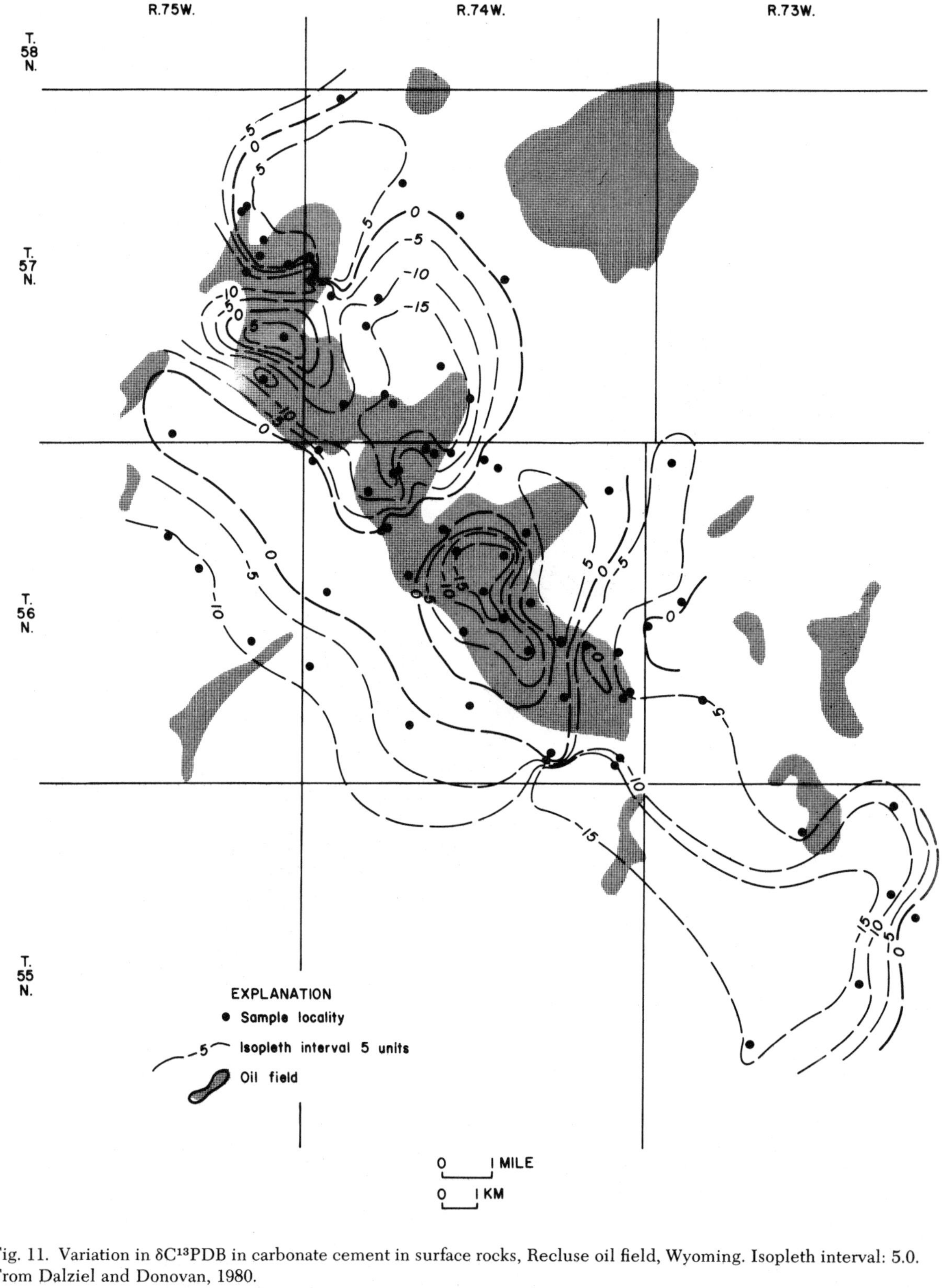

Fig. 11. Variation in δC^{13}PDB in carbonate cement in surface rocks, Recluse oil field, Wyoming. Isopleth interval: 5.0. From Dalziel and Donovan, 1980.

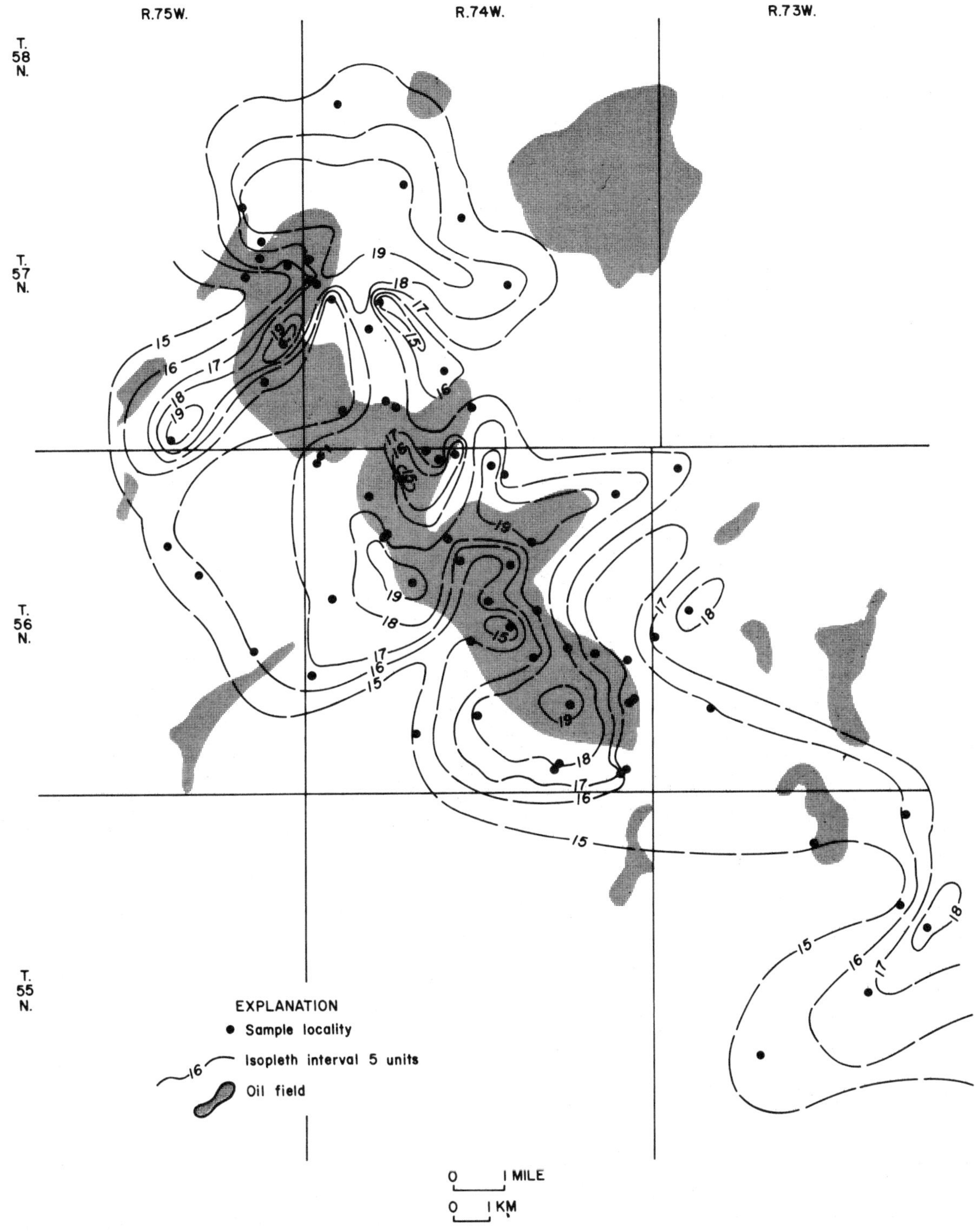

Fig. 12. Variation in δO^{18}SMOW in carbonate cement in surface rocks, Recluse oil field, Wyoming. Isopleth interval: 1.0. From Dalziel and Donovan, 1980. Shaded areas show oil and/or gas production limits at Recluse and in the immediate area.

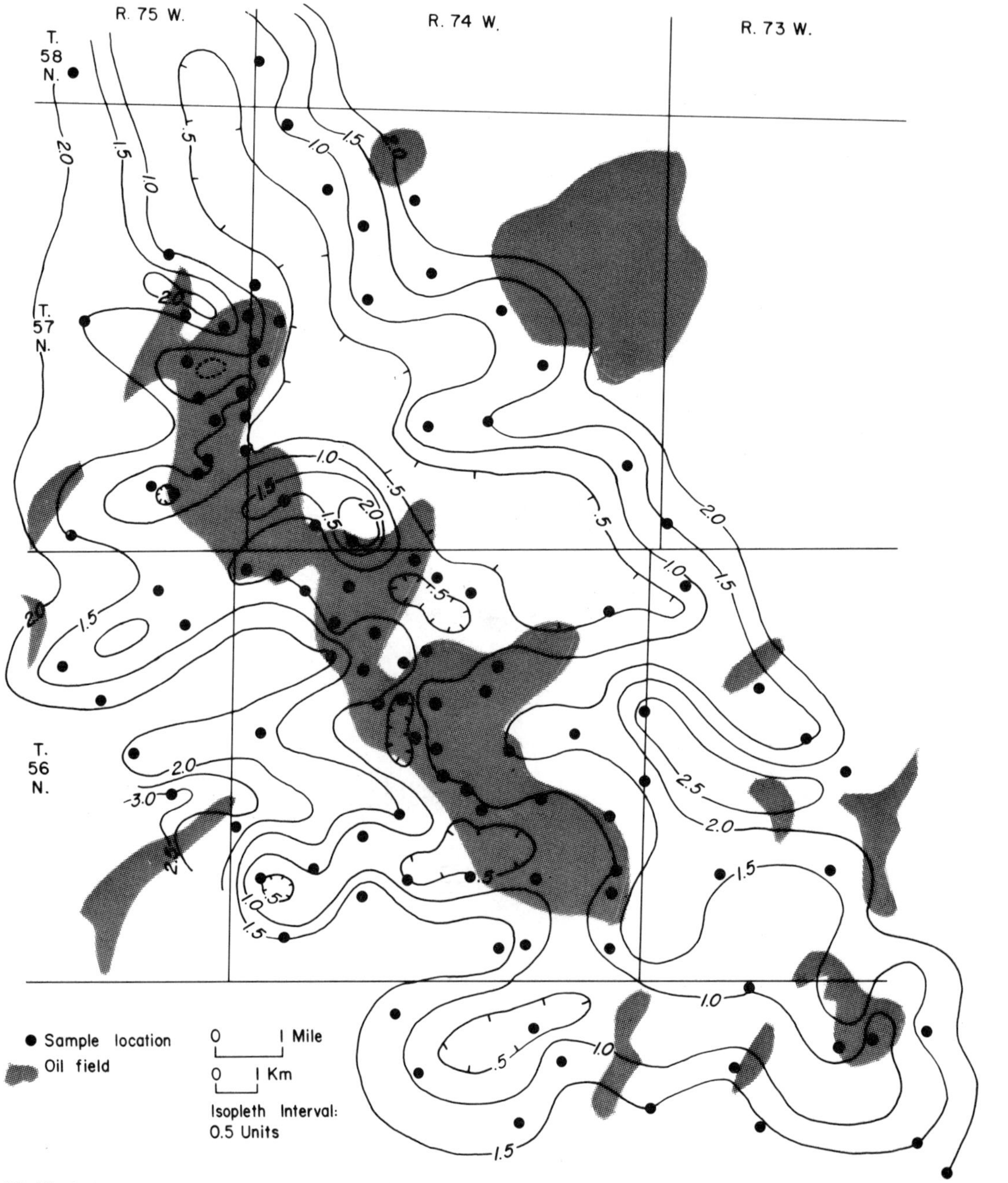

Fig. 13. Variation in manganese/iron ratios (×106) in soil, Recluse oil field, Wyoming. Isopleth interval: 0.5. From Dalziel and Donovan, 1980. Shaded areas show oil and/or gas production limits at Recluse and in the immediate area.

play systematic lower ratios over and along the producing trend, suggesting relative manganese depletion (Figure 13).

Anomalous manganese/iron ratios also occur in pine needles and sage leaves at Recluse; this likewise suggests the effects of petroleum microseepage on the plants (Figures 14 and 15). Hydrocarbons provided reducing conditions sufficient to enable divalent iron and manganese to be adsorbed on solids or organically complexed in the soils. In these reduced, bound and adsorbed states, iron and manganese uptake by the plant roots was accelerated and the elements were assimilated by

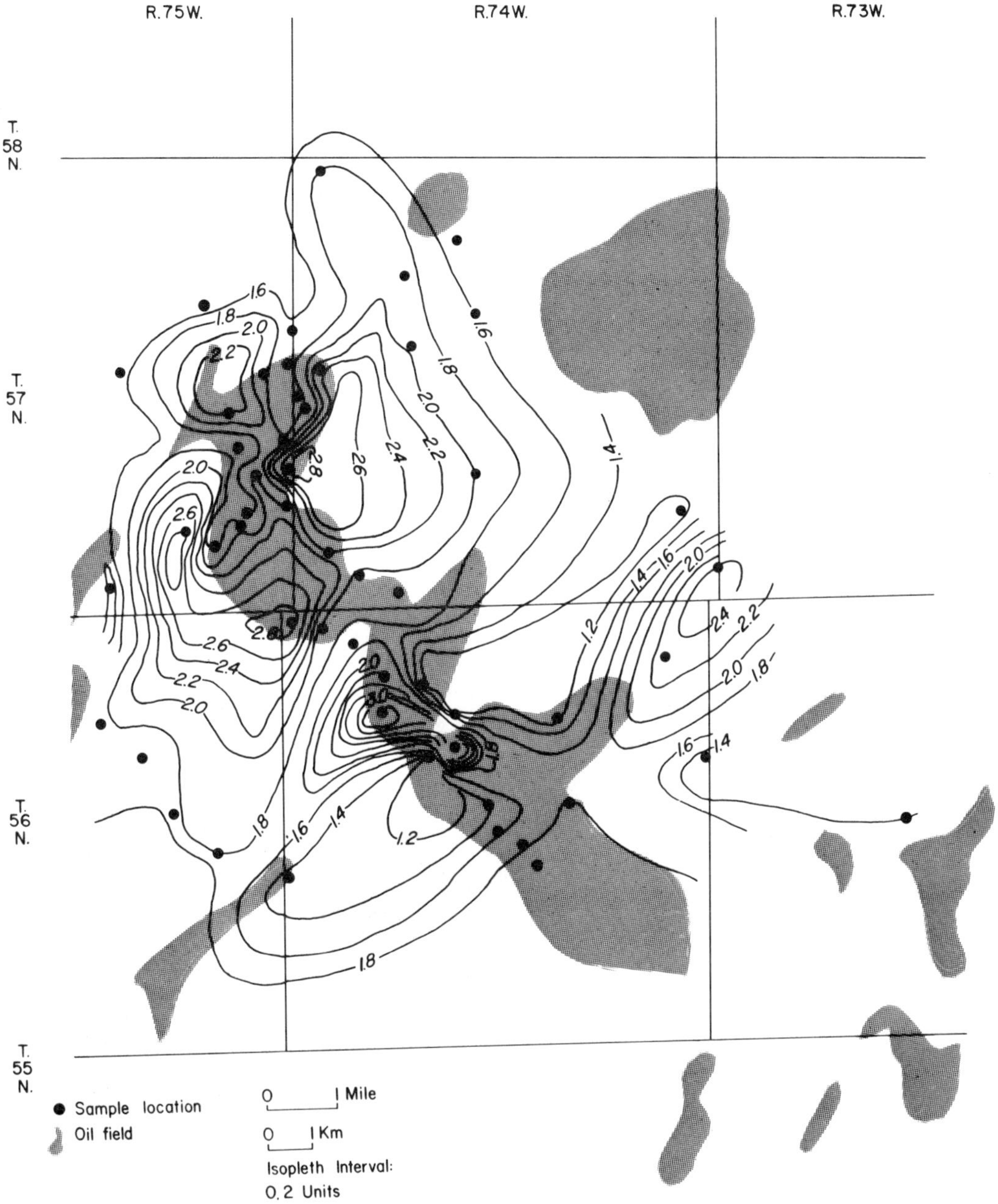

Fig. 14. Variation in log of manganese/iron ratios (×100) in pine needles, Recluse oil field, Wyoming. Isopleth interval: 0.2. From Dalziel and Donovan, 1980. Shaded areas show oil and/or gas production limits at Recluse and in the immediate area.

plants even though there is no measurable water-soluble pool of dissolved ions (Jenny and Overstreet, 1939; Hodgson, 1968; Chaney, Brown, and Tiffin, 1972).

Analysis of the iron and manganese content of airborne particulates, collected at low altitude (30 m) by a fixed-wing aircraft, suggests that the plants shed chemically anomalous particulates, probably from their leaf surfaces, and these particulates are borne upward by near-surface convective plumes—nonrotating columns of rising, warm air. Generation of con-

Fig. 15. Variation of log of manganese/iron ratios (×100) in sage leaves, Recluse oil field, Wyoming. Isopleth interval: 0.2. From Dalziel and Donovan, 1980. Shaded areas show oil and/or gas production limits at Recluse and in the immediate area.

vective plumes requires proper sunlight and wind conditions (Taylor, 1958; Kaimal and Businger, 1970). Release of elements to the atmosphere from evergreens has been documented (Curtin, King, and Mosier, 1974), and recent work indicates that the primary release is in particulate form (Beauford, Barber, and Barringer, 1975). Barringer (1977) considered it highly probable that heavy metals and trace elements are deposited at the leaf surface during the transpiration process, and some elements may be concentrated in tissue near

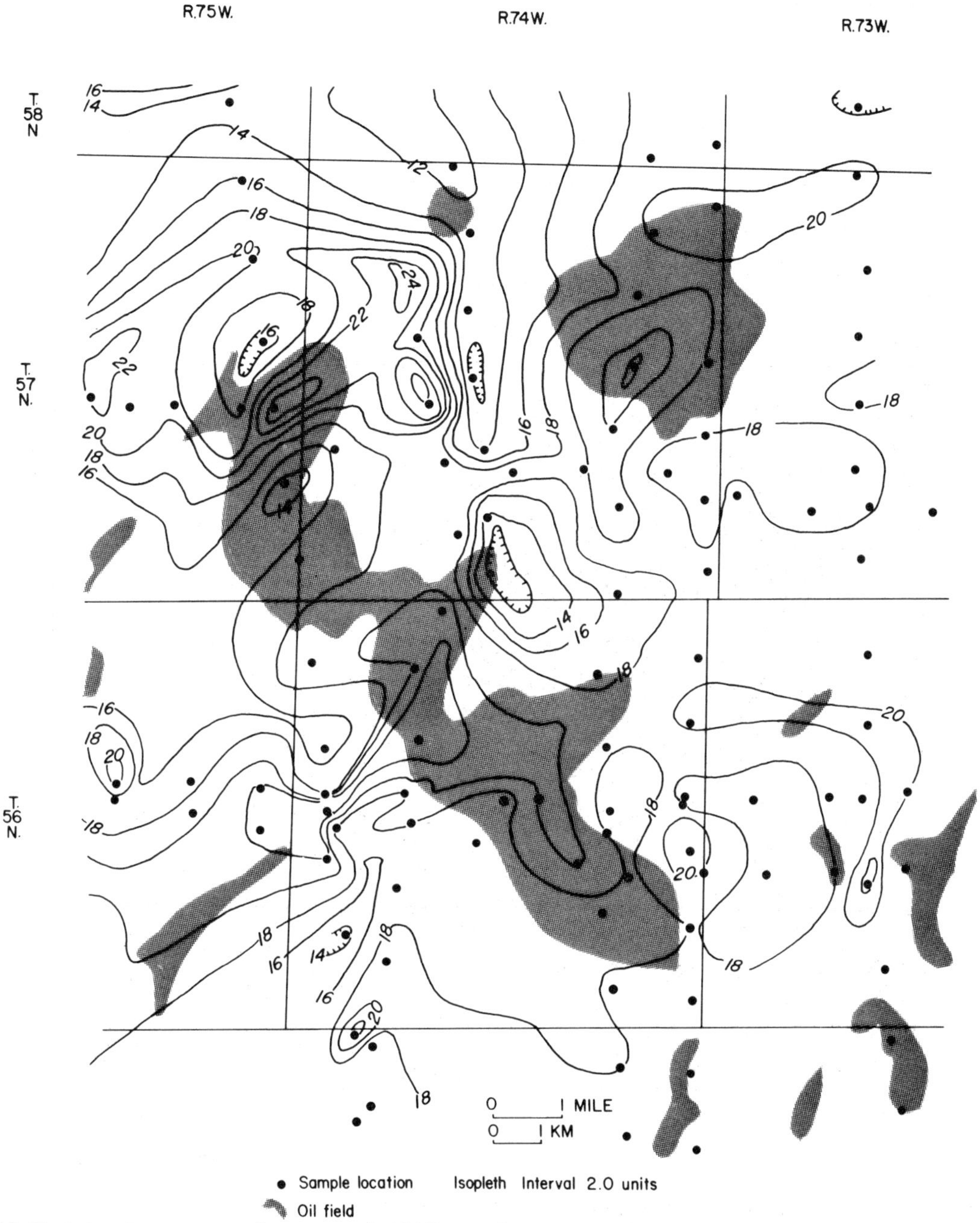

Fig. 16. Variation in manganese/iron ratios (×1000) in airborne particulates, Recluse oil field, Wyoming. Isopleth interval: 2.0. Shaded areas show oil and/or gas production limits at Recluse and in the immediate area.

the leaf surface. Particulates are shed continuously by abrasion and the rubbing of leaves against each other.

The upward flux of particulates during atmospheric mixing is greater than the downward sedimentation-controlled flux (Barringer, 1977), and the sampling equipment, used effectively, discriminates between predominantly local particulates and those of ambient background. Samples are collected on an advancing special tape. Tape advancement is controlled either by predetermined time settings or by mass loading and is correlated photographically and by time with the flight path.

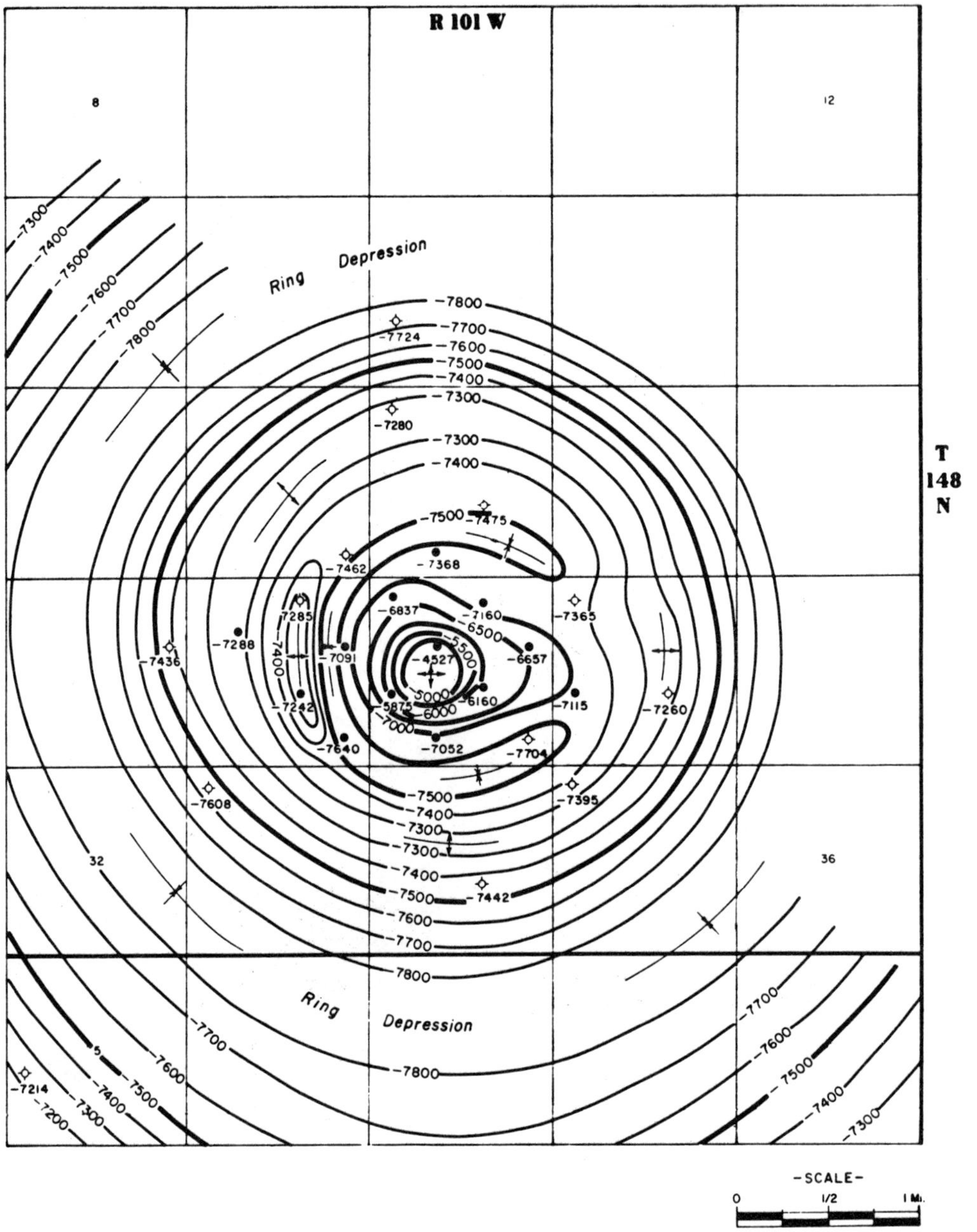

Fig. 17. Structure-contour map of the Red Wing Creek oil field, North Dakota. Structure contours on top of Mission Canyon Limestone (Mississippian). Contour interval: 100 ft (scale: ¾ in = 1 mi). From Brenan, Peterson, and Smith, 1975.

Particulates are subsequently analyzed for multielemental composition by using a laser to vaporize the material and channeling the vapors to an inductively coupled plasma-emission spectrometer. These experiments are continuing; some details of the theory, instrumentation, and procedures have been reported (Barringer, 1977; this volume).

Figure 16 is a preliminary map showing the areal variation of the manganese/iron ratios of airborne particulates sampled over the Recluse oil field in the fall of 1978. Samples collected during 30 sec. intervals along the flight path were analyzed, averaged, and plotted; each sample point on the map represents the mean ratio value for material collected 15 sec. fore

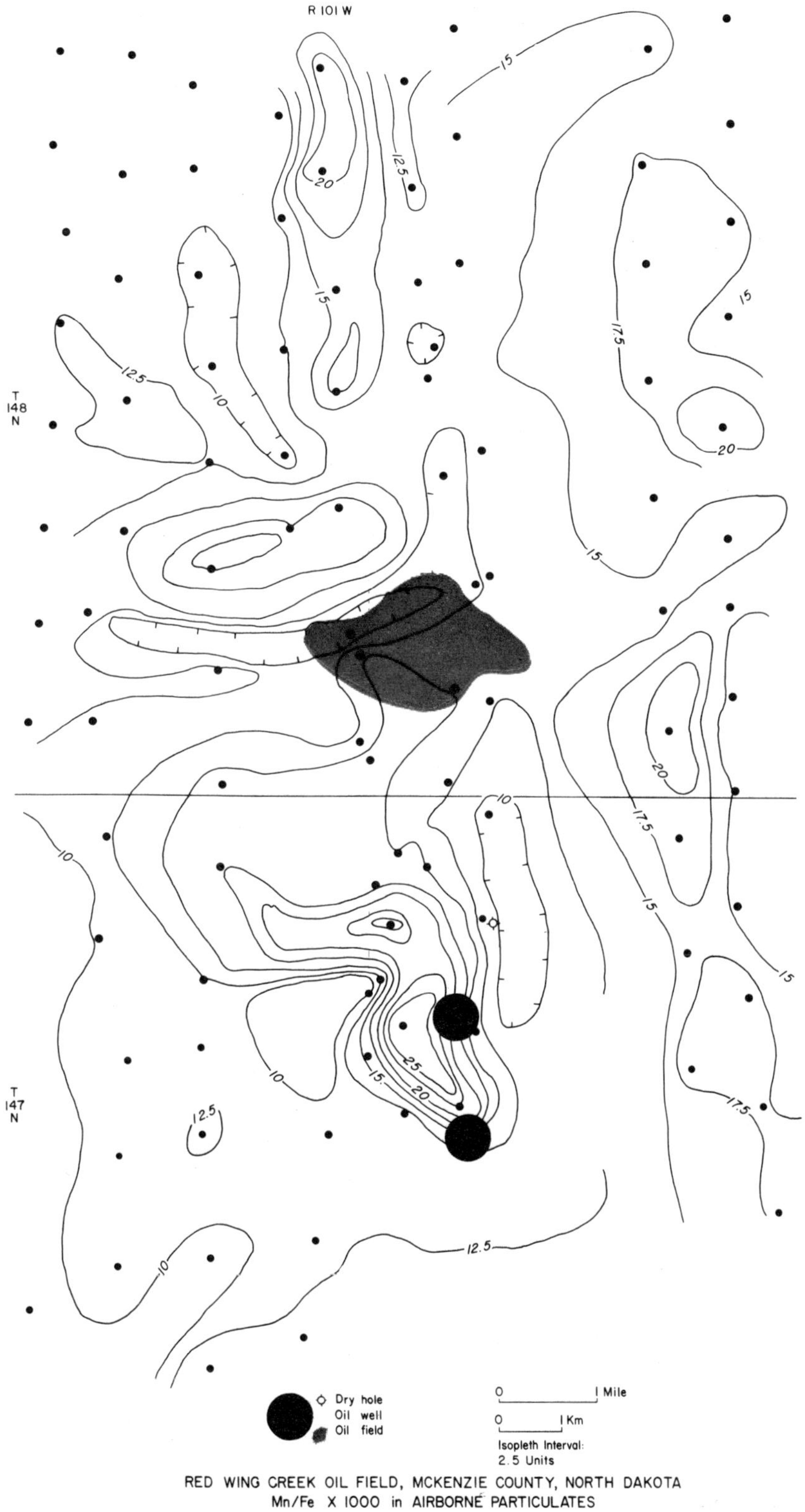

Fig. 18. Variation in manganese/iron ratios (×1000) in airborne particulates, Red Wing Creek oil field, North Dakota. Isopleth interval: 2.5. Shaded area shows oil production limits at Red Wing Creek.

and aft of the plotted point. The mapped pattern of manganese/iron variation closely corresponds with that of sage and pine, suggesting that the airborne particulates in the area originated from vegetation.

The Red Wing Creek structure is in the central Williston basin of western North Dakota in McKenzie County. The highly deformed, oil-bearing uplift, generally circular in plan, consists of a central uplift, ring depression, and outer rim (Figure 17), and has been interpreted as a meteoric impact crater (Brenan, Peterson, and Smith, 1975). The central uplift is chaotically thrust faulted throughout the Mississippian to Triassic rock interval. Deformation occurred in Triassic or Early Jurassic time, and oil is produced from fractured Mississippian Mission Canyon Limestone. The maximum oil column is 845 m thick. As of October 1975, 2,070,158 barrels of oil have been produced (Brenan et al., 1975).

The ground surface at Red Wing Creek is generally flat, but it gains relief in all directions; terrain becomes rolling to the north and northeast, but it is sharply dissected to the south and southwest. Crops cover most of the ground around and north of Red Wing Creek, whereas low native shrubs predominate to the south.

Interpretation of manganese/iron ratios in airborne particulates sampled over Red Wing Creek is incomplete because of a lack of ground-truth data. However, an area of moderately high manganese/iron ratios, separated from the producing area by an elongate zone of low ratios, occurs immediately to the northwest (Figure 18). A sigmoidal pattern of anomalously high values is centered about 5 km south of Red Wing Creek over more densely vegetated terrain. Recent wells drilled for oil near the southern and eastern edges of this anomaly resulted in the discovery of the Bull Moose field. Further interpretation of the Red Wing Creek data will require extensive ground-truth sampling.

SUMMARY

Preliminary evaluation of a variety of airborne and spaceborne remote-sensing techniques suggests their cost-effective utility in isolating evidence of epigenetic phenomena in rocks, soils, and plants over oil fields. This report has reviewed in a cursory way some of the more promising techniques for remote extraction of chemical information useful to explorationists. Refinements and additional experiments are now being designed in order to complete this evaluation.

References Cited

Al-Shaieb, Zuhair, Olmsted, R. W., Shelton, J. W., May, R. T., Owens, R. T., and Hanson, R. E., 1977, Uranium potential of Permian and Pennsylvanian sandstone in Oklahoma: American Association of Petroleum Geologists Bulletin, v. 61, p. 360-375.

Bailey, P. T., 1953, The development and history of the Garza field, Garza County, Texas: Lubbock, Texas, Texas Technological College, M.S. Thesis, 62 p.

Barringer, A. R., 1977, AIRTRACE—an airborne geochemical exploration technique: U.S. Geological Survey Professional Paper 1015, p. 231-251 (Proceedings, 1st Annual William T. Pecora Memorial Symposium, October 1975, Sioux Falls, South Dakota, P. W. Wall and W. A. Fisher, eds.)

Beauford, W., Barber, J., and Barringer, A. R., 1975, Heavy metal release from plants into the atmosphere: Nature, v. 256, July 3, 1975, p. 35-37.

Brenan, R. L., Peterson, B. L., and Smith, H. J., 1975, The origin of Red Wing Creek structure: McKenzie County, North Dakota: Wyoming Geological Association, Earth Science Bulletin, v. 8, no. 3, Sept. 1975, 41 p.

Chaney, R. L., Brown, J. C., and Tiffin, L. O., 1972, Obligatory reduction of ferric chelates in iron uptake by soybeans: Plant Physiology, v. 50, p. 208-213.

Chavez, P. S., Jr., 1975, Atmospheric, solar and MTF corrections for ERTS digital imagery (abs.): American Society of Photogrammetry Proceedings, p. 69-69A.

Condit, C. D., and Chavez, P. S., Jr., 1979, Basic concepts of computerized digital image processing for geologists: U.S. Geological Survey Bulletin 1462, 16 p.

Curtin, G. C., King, H. D., and Mosier, E. L., 1974, Movement of elements into the atmosphere from coniferous trees in subalpine forests of Colorado and Idaho: Journal of Geochemical Exploration, v. 3, p. 245-263.

Dalziel, M. C., and Donovan, T. J., 1980, Biogeochemical evidence for sub-surface hydrocarbon occurrence, Recluse oil field, Wyoming—preliminary results: U.S. Geological Survey Circular [in press].

Donovan, T. J., 1974, Petroleum microseepage at Cement, Oklahoma—evidence and mechanism: American Association of Petroleum Geologists Bulletin, v. 58, p. 429-446.

Donovan, T. J., Barringer, A. R., Foote, R. S., and Watson, R. D., 1975, Low altitude remote sensing experiments at Cement and Davenport oil fields, Oklahoma (abs.): Geophysics, v. 41, p. 350.

Donovan, T. J., and Dalziel, M. C., 1977, Late diagenetic indicators of buried oil and gas: U.S. Geological Survey Open-file Report 77-816, 44 p.

Donovan, T. J., Forgey, R. L., and Roberts, A. A., 1979, Aeromagnetic detection of diagenetic magnetite over oil fields: American Association of Petroleum Geologists Bulletin, v. 63, p. 245-248.

Donovan, T. J., Friedman, Irving, and Gleason, J. D., 1974, Recognition of petroleum-bearing traps by unusual isotopic compositions of carbonate-cemented surface rocks: Geology, v. 2, p. 351-354.

Donovan, T. J., Termain, P. A., and Henry, M. E., 1979, Late diagenetic indicators of buried oil and gas: II, Direction detection experiment at Cement and Garza oil fields, Oklahoma and Texas, using enhanced LANDSAT I and II images: U.S. Geological Survey Open-file Report 79-243, 45 p.

Eliason, E. M., Chavez, P. S., Jr., and Soderblom, L. A., 1974, Simulated "true color" images from ERTS data: Geology, v. 2, p. 231-234.

Gableman, J. W., 1977, Migration of uranium and thorium—exploration significance: American Association of Petroleum Geologists, Studies in Geology, no. 3, 168 p.

Goetz, A. F. H., 1975, Use of ERTS and other correlative data, *in* Goetz, A. F. H., and others, application of ERTS images and image processing to regional geologic problems and geologic mapping in northern Arizona: Jet Propulsion Laboratory, California Institute of Technology (National Aeronautics and Space Administration) Technical Report 32-1597, p. 4-12.

Halbouty, M. T., 1968, Giant oil and gas fields in the United States: American Association of Petroleum Geologists Bulletin, v. 52, p. 1115-1151.

Henry, M. E., and Donovan, T. J., 1978, Isotopic and chemical data from carbonate cements in surface rocks over and near four Oklahoma oil fields: U.S. Geological Survey Open-file Report 78-927, 17 p.

Hodgson, J. F., 1968, Theoretical approach for the contribution of chelates to the movement of iron to rocks: International Congress, Soil Science, Transactions, 9th, Adelaide, v. 2, p. 229-241.

Jenny, H., and Overstreet, R., 1939, Cation exchange between rocks and soil colloids: Soil Science, v. 47, p. 257-272.

Jones, V. L., and Lyons, P. L., 1964, Vertical-intensity magnetic map of Oklahoma: Oklahoma Geological Survey Map GM-6, scale 1:750,000.

Kaimal, J. S., and Businger, J. A., 1970, Case studies of a convective plume and a dust devil: Journal of Applied Meteorology, v. 9, p. 612-620.

Krauskopf, K. B., 1957, Separation of manganese from iron in sedimentary processes: Geochimica et Cosmochimica Acta, v. 12, p. 61-68.

McKay, E. J., and Hyden, H. J., 1956, Permian of north Texas and southern Oklahoma, *in* Geologic investigations of radioactive deposits; semiannual report, June 1 to November 30, 1956: U.S. Geological Survey Report TEI-640, p. 208-216.

Moffatt, H. H., 1973, Soil survey of Caddo County, Oklahoma: Washington, D.C., U.S. Department of Agriculture, Soil Conservation Service, 70 p.

Olmsted, R. W., and Al-Shaieb, Zuhair, 1975, Geochemical anomalies, uranium potential of south-central Oklahoma [abs.]: Geological Society of America, annual meeting, Salt Lake City, October 20-22, 1975, Abstracts with Programs, v. 7, no. 7, p. 1219.

Pandey, G. N., Tek, M. R., and Katz, D. L., 1974, Diffusion of fluids through porous media with implications in petroleum geology: American Association of Petroleum Geologists Bulletin, v. 58, p. 291-303.

Richardson, W. E., Grice, D. G., and Putnam, L. A., 1975, Soil survey of Garza County, Texas: Washington, U.S. Dept. of Agriculture, Soil Conservation Service, 84 p.

Rowan, L. C., Wetlaufer, P. H., Goetz, A. F. H., Billingsley, F. C., and Stewart, J. H., 1974, Discrimination of rock types and detection of hydrothermally altered areas in south-central Nevada by the use of computer-enhanced ERTS images: U.S. Geological Survey Professional Paper 883, 35 p.

Taranik, J. V., 1978, Principles of computer processing of LANDSAT data for geologic applications: U.S. Geological Survey Open-file Report 78-117, 50 p.

Taylor, R. J., 1958, Thermal structures in the lowest layers of the atmosphere: Australian Journal of Physics, v. 11, p. 168-176.

Termain, P. A., Donovan, T. J., and Chavez, P. C., Jr., 1980, Integration of geological, geochemical and geophysical spatial data of the Cement oil field, Oklahoma, Test Site [abs.]: Pecora Conference, 6th Annual, Sioux Falls, April 13-17, Program with Abstracts [in press].

Toth, J., 1962, A theory of groundwater motion in small drainage basins in central Alberta, Canada: Journal of Geophysical Research, v. 67, p. 4375-4387.

Toth, J., 1963, A theoretical analysis of ground water flow in small drainage basins: Journal of Geophysical Research, v. 68, p. 4795-4812.

Toth, J., 1972, Ground water motion in drainage basins, *in* Fairbridge, R. W., Encyclopedia of geochemistry and environmental sciences: New York, Van Nostrand Reinhold, p. 478-487.

Woncik, John, 1969, Recluse field, Campbell County, Wyoming: Mountain Geologist, v. 6, no. 4, p. 221-226.

EXPLORATION RADIOMETRICS: POSTSURVEY DRILLING RESULTS

Richard C. Weart and George Heimberg

SUNMARK EXPLORATION COMPANY

Sunmark Exploration Company initiated a program in 1977 to investigate hydrocarbon exploration techniques. This study was prompted by the desire to evaluate various exploration methods for possible application in frontier areas where the geological and physical conditions have impeded the successful utilization of conventional tools, and to lower front-end petroleum exploration costs. Radiometrics was one of the methods chosen for evaluation, based primarily on the plausibility of the theoretical considerations supporting the method, and the very low field survey costs.

Radiometric surveying, first used in the 1920s for hydrocarbon exploration, is considered to be a direct oil and gas finding method. With the exception of the drill bit, all direct hydrocarbon exploration methods are categorized as unconventional by the majority of geoscientists and accepted as viable by only a few. The primary reason for this lack of acceptance is failure of the explorationist to conceptualize and accept as valid the vertical migration process. This process, which most unconventional techniques depend upon, is predicated on the principle that formation fluids and gases do ascend vertically, or nearly so, through the sedimentary section, and that as a result of this migration surface manifestations of anomalous conditions at depth, such as hydrocarbon accumulations, are measurable and mappable. Geoscientists have largely been conditioned to think in terms of impermeable barriers in connection with trapping mechanisms, and of lateral rather than vertical migration with respect to porous and permeable aquifers. Many investigators, however, are marshaling a large amount of evidence which is providing compelling arguments for vertical migration and theoretical support for the empirical approach of case histories.

Assuming that vertical migration of formation fluids is an active, nearly continuous, and universal mechanism in sedimentary areas, it is also the necessary agent for the transport toward, and to, the earth's surface of all kinds of matter in solution. This includes the radioactive, soluble deterioration products of the igneous basement and the overlying sediments. The capability of hydrocarbons to adsorb radioactive materials, as well as the shielding effect that subsurface hydrocarbon accumulations have to vertical fluid migration, creates a low radiation anomaly above such an accumulation—the well-known chimney effect (Figure 1).*

The high radiation halo associated with such an anomaly usually approximates the outline of a probable oil and gas accumulation of unknown magnitude and indeterminate depth. Hydrodynamic conditions, fracture patterns, coning effects from diffusion, and other geologic conditions can often distort and shift the halo pattern. The halo of high radiation values is probably due to an increase in the vertical flow volume of formation fluids diverted around, and also expelled from, the hydrocarbon trap, as indicated by the arrows. The water expelled from the trap as it is replaced by hydrocarbons is most likely to contain greater amounts of radioactive material in solution than normal formation waters because of its proximity to increased concentration of radioactive elements normally found in association with hydrocarbons.

As the formation fluids with their dissolved components rise vertically through the overlying sedimentary burden, exposure to changing chemical and physical environments promotes the precipitation of various elements formerly

*See atlas for Fig. 1 and Figs. 4-14.

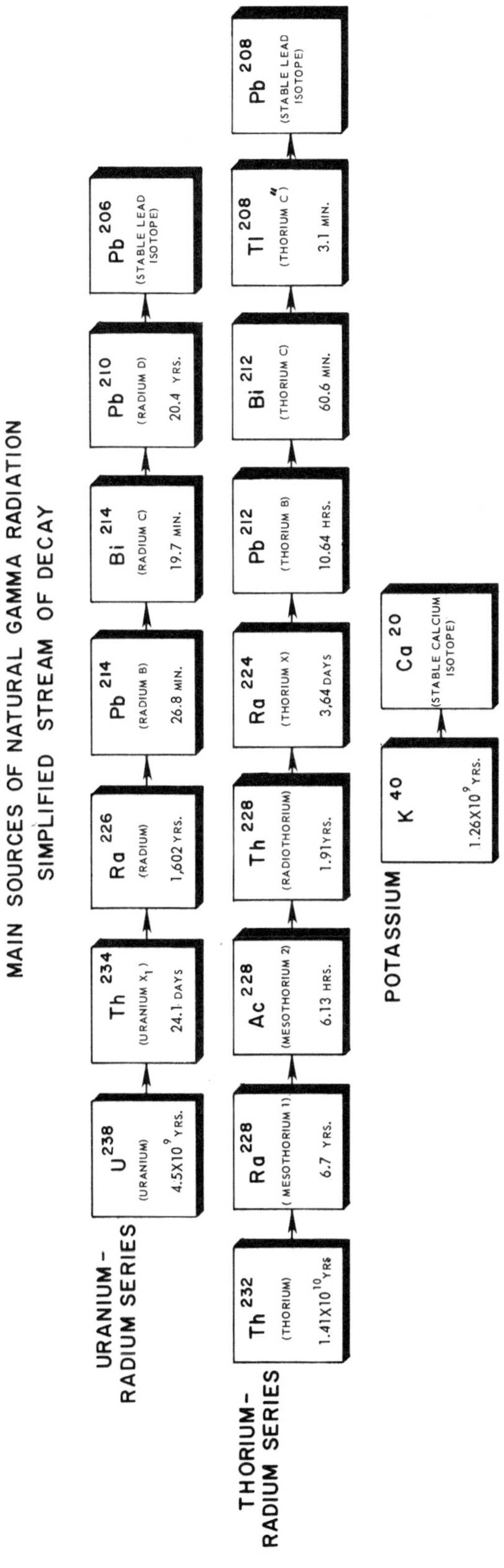

Fig. 2.

held in solution. Included among these elements are the gamma ray emitters uranium, thorium, and potassium-40. Each of these elements has a long half-life and tends to provide a permanent gamma ray source in the near-surface soils and rocks where it has been emplaced. Major members of the uranium and thorium decay series, several of which are high gamma contributors, are indicated in Figure 2. Uranium is by far the most soluble in water compared to thorium and potassium-40. It and its daughter products are, for this reason among others, the greatest contributors to the total gamma radiation intensity in most areas.

Total gamma radiation was measured in all field surveys. It is recognized that cosmic radiation and other sources contribute to the total measured intensity, which is in turn influenced by a number of different variables, such as stratigraphic-pedological changes, variations in moisture, temperature, and pressure, topography, faulting, and others. False anomalies can usually be detected by close observation of the terrain, response of the instruments, and size, shape, and character of the anomaly. Experience and knowledge of the geology in the survey area is very helpful if not necessary in this regard. Diurnal variations in radiation intensity occur and can be detected by instrument drift. The diurnal changes occur rather slowly and can be largely discounted because of the short line segments usually surveyed and the resetting of the recorder to a standard base line position at the end of each segment.

All of our surveys were conducted on a contract basis. A vehicle-mounted 5½-foot-long, 5,000 cu in ionization chamber was used to measure total gamma radiation. The direct current generated by the chamber was preamplified and then continuously recorded with a Hewlett-Packard 7028-A gas chromatograph recorder synchronized with the odometer of the vehicle. The recorded traces were manually and/or computer reduced to ten averaged radiometric values per survey mile, transferred to a base map, and used for contouring the radiometric signature of the surveyed area. Sites inaccessible to four-wheel-drive vehicles were surveyed with a 1,000 cu in ionization chamber that is portable by two men.

CLASSIFICATION OF RADIOMETRIC ANOMALIES

CLASS	SURVEY GRID	SIZE AND SHAPE.	INTERNAL VALUES	EDGE CONTRAST	HALO	REMARKS
I	GOOD DETAIL	FAVORABLE	CONSISTENTLY LOW	MOSTLY 10 UNITS OR MORE	WELL DEFINED	FULLY ACCEPTABLE
II	SEMI DETAIL	FAVORABLE	CONSISTENTLY LOW WITH FEW EXCEPTIONS	MOSTLY 10 UNITS OR MORE	EVIDENT ALONG MOST SURVEY LINES	SHOULD MOVE UP TO CLASS 1 WITH ADDITIONAL CONTROL
III	SPARSE	APPARENTLY FAVORABLE	VARIABLE	NORMALLY SLIGHTLY BELOW 10 UNITS	VAGUELY DEFINED	FURNISHES LEADS FOR ADDITIONAL WORK
IV	NORMALLY INSUFFICIENT BUT MAY VARY	NORMALLY UNFAVORABLE BUT MAY VARY	OFTEN HIGHLY VARIABLE	NORMALLY VERY LOW BUT MAY VARY	POOR OR ABSENT	NOT USABLE. ONE OR MORE UNACCEPTABLE PARAMETERS PRESENT

Fig. 3.

Reconnaissance and semidetail surveys were usually conducted utilizing a grid density of one mile or larger depending on the estimated areal extent of the objective. Interesting anomalies were then detailed with a finer grid. Independent interpretations were made by the contractor and Sunmark personnel.

The size, shape, and definition of radiometric anomalies are often a function of survey grid density as well as of interference by geological, meteorological, and cultural parameters, some of which have been mentioned previously. All of these can and do have a bearing on the significance of a given anomaly, and thus on its usefulness in exploration. In order to rank the mapped anomalies according to their significance, a classification system has been developed based on a critical analysis of the radiometric values, interpretation, and tightness of the grid survey. The fourfold classification (Figure 3), while obviously subjective, appears to be quite effective.

The four classes are based on the quality of five parameters. Tightness of the survey grid is important and should be dependent on the size of the objective expected in a particular area. Size, shape, and orientation of an anomaly are often the factors which, in conjunction with other geologic information, determine the economic viability and the validity of the prospect or lead. The lower and more consistent the radioactive values are within the halo, the more attractive the anomaly. The larger and sharper the difference in radioactive values between the halo, normal background, and the area within the halo, the greater the significance or validity of the anomaly, although it cannot yet be claimed that radiometric exploration is a quantitative or depth delineation tool. Finally, the more continuous the halo, in terms of areal definition, the more significant the anomaly.

Sunmark currently has in excess of seventy radiometric surveys on file. Distribution is as indicated in Figure 4 in the contiguous Lower-48 states. In some cases, one dot represents more than one survey. Well over 50% of the surveys were conducted by or for Sunmark. Sites were chosen with various objectives in mind in order to determine and compare radiometric response to oil, gas, and gas-oil reservoirs, shallow, intermediate, and deep hydrocarbon accumulations, and such other factors as soil, moisture, terrain, coal beds, and lithology.

Surveys were tied into known production where possible in order to determine radioactive response to a local field as compared to normal background. An attempt was made to coordinate surveys with future drilling, wildcat or development, by Sunmark or others in the industry in order to "call the shots" as much as possible.

As in most unconventional methods, repeatability of data is much desired and needed for authenticity purposes. The data of portions of many of these survey sites have been repeated with good results. Data from sites revisited, even after a few years' hiatus, have shown a high degree of repeatability.

Figure 5a shows the track of one of the radiometric survey lines run in 1973 across the approximately one-mile-wide Albion-Scipio Field in southern Michigan, shown in yellow, and rerun in 1978. Both the 1973 and 1978 recorder traces show the field anomaly and the somewhat unimpressive halos in a similar manner (Figure 5b). In Figure 5c the 1978 trace is superimposed on the 1973 trace. The good congruence is obvious. Notice that the *only* radiometric anomaly along this line segment occurs over Albion-Scipio Field.

WEST FELDA FIELD, FLORIDA

In 1971, when it was technically very difficult or even impossible to carry out seismic exploration in southern Florida, Sun Oil Company became interested in the idea of conducting a radiometric reconnaissance survey covering the updip portion of the Sunniland producing trend. Such a survey appeared even more feasible because of the then ongoing "100-year" drought which had lowered water levels by some ten feet. By that time, only the southeastern portion of West Felda Field had been developed, and radiometrics, it was hoped, would provide clues concerning the ultimate size and shape of the field.

Prior to final approval of the survey, a field check of the "halo" or "chimney" concept was suggested based on the proposition that a "chimney" should be recognizable at any given subsurface level above the oil body. Since Sun's wells in the West Felda Field had been drilled and completed in an identical manner and logged with the very same logging tools, and since the Lower Cretaceous strata in this vicinity consist of chemically precipitated carbonates and evaporites on a tectonically quiescent shelf, exceptional uniformity from well to well could be anticipated and conditions for such a check appeared very favorable.

The gamma ray curves from eight wells along a line running roughly north-south across the east end of the West Felda Field were used for this purpose (Figure 6a). Seventeen individual gamma curve points between the top of the Lower Cretaceous at 6,000 ft and the Middle Sunniland pay at 11,500 ft were found to be correlatable in all eight wells. The intensities of these points were measured in API units and plotted on a section (Figure 6b). Their combined averages of all seventeen points are shown diagrammatically on the profile.

A definite radiometric deficiency, shown in yellow, is evident above the field, although well 28-3 registers somewhat higher for an unknown reason. A halo, shown in red, is also indicated. The pay thickness in these wells is shown in green at the bottom of the diagram. Although one may want to construe a quantitative relationship between pay thickness and radiometric values, no such conclusion can or should be drawn at this time.

Postsurvey drilling results (Figure 7) are indicated by the large dots in West Felda Field and its immediate vicinity. Green dots are producers. The pink dots are dry holes. The pink line indicates the position of the previously discussed radiometric profile.

In the color scheme, yellow shows radiometrically deficient areas, while red means higher than background values. Thus, fields and radiometric prospects should appear in yellow, halos and other unfavorable anomalies in red. Small dots indicate presurvey wells, while large dots mark those wells that were drilled after the completion of the radiometric survey. No classification or ranking of radiometric anomalies, with one exception, is shown in this and the following illustrations.

Figure 7 shows only a portion of the survey. Felda Field on the right, developed prior to the survey, is radiometrically well defined with a halo indicated along most of the periphery. The larger West Felda anomaly on the left shows an even better developed, more continuous halo. Postsurvey drilling at West Felda agrees well with the radiometric data.

The dry hole on the large anomaly in T46S, R27E is the Sun 14-4 Turner, which had oil and gas shows in cores and tested a very small amount of free oil. Northwest of Felda Field, Tribal drilled two wells on Sun farmout on a small radiometric anomaly. Both wells had good shows in cores and recovered free oil, up to 120 ft on drill stem test. The wells were plugged as noncommercial based on 1971 prices. This is a good case in point showing that radometrics can only indicate the presence or absence, but not the quantity of hydrocarbons in a reservoir. All dry holes should be checked for bona fide shows in a statistical evaluation for radiometric data.

The somewhat limited statistical base of forty-eight postsurvey wells indicates that 81% of the wells completed as producers fall on a radiometrically favorable anomaly, while two-thirds of the dry holes were predictable. The combined percentage shows that seven out of ten wells were predicted correctly. Ambiguous wells and wells over a mile from an actual radiometric survey line are excluded from the statistics.

LITTLE CREEK FIELD, MISSISSIPPI

Sun's earliest association with radiometric surveying occurred in 1959 in conjunction with a survey conducted for independents Willie Hughes and John Noble. The discovery well for McComb Field was drilled on a Sun farmout solely on the strength of this survey. Details are given in an article by George Miller published in the February 13, 1961, issue of the *Oil and Gas Journal.* At the time of the survey there were no wells in the county. Figures 8a and 8b are fashioned after an illustration in Miller's articles. Figure 8a shows the postsurvey drilling results in the Pike County portion of the Little Creek Field as of 1961, the year of publication. The congruence is obvious. The published illustration shows a radiometric prospect west of the field, presumably separated by a narrow unproductive area.

In Figure 8b the large dots show all the wells drilled *after* 1961 at Little Creek. Obviously, the radiometric prospect held up well. The statistical results, based on 345 postsurvey control wells, show the radiometric predictions to be 82% correct for the entire survey area, which includes McComb Field to the southwest.

SOUTHWEST HINTON AREA, OKLAHOMA

Beginning in 1977 Sunmark became progressively more involved with radiometric surveying. One of the earlier surveys was conducted in Caddo County, Oklahoma (Figure 9), while Sun was in the process of developing Lower Pennsylvanian Marchand stratigraphic sand production in Northwest Sickles Field. Some wells had been completed, others were making hole, and some were locations. The nature, trend, and relationships of the various sand bodies were not very clear.

Radiometrics indicated several northwesterly oriented, weakly connected prospective trends. Subsequent drilling results generally agree with this. The three nonproducers aligned west-east near the south end of the Northwest Sickles Field within the radiometric anomaly in T11N, R13W encountered 7, 1, and 4 ft of pay, respectively. Lengthy completion attempts were made in two of them before abandonment. Wildcat and development drilling is currently taking place a short distance south of, and on trend with, the untested interesting radiometric anomalies in the southwest corner of the map.

Because of the rather small number of control wells available, both presurvey and postsurvey wells are included in the statistics. The percentage of agreement with radiometric data is very high, approaching 85%.

SHANNON TREND, CAMPBELL COUNTY, WYOMING

Several radiometric projects were carried out by Sunmark in 1977 along the Cretaceous Shannon, Ferguson, and Muddy trends in Wyoming's Powder River Basin. The only trend with statistically sufficient postsurvey drilling is the stratigraphic Shannon sand trend (Figure 10) in Campbell County and its continuation into northern Converse County, where it mingles with the oil and gas production from Cretaceous Dakota sands in Powell Field.

Hartzog Draw Field, which had been discovered a short time prior to the survey, had seen some drilling essentially as far southeast as T45N. Drilling in the remainder of the map area was fairly sparse and scattered at the time. Since the survey, 141 wells have been added in the area.

Interestingly—but possibly a matter of coincidence—the pattern of the radiometric anomalies in T41-42N agrees with the deterioration of the Shannon sand condition in this vicinity.

Statistics, based on the 141 postsurvey wells, again show about 80% correct calls for the radiometric survey.

BEECHER ISLAND AREA, LAS ANIMAS ARCH

The Beecher Island area is one of nine areas in eastern Colorado and adjacent parts of Kansas and Nebraska that were radiometrically surveyed as part of an experimental unconventional exploration project carried out by Sun in 1978. Figure 11 shows the actual appearance of one of our radiometric maps after evaluation and ranking of the anomalies. Class I through IV anomalies are color coded as in Figure 3.

The Beecher Island Field anomaly is in the center. It is the largest Niobrara gas field in the area and demonstrates a classic radiometric anomaly. Most Niobrara production in the area is structurally controlled. The several undrilled east-west oriented anomalies north of Beecher Island should be interpreted as one single prospect, since the high radiometric values within this prospect are probably caused by major faulting.

The significant regional Goodland fault lineament runs diagonally SE-NW through the area. It follows essentially a zone of unfavorable radiometric values that separates the Beecher Island area anomalies from a belt of similar class I anomalies southeast of the lineament. The cluster of green dots near the center of the west edge of the map are postsurvey wells in Mildred Field. Mildred Field was not surveyed and thus does not appear as a defined anomaly.

The large anomalies in the southeast corner of the map were recommended as good prospects. Subsequently, Murfin drilled a Niobrara gas discovery, shown by the green dot, in December, 1978. Since then, additional oil and gas offset wells have been completed north of the discovery.

The radiometric survey was 68% correct based on 28 post-survey wells (large dots in Figure 11), and of 197 presurvey wells 71% are correctly called by the survey. The difference in accuracy between presurvey and postsurvey drilling is slight.

PACHUTA CREEK AREA, MISSISSIPPI

A semidetailed survey in the Pachuta Creek area, Clarke County, Mississippi (Figure 12), shown here by permission of Rush Oil Company, was made in 1966 prior to any production in Clarke County. The high incidence of discoveries to radiometric anomalies that were made in the eight years subsequent to the survey in the Pachuta Creek area is striking.

The radiometric anomalies of the Nancy trend agree quite well with fault blocks at the Smackover level. The nearly undrilled north half of the Pachuta Creek anomaly lies in a large east-west trending graben.

The results from 147 wells drilled postsurvey indicate that about 73% were called correctly by radiometrics.

COYOTE CREEK AND MEDICINE POLE HILLS FIELDS, NORTH DAKOTA

A 1971 radiometric survey in Bowman County, North Dakota (Figure 13), within the Williston Basin, was made for two independents, Youngblood and Wilhite. The three wells that are circled had been drilled. The survey was made to help determine whether the two discoveries would develop into a single field or would remain separated. Radiometrics apparently predicted correctly the separation between Coyote Creek Field in the north and Medicine Pole Hills Field in the south.

The thirty postsurvey wells show a statistical average of 66%, two out of three correct.

WEST MORK LAKE AREA, NORTH DAKOTA

A final example (Figure 14) is a survey in the West Mork Lake area, Ward County, North Dakota, made by an independent contractor for Youngblood and Wilhite in 1971. The Mississippian sand reservoir is linear and stratigraphic in nature. Good halo development is evident. The large dots designate the locations of the discovery wells for the two fields drilled prior to the survey. Radiometrics indicate a nonproductive zone or barrier between the two fields.

The statistical figures for this particular case include both presurvey and postsurvey wells. The correct call rate of 86% is exceptionally high.

STATISTICAL SUMMARY

Figure 15 tabulates the statistical results of the examples previously shown, with the exception of the Clarke County, Mississippi, area. In order to broaden the data base, however, the Las Animas Arch statistics include the postsurvey wells of *all* nine areas of Sunmark's 1978 experimental project.

The yellow bars indicate the percentage of producers drilled in radiometrically favorable areas, the red bars the dry holes drilled in radiometrically unfavorable areas, and the green bars the combination of the two. The appropriate percentage figures are given in the last column.

As is shown, the radiometric results are consistently over 50% correct and are often in the 70% or better range. The same holds true for almost all other areas that we have approached statistically. It does not seem to matter whether only postsurvey wells or all wells are included.

The three bars on the bottom line give the total tally which says that, on the average, the radiometric method may be correct three out of four times. This figure closely agrees with domestic drilling statistics, compiled by various sources, which indicate approximately 75%

to 80% of all wells drilled result in producers.

A large number of radiometric surveys completed and in Sunmark's file are of recent vintage. Postsurvey drilling statistics are not yet available. Our experience to date strongly implies that the proper utilization of this type survey in an integrated exploration program can be effective in the reconnaissance phase for developing leads and prospects, and can provide supplementary evidence in localizing and evaluating prospects. Its place in hydrocarbon exploration as a cost-effective front-end tool and a risk-reducer, as well as a potential substitute for conventional exploration methods under certain geological conditions, should be seriously considered by the explorationist.

HELIUM EMANOMETRY IN EXPLORING FOR HYDROCARBONS: PART I

Louis A. Pogorski and G. Stewart Quirt

CHEMICAL PROJECTS LIMITED

ABSTRACT

Helium emanometry is a new exploration tool that can be applied in the search not only for deeply buried uranium deposits but also for petroleum reservoirs. It has been tested over known oil and gas pools in regions including Alaska, Alberta, Colorado, Guatemala, Kansas, Michigan, Oklahoma, Ontario, South Dakota, and Texas.

Helium produced by the decay of uranium and thorium in the earth is swept into petroleum reservoirs where it accumulates and produces microseeps that can be detected at the surface. This gas is an ideal geochemical indicator of petroleum in that it is inert, stable, slightly soluble in ground fluids, practically nonadsorbable, and highly mobile.

This technique can be utilized by collecting soil, soil gas, water, or bottom sediment samples in reconnaissance, semidetailed or detailed sampling patterns under a wide range of climatic conditions varying from subtropical to arctic. The helium analyses are performed by gas source mass spectrometry. In addition to the helium determinations, a number of other parameters must be measured. Data interpretation is accomplished with the aid of computerized methods.

Examples of results from tests over known petroliferous research sites, which were conducted in cooperation with the U.S. Geological Survey, are presented in "Helium Emanometry in Exploring for Hydrocarbons, Part II."

INTRODUCTION

Our society enjoys a standard of living that is much higher than that experienced by most of the world population. The key to sustaining this life style is obtaining a continuing supply of energy. We have now reached the point in our history at which alternative energy sources, such as solar, nuclear, and geothermal, must be developed. Fossil fuels, particularly petroleum, have provided and are providing us with a great variety of products, ranging from plastics to gasoline. These commodities which we have all taken for granted cannot be easily replaced. Consequently, within the immediate future the demand for petroleum will continue to be great. In order to meet this demand, we must now search for deposits, such as stratigraphic traps and small reservoirs, which are most difficult to detect. This situation has created a need to develop new exploration approaches and techniques which will prove to be successful and economical in locating these types of pools. One of these new approaches involves the application of helium emanometry.

PROPERTIES AND ORIGIN OF HELIUM

Helium tends to be associated with most hydrocarbon reservoirs, as shown in the table below. It therefore can be considered as being a primary indicator of oil and gas.

TABLE I

Helium Content of Natural Gas from Various Regions of North America

LOCATION	HELIUM CONCENTRATION RANGE (ppm by volume)	REFERENCE
U.S.A.	100-80,000	Lipper and Wilcox (1960)
Texas	100-17,000	Pierce et al. (1964)
Alberta	100-3,820	Elworthy (1926) Anonymous (1972)
Saskatchewan	200-1,900	Elworthy (1926)
Ontario	100-8,000	Elworthy (1926) Anonymous (1970)

Helium is an almost ideal geochemical indicator in that it is chemically inert; physically stable; quite soluble in hydrocarbon and ground fluids at depth, but sparingly soluble in water under ambient conditions; practically nonadsorbable and having a low temperature solubility coefficient within moderate temperature ranges; highly mobile and therefore able to penetrate thousands of feet of overburden; and present in the atmosphere at a spatially and temporally low constant abundance level of 5.239 parts per million (ppm) by volume (Gluekauf, 1946). Helium has two stable

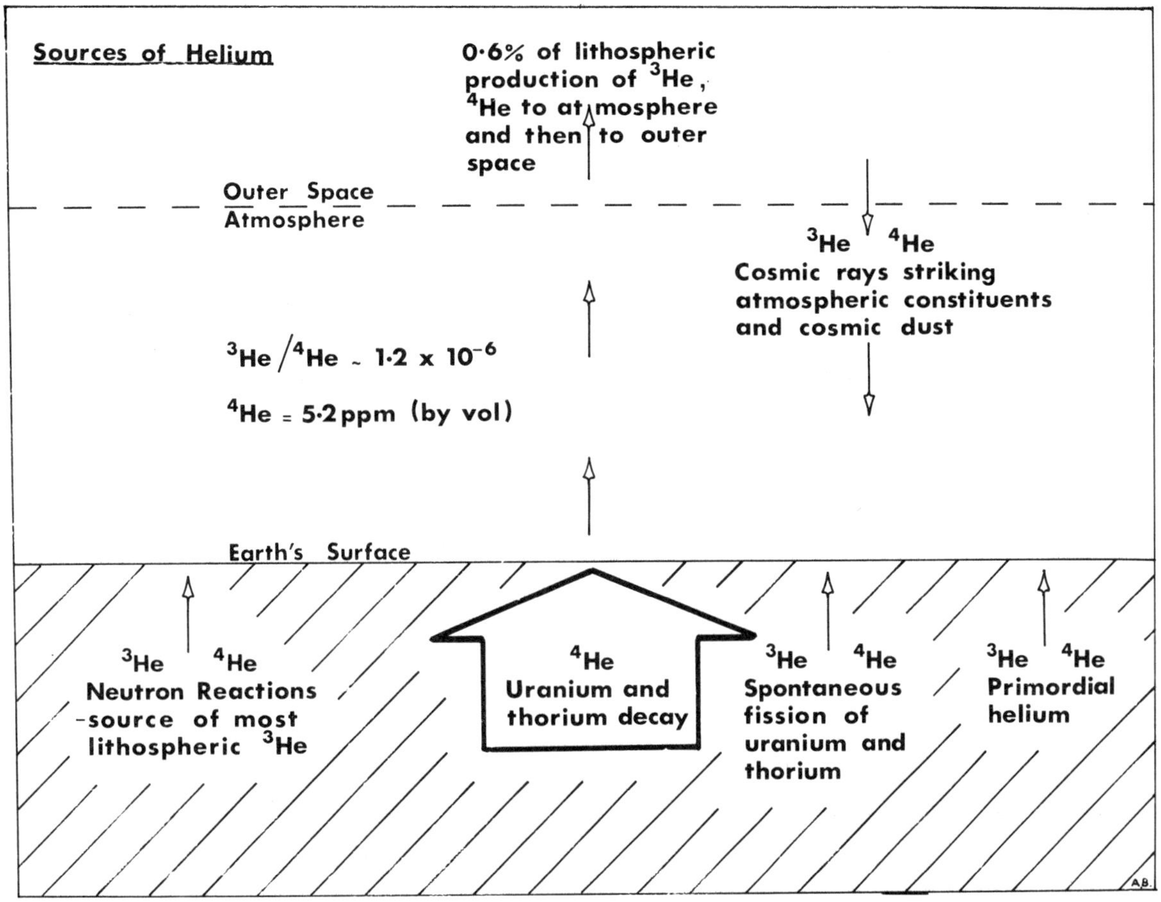

Fig. 1. The terrestrial helium balance illustrating the sources of helium.

isotopes: helium-3 and helium-4. However, as the latter is approximately 1.2 to 1.4 million times more abundant than the former (Aldrich and Nier, 1948; Mamyrin et al., 1970), helium is essentially helium-4. Detailed accounts of the properties of helium have been documented by Cook (1961), Asimov (1966), and Yakutseni (1968). The terrestrial material balance of helium is shown in Figure 1. Effectively all of the helium-4 present in the earth has been produced as a decay product of uranium-238, uranium-235, and thorium-232 (Asimov, 1966; Pogorski and Quirt, 1979), according to the following simplified decay schemes (Moore and Esfandiari, 1971):

$$(\text{uranium-238}) \longrightarrow (\text{lead-206}) + 8\ (\text{helium-4})$$
$$(\text{uranium-235}) \longrightarrow (\text{lead-207}) + 7\ (\text{helium-4})$$
$$(\text{thorium-232}) \longrightarrow (\text{lead-208}) + 6\ (\text{helium-4})$$

At present, the decay of uranium and thorium in the earth annually produces in the order of 10^{10} liters of helium-4, approximately half of which is formed in the crust while the remainder originates in the mantle and the core (Damon and Kulp, 1958; Vinogradov, 1964). It should be noted, as illustrated in Figure 2, that approximately 97% of the terrestrial helium is produced beneath the sedimentary layer (Eremeev et al., 1973).

HELIUM MIGRATION AND FLUX

Because the concentration of helium in the crust exceeds that of the atmosphere, this inert gas migrates outward and replaces that lost to outer space in the upper regions of the atmosphere (Damon and Kulp, 1958). This emanation or flux of helium is achieved principally by filtration or fluid transport (Golubev et al.,

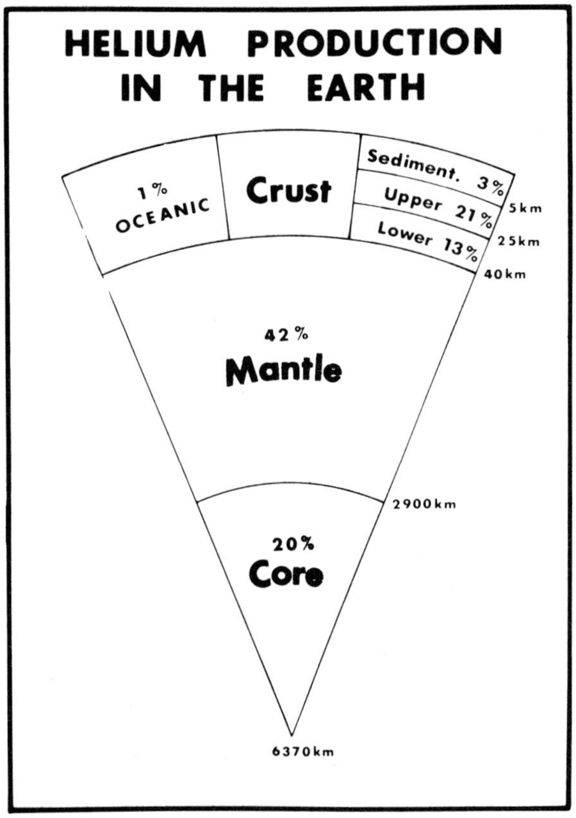

Fig. 2. Helium production in the various zones of the earth.

1970, 1974) and results in an average helium flux at the surface of the order of 1 liter/km²/year (Asimov, 1966; Naughton et al., 1973; Tugarinov and Osipov, 1974). Diffusive movement of helium is of importance at the level of the rock matrix (Moore and Esfandiari, 1971) and in the zone of aeration; however, it is generally slower than the sedimentation rate (Newton and Round, 1961).

The actual surface helium flux is not uniform (Bulashevich and Bashorin, 1970; Golubev et al., 1970, 1974; Eremeev et al., 1973; Goldak, 1974; Roberts, 1975; Roberts et al., 1975; Dyck, 1976; Pogorski and Quirt, 1976a, 1976b; Reimer, 1976; Reimer and Otton, 1976; Martin and Bergquist, 1977; Martin et al., 1978; Pogorski and Quirt, 1978, 1979; Roberts, this volume). It is modified by the presence of: (a) faults which create an easier path for helium migration (Figure 3), (b) radioactive ore bodies which act as local helium sources (Figure 4), (c) petroleum and geothermal reservoirs which accumulate and re-emit helium (Figures 5 and 6).

The objective of helium emanometry is to measure and interpret the near-surface variations in the helium flux. The easiest and the most tempting approach would be to measure the variations in the helium concentration in near-surface air. Unfortunately, this approach is not practical. Because the rate at which helium emanates into the air is very slow while the rate at which it mixes with the air is very fast, the possibility of finding meaningful variations is eliminated (Gluekauf, 1946; Gluekauf and Paneth, 1946).

This mixing effect is much less pronounced, however, in the subsurface environment. Therefore, it is possible to detect meaningful variations of the helium flux in subsurface media. Helium is present in the near-surface environment in the following modes, some of which are illustrated in Figure 7: (a) in the macropores or intercrumb pores of soil, (b) in the micropores or crumb pores of soil, (c) in permafrost, (d) in ground water, (e) in saturated sediments both onshore and offshore, and (f) in the near bottom waters both onshore and offshore.

SAMPLING TECHNIQUES

A variety of sampling techniques are needed to sample the various types of media in which helium is present. Soil samples are normally collected with the aid of hand or power augers at a depth of approximately 60 cm (Figure 8) and are hermetically sealed in specially designed 200 cc capacity aluminum sample containers with the aid of portable sealing units (Pogorski and Quirt, 1976a, 1976b, 1978, 1979) (Figure 9).

Soil gas is extracted from a depth of approximately 90 cm with the assistance of a low dead volume, manually operated, nonclogging, steel probes of the type shown in Figure 10 (Pogorski, 1974). In collecting soil gas samples with probes, samples of only a few cubic centimeters should be taken so as not to disturb severely the environment of the soil macropores. Such a disturbance can cause dilution of the sample by atmospheric air or desorbed soil gases. Soil gas samples are withdrawn from the

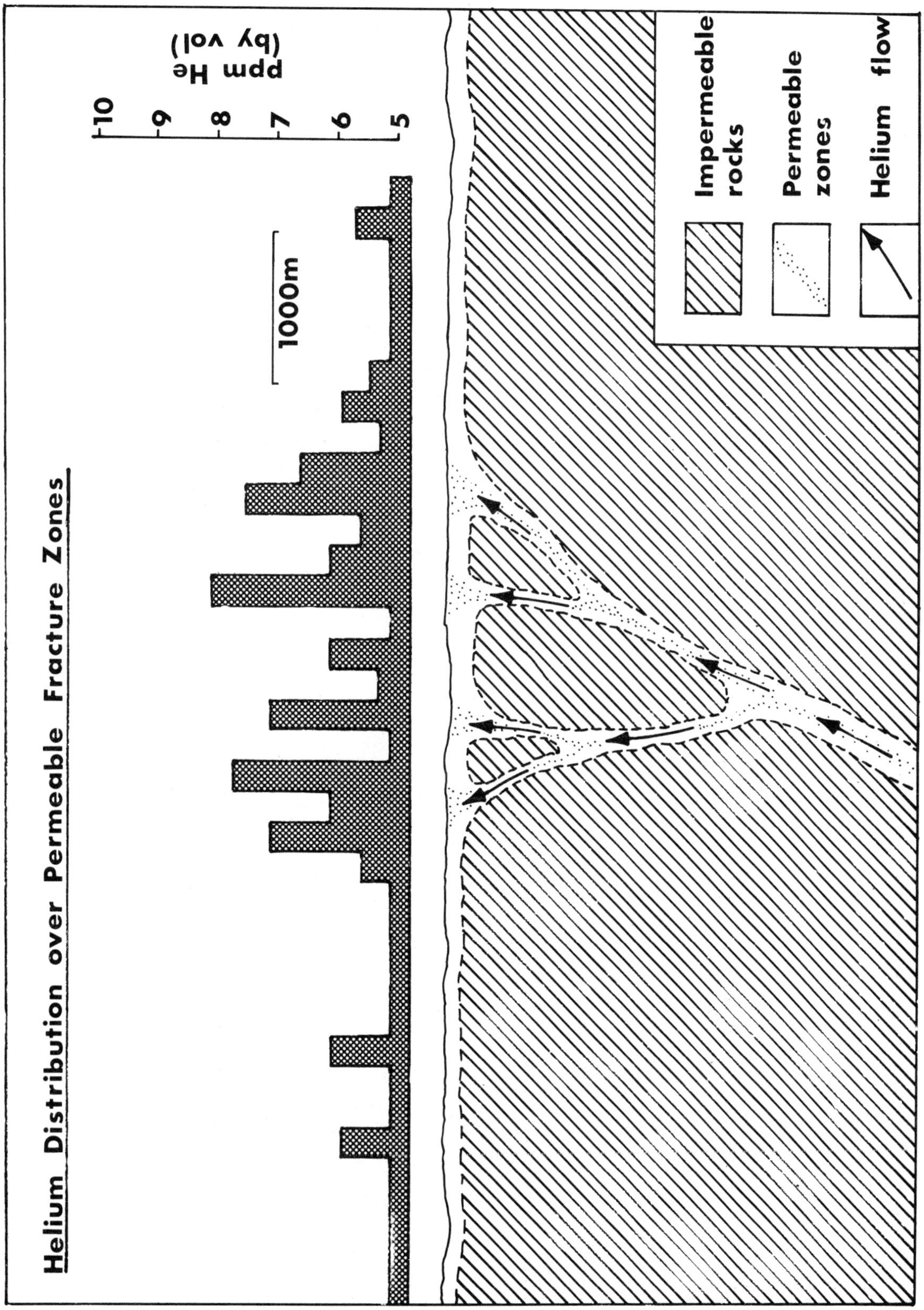

Fig. 3. A schematic diagram illustrating that faults provide an easy path for helium migration from depth.

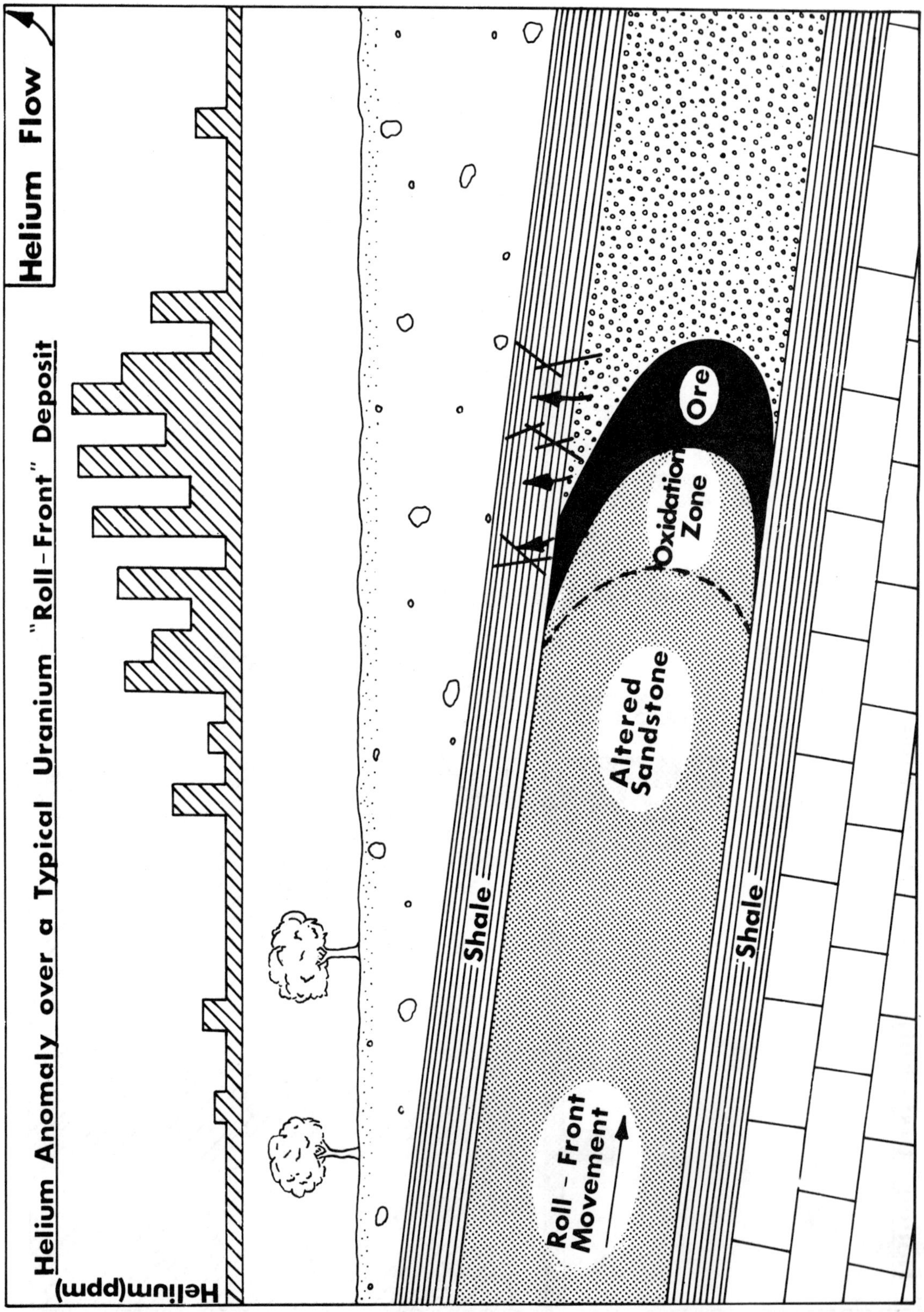

Fig. 4. A schematic diagram illustrating that ore deposits act as local helium sources.

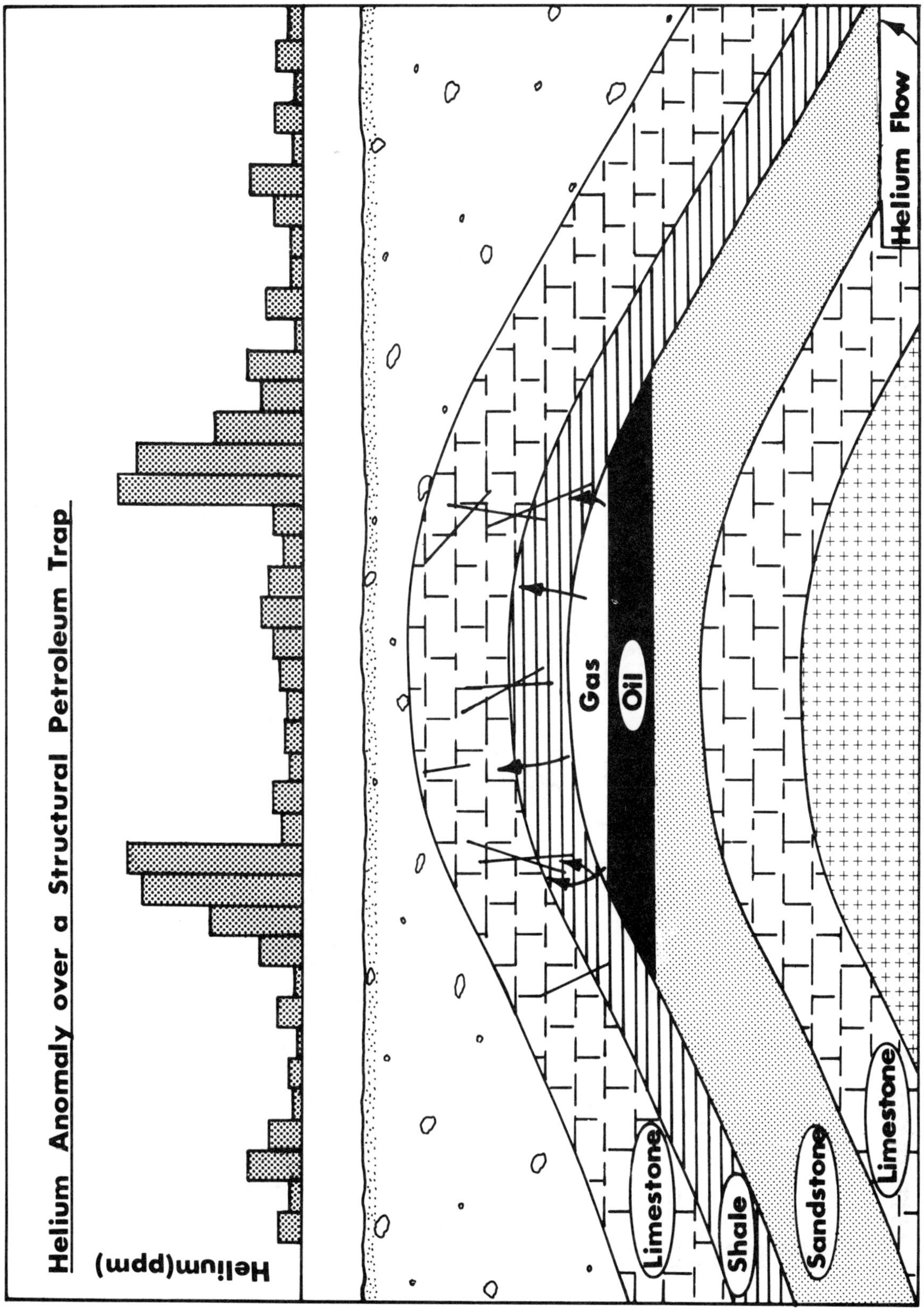

Fig. 5. A schematic diagram showing that an anticlinal petroleum reservoir acts as a local helium source. Note that the helium anomaly is a halo.

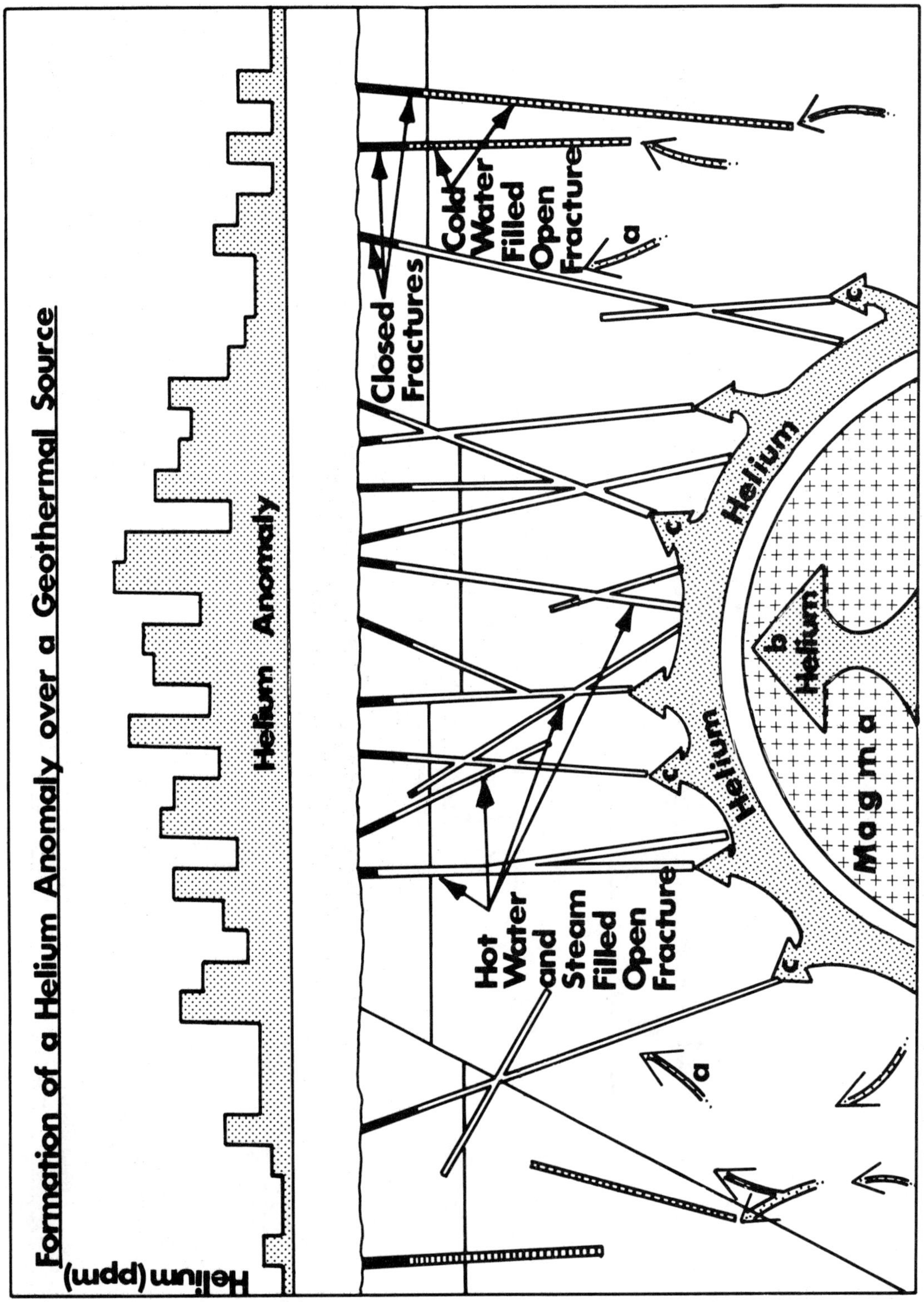

Fig. 6. A schematic diagram illustrating that a geothermal reservoir acts as a local source of helium. In this diagram there are three sources of helium: (a) helium from the background flux; (b) helium from the magma heat source; and (c) helium scavenged by the hot geothermal waters.

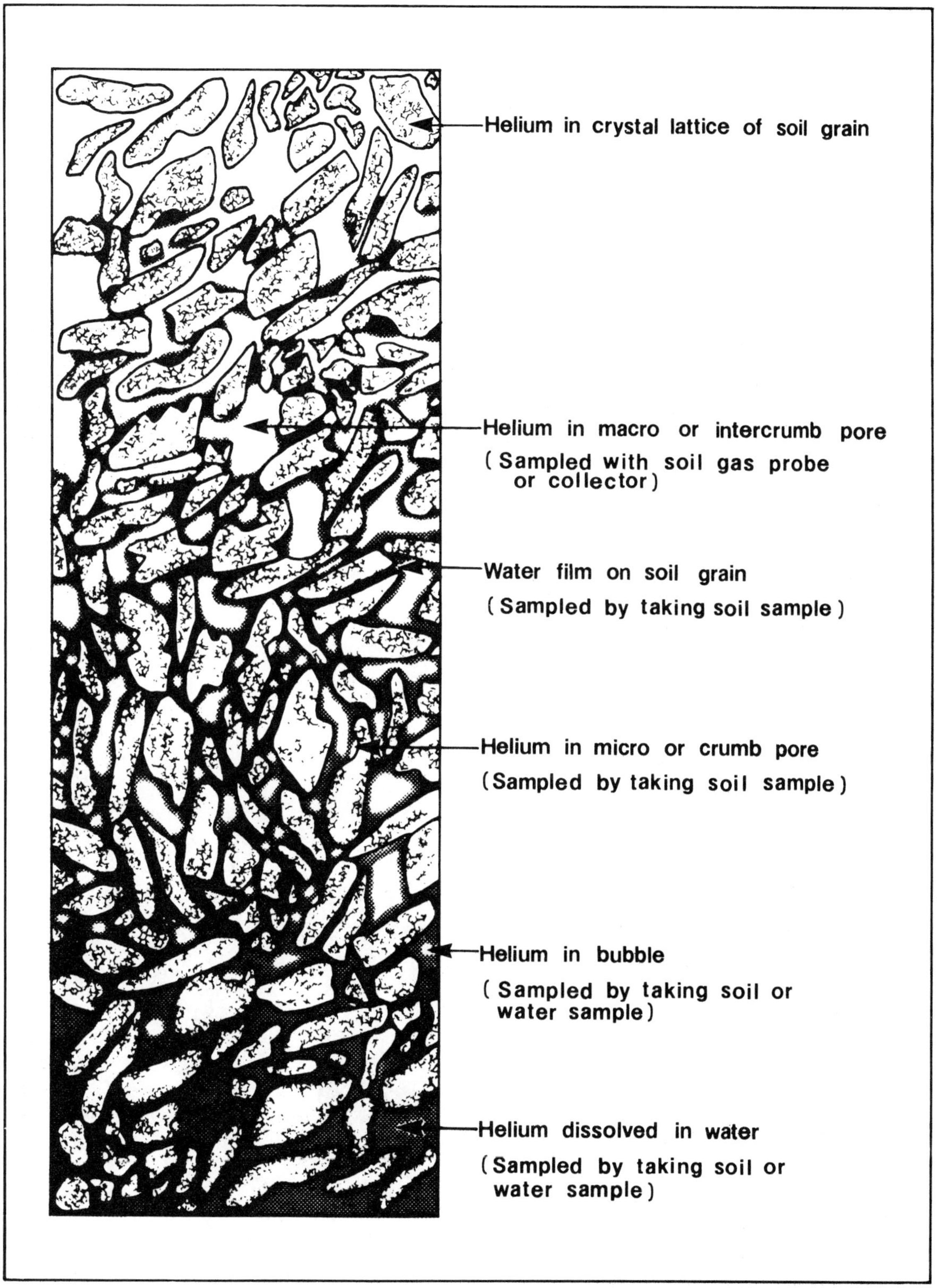

Fig. 7. A schematic diagram showing how helium occurs in soil.

Fig. 8. Collecting a soil sample with a hand auger.

Fig. 9. Geologist with a portable sample container sealer.

probe with a syringe and are subsequently stored in hermetically sealed Bistable gas samplers (Figure 11) (Pogorski and Reimer, 1972).

Onshore water samples can be taken from farm wells, windmills, or lake bottoms. After collection, the water samples are sealed in the same type of aluminum sample container that is used for the soil bottom sediment samples. This simplifies the logistics of the exploration programs.

Offshore near-bottom water and saturated bottom sediments can be collected economically and efficiently from boats or helicopters. All of these techniques are rapid and economical and can be carried out without hazard to the

Fig. 10. Geologist holding a manually operated nonclogging steel probe that is used in collecting soil gas samples.

Fig. 11. Geologist extracting a soil gas sample from a probe with the aid of a disposable syringe. He is holding a Bistable gas sampler in his left hand.

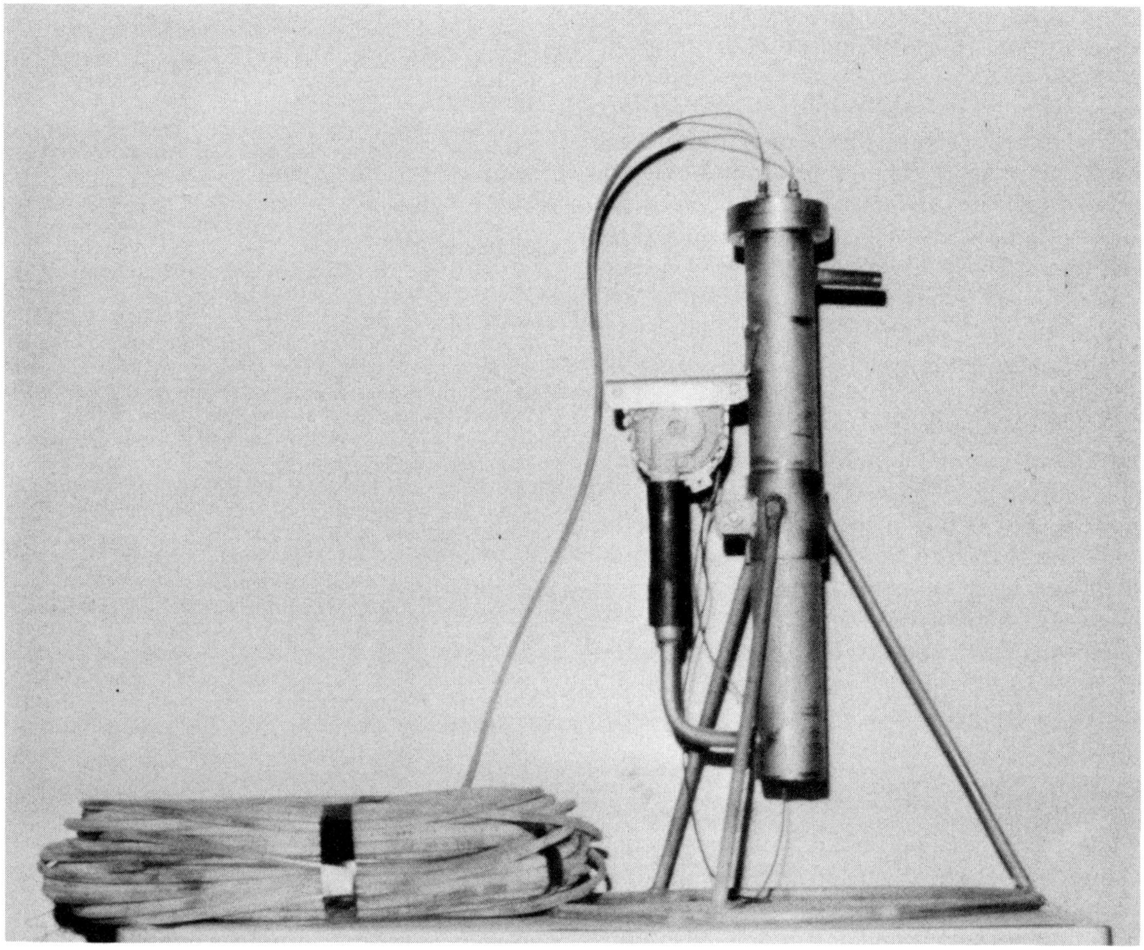

Fig. 12. The gill-type aquatic gas sampler.

environment. A new experimental approach to measuring the helium concentration in subsurface water involves the application of a gill-type device which can be attached to a fish and towed behind a ship. This device extracts helium from water in situ (Figure 12).

Several sampling patterns can be used. In general, reconnaissance surveys can be conducted by taking water samples at a density of about one sample per square kilometer and soil and soil gas samples at a density of about five samples per square kilometer. In carrying out detailed surveys in arid regions, soil gas samples can be extracted from the soil macropores with the aid of probes, while in areas in which the soil moisture content is higher, soil samples should be collected. Detailed sampling of lake bottom water or sediment may also be employed. For detailed sampling, the sampling density will be a function of the anticipated size of the target. The recommended density for detailed exploration for small targets is, for example, approximately ten samples per square kilometer.

ANALYTICAL TECHNIQUES

Small quantities of helium-4 are commonly determined by gas source mass spectrometry. Both laboratory and semiportable field units can be employed for this purpose. There are advantages and disadvantages to utilizing either type of unit.

During each analysis, the concentration of helium-4 is compared to that of a reference sample. In analyzing soil and water samples, it is necessary to determine a number of other parameters as well as the helium concentration. Examples of some of the parameters that must be determined for soil samples are the content of hydrocarbons, air components, moisture level, and porosity. Therefore, a number of additional instruments are required, such as gas chromatographs and a complete soil and water laboratory. These additonal parameters are essential for achieving a meaningful interpretation of the soil and water results. The mass spectrometer readings alone would usually be of little value.

In determining the quantity of helium-4 present in the soil gas samples, the practical overall precision is of about ± 40 parts per billion (ppb); however, under optimum conditions the analytical error can be reduced to ± 10 ppb. The reproducibility of the helium-4 analyses of the water samples varies from ± 5% to ± 20%, depending on the range. The total error in the calculated soil helium-4 levels varies from 5% to 10% of the value.

INTERPRETIVE METHODS

The calculated corrected soil, water, and soil-gas helium results are analyzed employing computerized statistical techniques in order to separate the background and anomalous populations. The anomalous population is normally divided into several subsets for purposes of data presentation. The data are then displayed in the form of histograms, dot maps, line and anomaly contour density maps, or as helium surface block diagrams. Trend surface analysis is often of value in interpreting the helium data.

Helium emanometry has already found widespread application in North America and in the Soviet Union as a tool for locating uranium, major faults, and geothermal reservoirs. Recent work, conducted jointly by Chemical Projects Limited and its American affiliate, Helium Surveys Inc., with the U.S. Geological Survey, has demonstrated that the method has a considerable potential as a petroleum exploration technique. In Part II of this paper Dr. Alan Roberts of the U.S. Geological Survey has presented some of the results obtained over several known and prospective petroliferous areas.

REFERENCES

Aldrich, L. T., and Nier, A. O., 1948, The occurrence of ^{3}He in natural sources of helium: Physical Review, v. 74; p. 1590-1594.

Anonymous, 1970, Gas analysis Ontario gas wells: Ontario Department of Mines and Northern Affairs, Petroleum Resources Paper, 70-2, 194 p.

Anonymous, 1972, The helium content of Alberta natural gases: Energy Resources Conservation Board, Province of Alberta, 9 p.

Asimov, I., 1966, The noble gases: Basic Books, New York, 171 p.

Bulashevich, B. P., and Bashorin, V. N., 1970, Geliy y podzemnykh vodakh na profile glubinnykh seysmicheskikh zondinovaniy v Zaural'ye (Helium in ground water along a deep seismic sounding profile east of the Urals): Doklady Akademii Nauk SSSR, v. 193, p. 573-575.

Cook, G. A., 1961, Argon, helium and the noble gases: Interscience, New York, 394 p.

Damon P. E., and Kulp, J. L., 1958, Inert gases and the evolution of the atmosphere: Geochimica et Cosmochimica Acta, v. 13, p. 280-292.

Dyck, W., 1976, The use of helium in mineral exploration: Journal of Geochemical Exploration, v. 5, p. 3-20.

Elworthy, R. T., 1926, Helium in Canada: Department of Mines, Mines Branch Report no. 679, 64 p.

Eremeev, A. N., Sokolov, V. A., Solovov, A. P., and Yanitskii, I. N., 1973, Application of helium surveying to structural mapping and ore deposit forecasting, *in* Geochemical Exploration, 1972: The Institution of Mining and Metallurgy, London, p. 183-193.

Glueckauf, E., 1946, A micro-analysis of the helium and neon contents of air: Proceedings of the Royal Society of London, v. 185, p. 98-119.

———, and Paneth, F. A., 1946, The helium content of atmospheric air: Proceedings of the Royal Society of London, v. 185, p. 89-98.

Goldak, G. E., 1974, Helium-4 mass spectrometry for uranium exploration: American Institute of Mining Engineers, preprint 74-L-44, 12 p.

Golubev, V. S., Osipov, Yu. G., and Yanitskii, I. N., 1970, Some characteristics of migration of helium in the permeable systems of the upper part of the earth's crust: Geokhimiya, v. 11, p. 1341-1348.

———, Yeremeyev, A. N., and Yanitskii, I. N., 1974, Analysis of some models of helium migration in the lithosphere: Geochemistry International, v. 11, p. 734-742.

Lipper, H. W., and Wilcox, A. L., 1960, Helium: U. S. Bureau of Mines Publication, 36 p.

Mamyrin, B. A., Anufriyev, G. S., Kamenskiiy, I. L., and Tolstikhin, I. N., 1970, Determination of the isotopic composition of helium in the atmosphere: Geokhimiya, v. 6, p. 721-730, translated in Geochemistry International, v. 7, p. 498-505.

Martin, J. P., and Bergquist, L. E., 1977, Study of the applicability of ^{3}He/^{4}He Ratio for uranium prospecting: Final Report by Martin Marietta Corporation to the U. S. Energy Research Development Administration, 95 p.

———, Devoto, R. H., and Mead, R. H., 1978, The potential of helium as a guide to uranium ore, pt 1: Final project report to the Electric Power Research Institute (EPRI), EPRI Report EA-813, 168 p.

Moore, C. A., and Esfandiari, B., 1971, Geochemistry and geology of helium, *in* Lansberg, H. E., and Van Mieghem, J., eds., Advances in geophysics: Academic Press, New York, v. 15, p. 1-57.

Naughton, J. J., Lee, J. H., Keeling, D., Finlayson, J. B., and Dority, G., 1973, Helium flux from the earth's mantle as estimated from Hawaiian fumarolic degassing: Science, v. 180, p. 55-57.

Newton, R., and Round, H. F., 1961, The diffusion of helium through sedimentary rocks: Geochimica et Cosmochimica Acta, v. 22, p. 106-132.

Pierce, A. P., Gott, G. B., and Mytton, F. W., 1964, Uranium and helium in the Panhandle gas field, Texas and adjacent areas: U. S. Geological Survey Professional Paper 454-G, 47 p.

Pogorski, L. A., 1974, Geochemical exploration method: U. S. Patent 3,835,710.

———, and Reimer, E. M., 1972, Fluid sampling device: U. S. Patent 3,662,928.

———, and Quirt, G. S., 1976a, Helium surveys: an exploration tool for locating uranium, petroleum and geothermal energy sources: Ontario Petroleum Institute Technical Paper no. 9, v. 15, 35 p.

———, 1976b, Helium surveys emerging in exploration: The Northern Miner, Nov. 25, 1976.

———, 1978, The potential of helium as a guide to uranium ore, pt. 2: Supplemental report on a helium survey in McKinley County, N. M., EPRI Report EA-813, 32 p.

———, 1979, Helium emanometry as an indicator of deeply buried uranium deposits: Final report to Bendix Field Engineering Corporation for work performed under Subcontract no. 77-078-S and Bendix contract EY-76-C-13 1664, 362 p.

Reimer, G. M., 1976, Design and assembly of a portable helium detector for evaluation as a uranium exploration instrument: U. S. Geological Survey Open-File Report 76-398, 18 p.

———, and Otton, J. K., 1976, Helium in soil gas and well water in the vicinity of a uranium deposit, Weld County, Colorado: U. S. Geological Survey Open-File Report 76-699, 10 p.

Roberts, A. A., 1975, Helium surveys over known geothermal resource areas in the Imperial Valley, California: U. S. Geological Survey Open-File Report 75-427, 6 p.

———, Friedman, I., Donovan, T. J., and Denton, E. H., 1975, Helium survey, a possible technique for locating geothermal reservoirs: Geophysical Research Letters, v. 2, p. 209-210.

Tugarinov, A. I., and Osipov, Yu. G., 1974, Flow of helium through granite massifs and zones containing radioactive ores: Geokhimiya, v. 9, p. 1424-1429, translated in Geochemistry International, v. 11, p. 981-986.

Vinogradov, A. P., 1964, Gas regime of the earth, in Chemistry of the Earth's Crust, v. 2, Vinogradov, A. P., ed., p. 1-19, translation by Israel Program for Scientific Translations Ltd., Jerusalem, 1967.

Yakutseni, V. P., 1968, Geologiya geliya (Geology of helium): Nedra, Leningrad, 232 p.

HELIUM EMANOMETRY IN EXPLORING FOR HYDROCARBONS: PART II

Alan A. Roberts

UNITED STATES GEOLOGICAL SURVEY

INTRODUCTION

Direct geochemical prospecting techniques for petroleum involve the detection of near-surface manifestations of seepage and subsequent migration of compounds from a petroleum reservoir. Most such techniques rely on the detection of petroleum hydrocarbons in the top 3 m of the soil horizon. Complications can arise because many of these compounds may also have biogenic origins, and these biogenically produced hydrocarbons may mask anomalous concentrations that are the result of migration from a fossil fuel source. Attempts have been made to differentiate between fossil hydrocarbons and those biogenically produced, with varying degrees of success (Hunt, 1979; Smith and Ellis, 1963; Horvitz, 1972; Debnam, 1969).

This paper deals with a different approach to the problem. Most petroleum reservoirs in the United States have significant concentrations—greater than 100 parts per million (ppm)—of the noble gas helium (Moore, 1976; Dyck, 1976). Helium is a very mobile molecule because of its small mass (4 amu), size, and diffusivity. It will therefore seep and migrate much more rapidly than any other component of petroleum. In addition, it cannot be produced biogenically, and so the complication of having it confused with decaying organic matter is avoided. Helium also cannot be destroyed or altered by chemical reactions in a natural environment. Thus, detection of anomalously high concentrations of helium in the soil can be more directly correlated with a deep source, such as a petroleum reservoir, than can hydrocarbon in soil surveys.

In addition to petroleum reservoirs, three other sources of high helium concentrations do exist. The ultimate origin of all nonprimordial helium in the environment is the alpha decay of uranium, thorium, and daughters within their decay series. Thus, high helium concentrations in the soil may be related to relatively rich uranium and thorium deposits (Dyck, 1976; Clarke et al., 1977; Reimer et al., 1979; Reimer and Adkisson, 1977; Torgerson and Clarke, 1978). Geothermal brines are observed to contain high concentrations of helium (Mazor, 1977; Mazor and Verhagen, 1976; Mazor and Fournier, 1973), and it has been suggested (Roberts, 1975; Roberts et al., 1975) that helium surveys may be useful in locating hot water geothermal reservoirs.

A fourth major cause of high concentrations of helium in soils is proximity to a permeable fault or fault system (Bulashevich and Bashorin, 1973; Plyusnin et al., 1972). In the Soviet Union, helium surveys are reported to be useful in locating secondary mineralization along such fault systems (Ovchinnikov et al., 1972; Eremeev et al., 1972; Dikun et al., 1976). Thus, it is important that interpretation of helium surveys be coupled with good geophysical and geological investigations in order to determine the source of the anomalously high helium concentrations.

The preceding paper (Pogorski and Quirt, this volume) has outlined the theoretical basis for using helium in geochemical prospecting for petroleum and has described sample collection and analytical procedures developed by Pogorski that were employed in many of our studies. This paper will concentrate on the application of these techniques and of other techniques developed by the U.S. Geological Survey and will test the possibilities of using them to prospect for petroleum.

EXPERIMENTAL METHODS

Two techniques for collecting and analyzing helium in soils have been employed in this study. In areas where the soil is wet or frozen the techniques described in the preceding paper (Pogorski and Quirt, this volume) are utilized. A soil sample is collected at a depth of about 1/2 m, hermetically sealed in an aluminum can, and shipped back to the laboratory for analysis of the helium content.

In arid or semiarid environments, techniques developed at the U.S. Geological Survey are used. A hollow, stainless steel probe is driven into the ground to a depth between 1/2 m and 2 m. The depth utilized, which is constant within any given survey, is determined by the permeability of the soil in the survey area. In a highly permeable soil, greater sample depth is necessary to avoid complete dilution of the helium anomaly by mixing with the atmosphere. A soil gas sample is drawn through the probe into a hypodermic syringe and discarded in order to purge the probe of air in its dead volume. A second 5-cm^3 sample is then collected and stored in the syringe. No significant leakage from these syringes has been observed in the few hours that may exist between collection and analysis. These samples are analyzed in the field using a portable mass spectrometer mounted in a truck (Friedman and Denton, 1975; Reimer, 1976; and Reimer and Denton, 1978). The soil gas sample is then compared to known helium-in-air standards and the helium content is determined with a precision of ± 10 parts per billion (ppb). Fifty to one hundred samples per day can be collected and analyzed with this equipment, which allows simultaneous field interpretation and sample collection.

These two techniques have been run simultaneously in six of the survey areas studied. In areas where the soil is very wet or frozen it is often impossible to obtain a soil gas sample through a probe, and it becomes necessary to collect a sample of the soil itself and seal it in a can. The sample is then allowed to reach equilibrium with a small volume of air of known helium concentration. The excess helium in this head-space gas is then analyzed, allowing a calculation of the quantity of helium in the original sample. In the two surveys where the soil was cohesive enough to maintain its integrity as a core, or as cuttings from an auger, and was also dry enough to allow collection of a soil gas sample through a probe, a good correlation was found between the patterns observed from both techniques. In the four surveys where the soil sample was very dry, however, little or no correlation was observed, probably because the very dry soils tended to crumble into small pieces upon collection, thereby allowing variable and large amounts of the helium contained in the pore spaces to escape to the atmosphere. Thus, the direct soil gas sampling technique will produce a more representative sample of the actual helium concentration in the soil.

RESULTS AND DISCUSSION

In order to determine the concentration range for a normal background, helium surveys were run over areas where there was no known or suspected source of excess helium. More than a thousand background samples of soil gas collected through a probe have been analyzed. Fewer than 10% of these samples had concentrations as much as 40 ppb above background, although isolated values reached as high as 100 ppb. For this reason, isolated, individual high-helium values should not be interpreted to indicate a high helium source but rather represent an unusual, very localized condition. Furthermore, in areas surveyed where high helium concentrations are observed, a few low helium values are also observed, even within a distance of just a meter or two from a high soil gas value. Therefore, it is necessary to collect a large number of samples in any given survey and to base interpretations on general patterns observed rather than on isolated high or low values.

Helium surveys have also been run over several known petroleum accumulations. Two distinctly different patterns of anomalously high values emerge. These two types of seepage patterns have been previously reported (Horvitz, 1969; Horvitz, 1954; Donovan et al., 1975; Donovan and Dalziel, 1977; Donovan, 1974; Debnam, 1969). The first type of pattern is the result of microseepage through the caprock with subsequent vertical migration of helium to the earth's surface. In this case an area of anomalously high helium concentrations is observed directly over the producing area and is called an apical anomaly.

The second pattern observed is a halo or

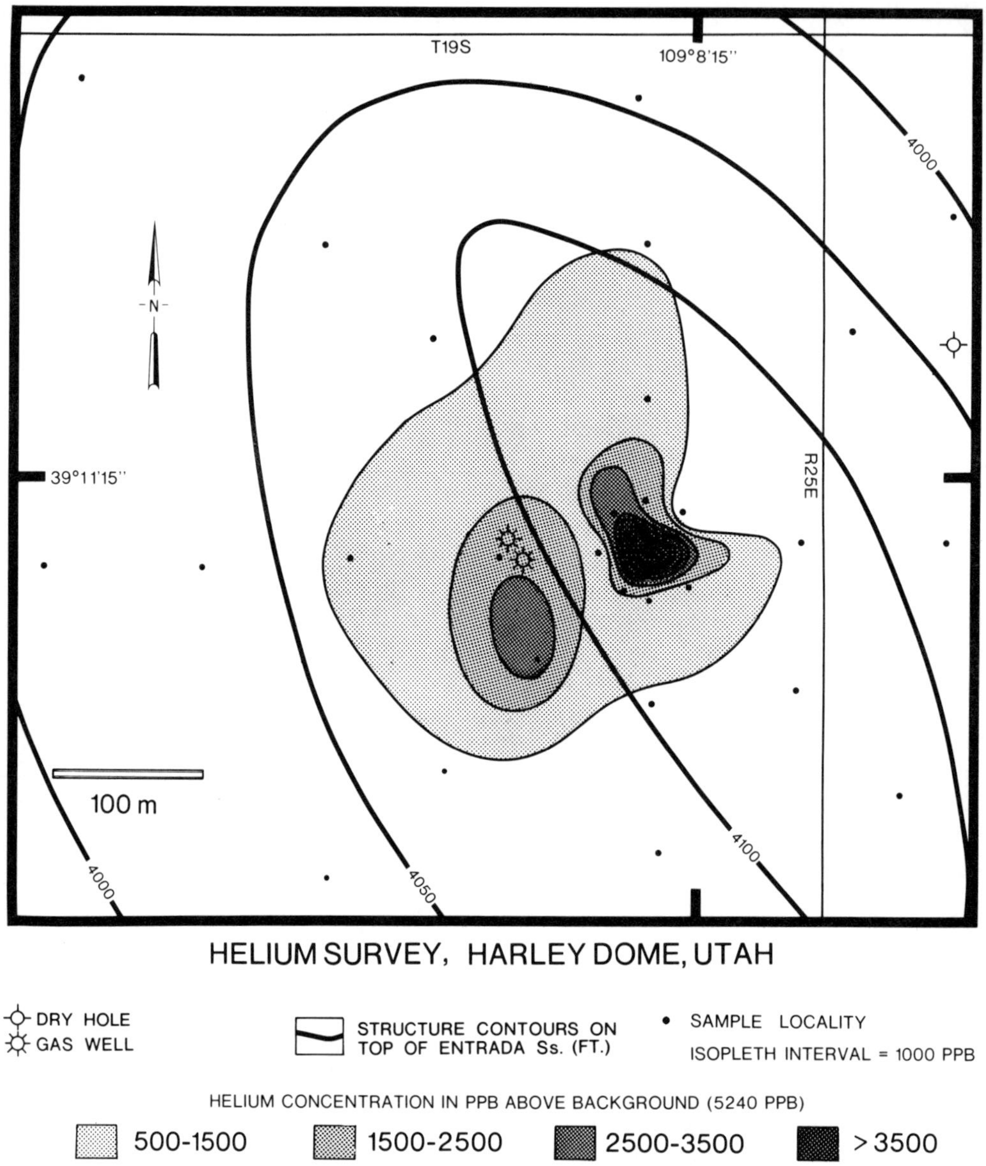

Fig. 1.

crescent of high helium values surrounding or partially surrounding the producing area. In general the reasons for halo-type anomalies are not well understood. They may at least sometimes be the result of calcite cementation of strata overlying the reservoir. As hydrocarbons from the reservoir migrate to aerobic, near-surface strata, they may be oxidized to carbon dioxide, which can be dissolved by the pore water and subsequently precipitated out with calcium as calcite cement (Donovan, 1974; Donovan and Dalziel, 1977). As this cement builds up over geologic time the strata become more impermeable to migration of the seeping gases. As a result, the gases would find the migration pathway of least resistance to be out and around this calcite chimney. The surface manifestation of present-day helium and hydrocarbon seepage would be in the form of a halo encircling this calcite chimney and hence encircling the producing area.

In areas where no source of excess helium

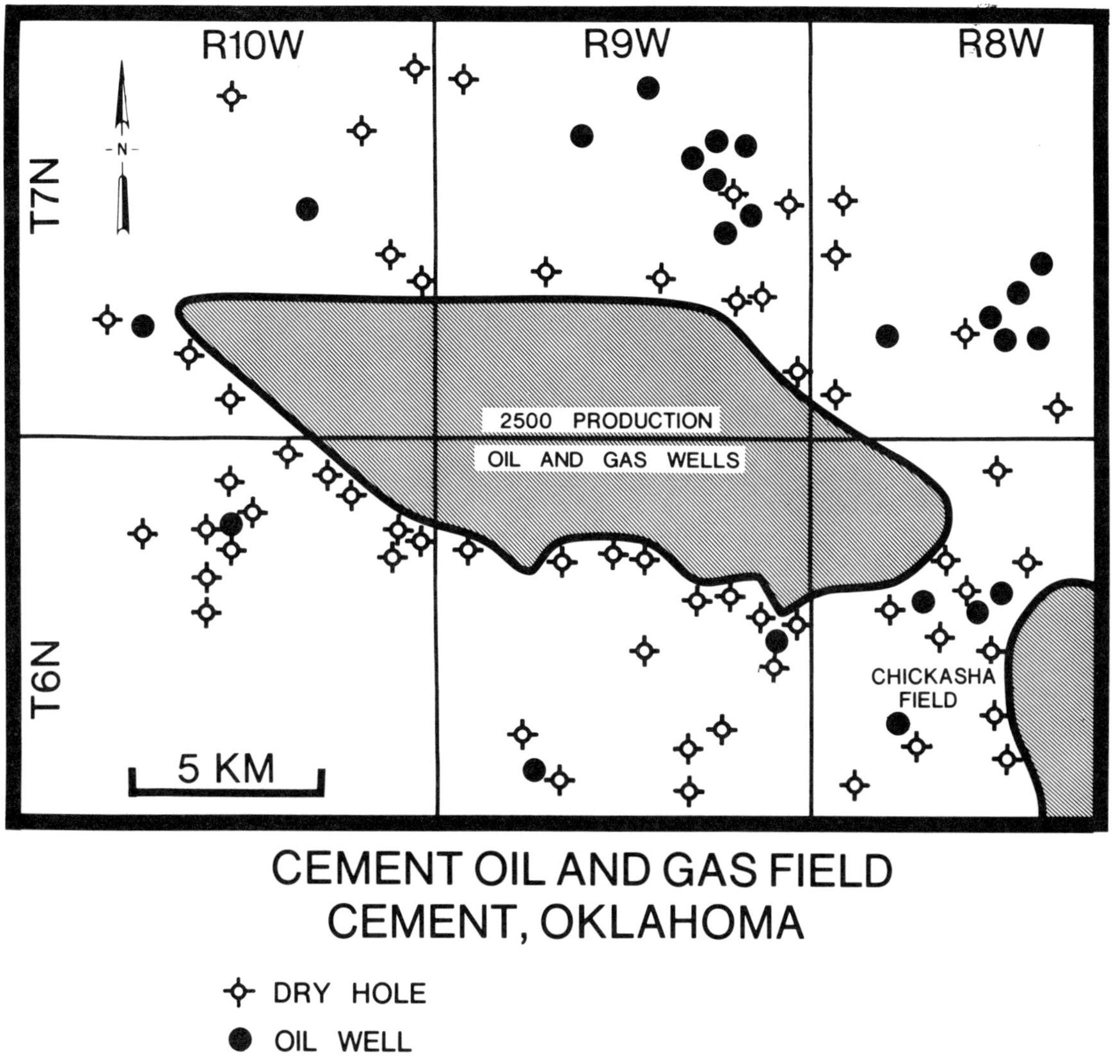

Fig. 2.

exists or is postulated, helium surveys reveal only a few, seemingly random and unconnected points of high helium concentrations, and the average value is close to that of background.

Preliminary field tests were run over the Harley Dome gas reservoir in Utah. The structure is a northwest-trending anticline on the southern shelf of the Uinta Basin. The sedimentary column at Harley Dome is about 550 m thick and consists of outcropping Cretaceous sandstones and shales underlain by the Morrison and Summerville Formations, Entrada Sandstone of Jurassic age, the Kayenta Formation, Wingate Sandstone, and the Chinle Formation of Triassic age. There is a major unconformity at the base of the Chinle Formation. There is no faulting observed or reported within the study area. The main gas reservoir is in the Entrada Sandstone at an average depth of about 260 m. The gas there has an unusually high helium content (7%) and contains 85% nitrogen. All wells drilled into the structure have been plugged and have never been produced, so the reservoir pressure is probably close to that existing at the time of discovery (1200 kPa).

Figure 1 displays the results of a helium survey in the Harley Dome area. Helium concentrations equal to normal atmospheric background (5.2 ppm) are observed in the soil in the area surrounding Harley Dome. These concentrations increase systematically toward the crest of the structure and reach a maximum of 15.6 ppm (9400 ppb above background) directly on top of the structure.

Figure 2 shows the producing area of the Cement oil field, located in the southeastern part of the Anadarko Basin in Oklahoma. The Cement oil field was discovered by surface mapping in 1916. The general area is characterized by outcropping reddish brown sandstones of Permian age that exhibit a marked color change to gray or white over the oil field itself (Donovan, 1974). Cement is a giant oil and gas accumulation with multizone reservoirs of Early Pennsylvanian to Early Permian age ranging from 600 m to 2,230 m depth. Structure is a slightly asymmetrical, NW-SE-striking anticline with two pronounced domes along the axis. There is an angular unconformity separating the overlying unfaulted Permian rocks from the highly faulted, very tightly folded Pennsylvanian sequence. Production from the shallow Permian reservoirs occurs along most of the productive length of the anticline, whereas the Pennsylvanian reservoirs are generally limited to the structurally highest parts of the two domes.

Many near-surface geochemical manifestations of petroleum microseepage have been reported at Cement (Donovan, 1974; Donovan and Dalziel, 1977). Figure 3 shows the variation in concentration of helium in the soil gas over the field. The 40 ppb (above background) isopleth encloses the area of high helium concentrations. There is a remarkably good correlation between this apical helium anomaly and the known producing area. The fact that the helium anomaly extends along the entire productive length of the field would seem to indicate that leakage of helium is not solely coming from reservoirs in the highly faulted Pennsylvanian rocks, but is additionally being derived from reservoirs in the overlying, unfaulted Permian sandstones.

A third example of a helium survey over a known oil reservoir is shown in Figure 4. The Rubelsanto oil field, Guatemala, lies in the southwestern part of the Chapayal Basin (also known as the Peten basin) along the Chiapas-Guatemala fold belt. The Rubelsanto anticline is one of a series of elongate, east-west-trending en echelon to subparallel structures, characterized by steep flanks, high-angle bounding faults, and a pronounced surface expression. Surface rocks in the region range from upper Tertiary claystone with interbedded sandstone lenses to Lower Cretaceous limestone and dolomite.

Little detail about the field has been published. Numerous Lower Cretaceous (Coban C. Formation), thin, fractured carbonate rocks embedded in anhydrite and increasing in thickness with depth produce oil at Rubelsanto (Buis and Evenson, 1977) at depths ranging from 1,500 m to 2,300 m. At the time of our survey there were four producing wells with a typical gas-to-oil ratio of 120 to 1 (Buis and Evenson, 1977), and a wildcat was also being drilled on the larger separate Chisec structure about 13 km south of Rubelsanto.

The survey over Rubelsanto was limited to a traverse along existent roads and footpaths, because of the hostility of the Guatemalan jungle. A mud sample was collected from the bottom of a 1/2 m hole drilled with a hand auger at each sample site. This sample was then sealed in a can and analyzed by Pogorski's (this volume) technique. These samples were so wet that there was essentially no dry pore space, and so all helium present must either be in aqueous solution or adsorbed on the soil. As such, helium values are reported as cm^3 of helium per cm^3 of mud $\times 10^{-8}$. The fence diagram in Figure 4 demonstrates the observation that all samples with high helium content occur over the known producing area.

Samples taken on a traverse south to the Chisec wildcat were also analyzed, and no samples with high helium values were found in the vicinity of the Chisec well. Since there were a very limited number of samples in this study, the results must be viewed with caution. Nevertheless, the lack of any high helium samples suggests a low potential for oil or gas production. In fact, no producible oil or gas was encountered in this well (Oil and Gas Journal, 1977).

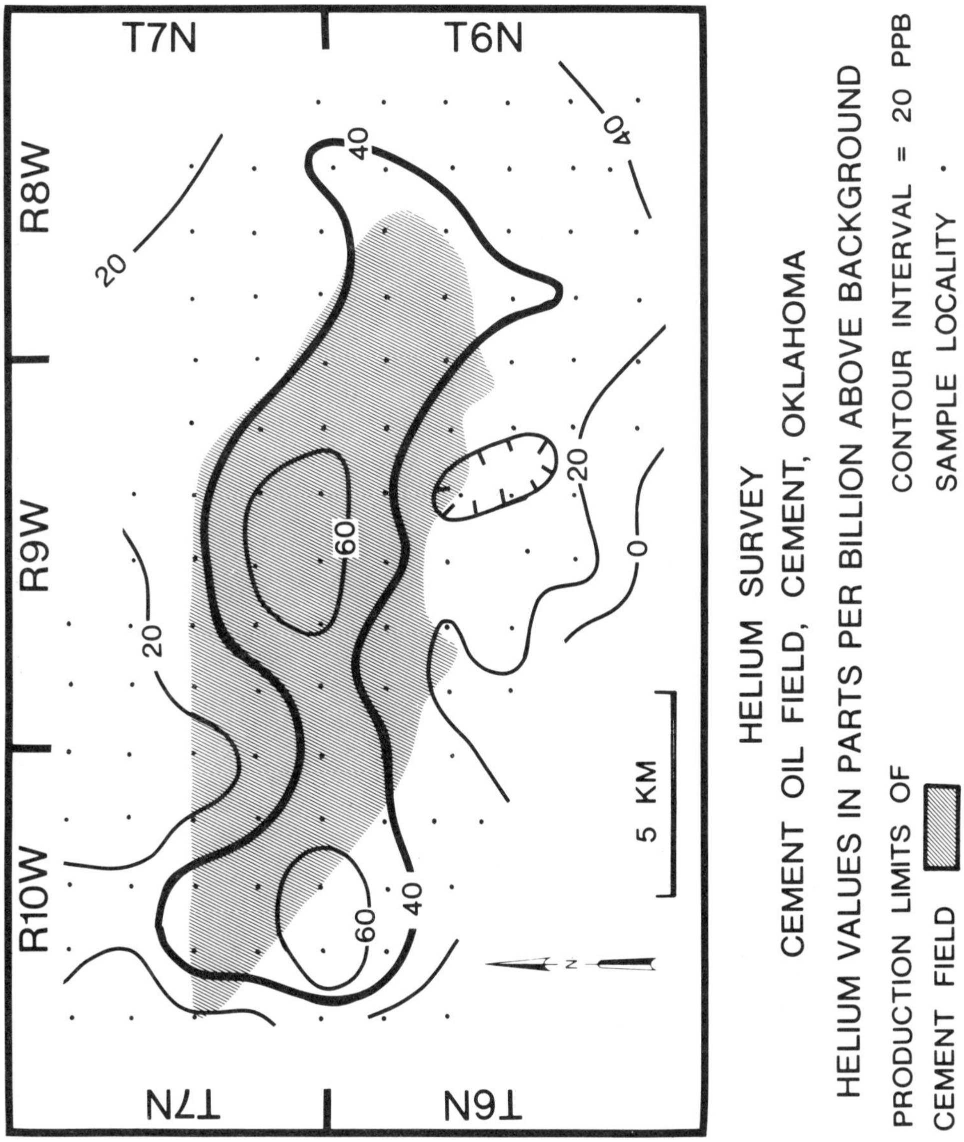
T7N
T6N
R10W
R9W
R8W
5 KM
HELIUM SURVEY
CEMENT OIL FIELD, CEMENT, OKLAHOMA
HELIUM VALUES IN PARTS PER BILLION ABOVE BACKGROUND
PRODUCTION LIMITS OF CEMENT FIELD
CONTOUR INTERVAL = 20 PPB
SAMPLE LOCALITY

Fig. 3.

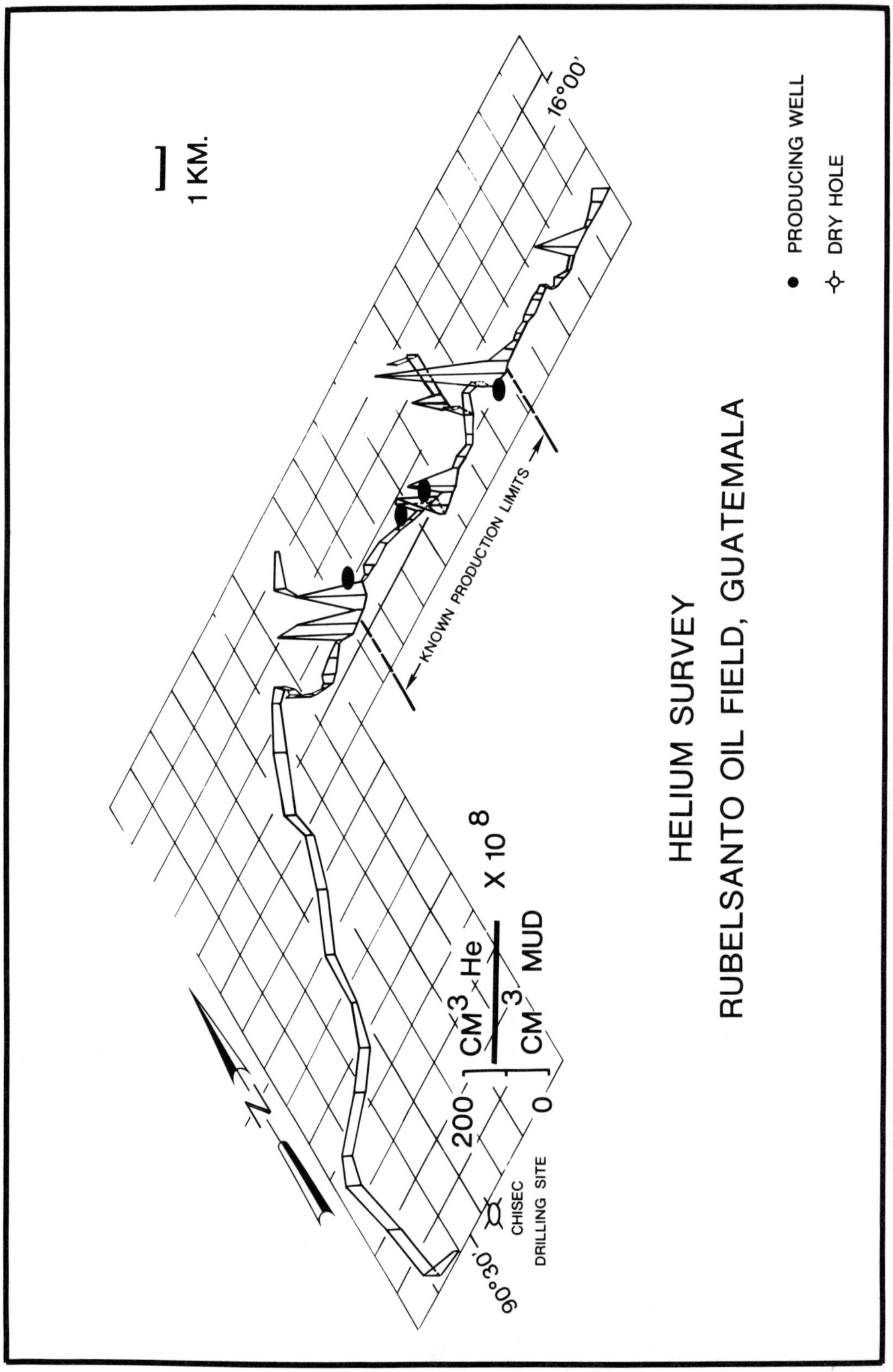
HELIUM SURVEY
RUBELSANTO OIL FIELD, GUATEMALA
1 KM.
16°00'
90°30'
N
200
0
CM3 He / CM3 MUD X 10 8
CHISEC DRILLING SITE
KNOWN PRODUCTION LIMITS
PRODUCING WELL
DRY HOLE

Fig. 4.

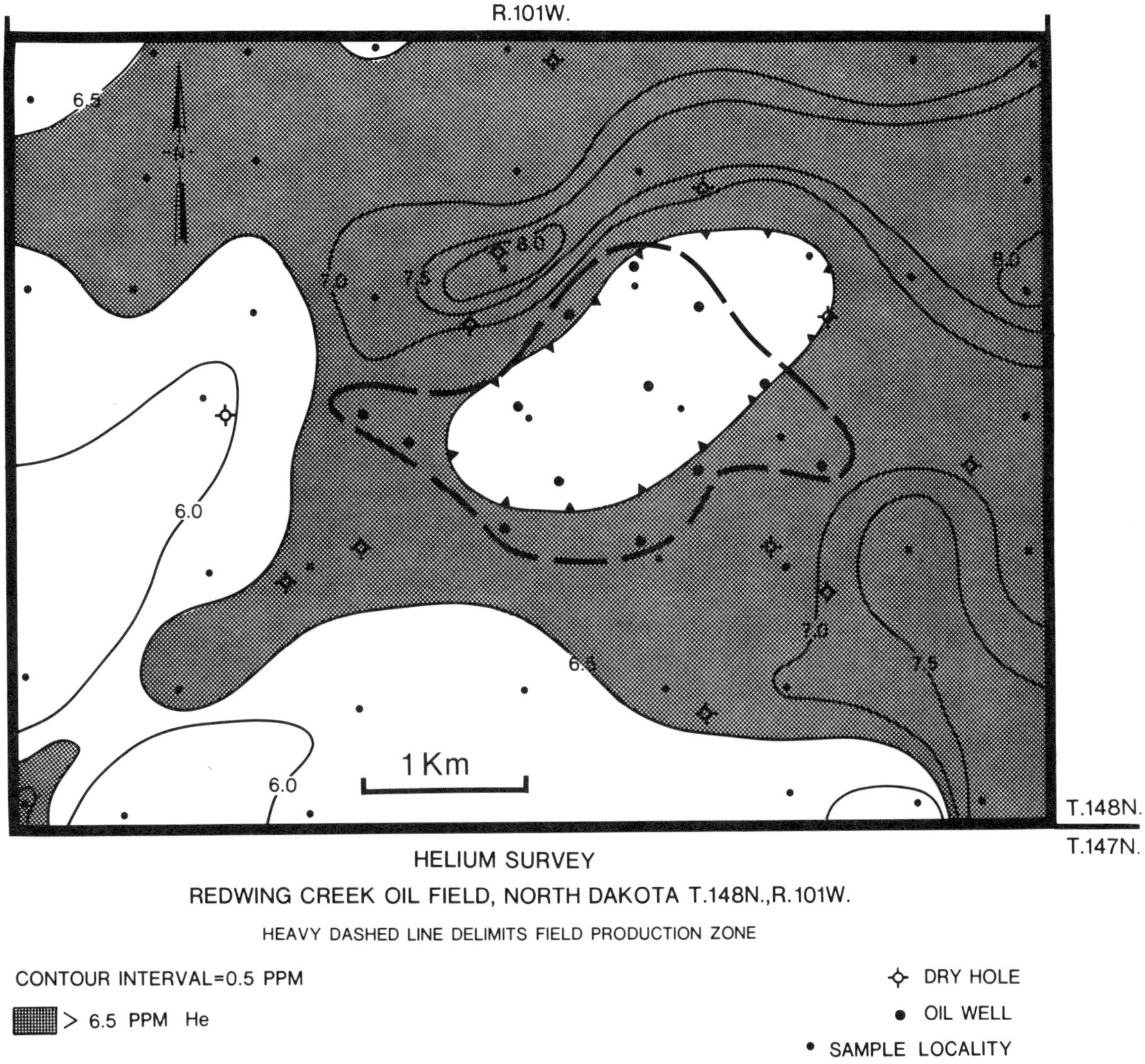

Fig. 5.

Another area surveyed is the Red Wing Creek oil field in North Dakota. For an excellent discussion of the Red Wing Creek structure see Brenan et al. (1975), on which the following information is based. The structure here is believed to be the result of an uplift of Mississippian carbonates and evaporites at the center of a meteoric impact crater. The impact is believed to have created a very complex, highly faulted, and fractured zone in the Mississippian sequence with an uplift in the center surrounded by a ring depression. This event was followed by normal Jurassic and Cretaceous sedimentation, leaving the Red Wing Creek structure at depth. Five and one-half million barrels of oil have been produced there since 1972. The producing horizon varies in depth from about 1,370 m to about 2,160 m and 785 m of oil column was encountered in the discovery well. Figure 5 shows the results of the helium survey run over the Red Wing Creek oil field. Here the survey revealed a halo of anomalously high helium content in the soil surrounding the main producing area. This high-helium halo is interpreted to be related to the oil accumulation rather than to the intense faulting. If the permeability of this fault system were the cause of high-helium values, these would be observed over the area of most intense faulting. The opposite is the case.

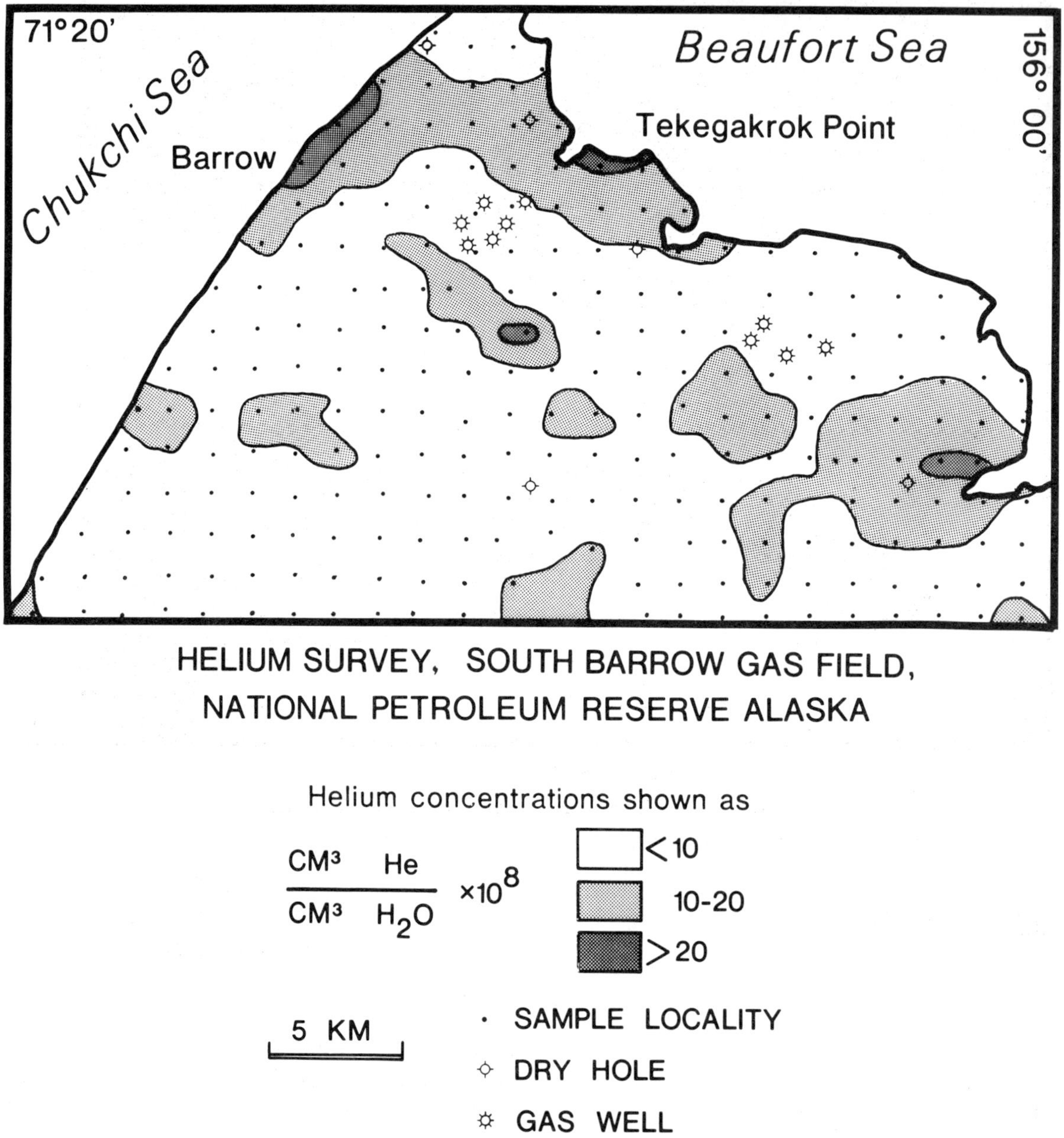

Fig. 6.

Helium surveys have also been used in permafrost areas on the North Slope of Alaska. Permafrost samples collected from a depth of about 0.7 m were hermetically sealed in cans and subsequently analyzed for helium content. Figure 6 shows the results of a survey run in the area around the South Barrow gas fields. The structural picture here is quite complex. Gas is produced for the Barrow Sand in the Kingak Shale of Jurassic age and also presumably from fractures in the argillite basement (Collins, 1961). There are at least three gas reservoirs in the area. The South Barrow East Area field is about 10 km east of the South Barrow field. The apparently unproductive area between these two fields is known as the disturbed zone. This zone shows up in the seismic surveys as "a nearly circular area of no seismic reflections,

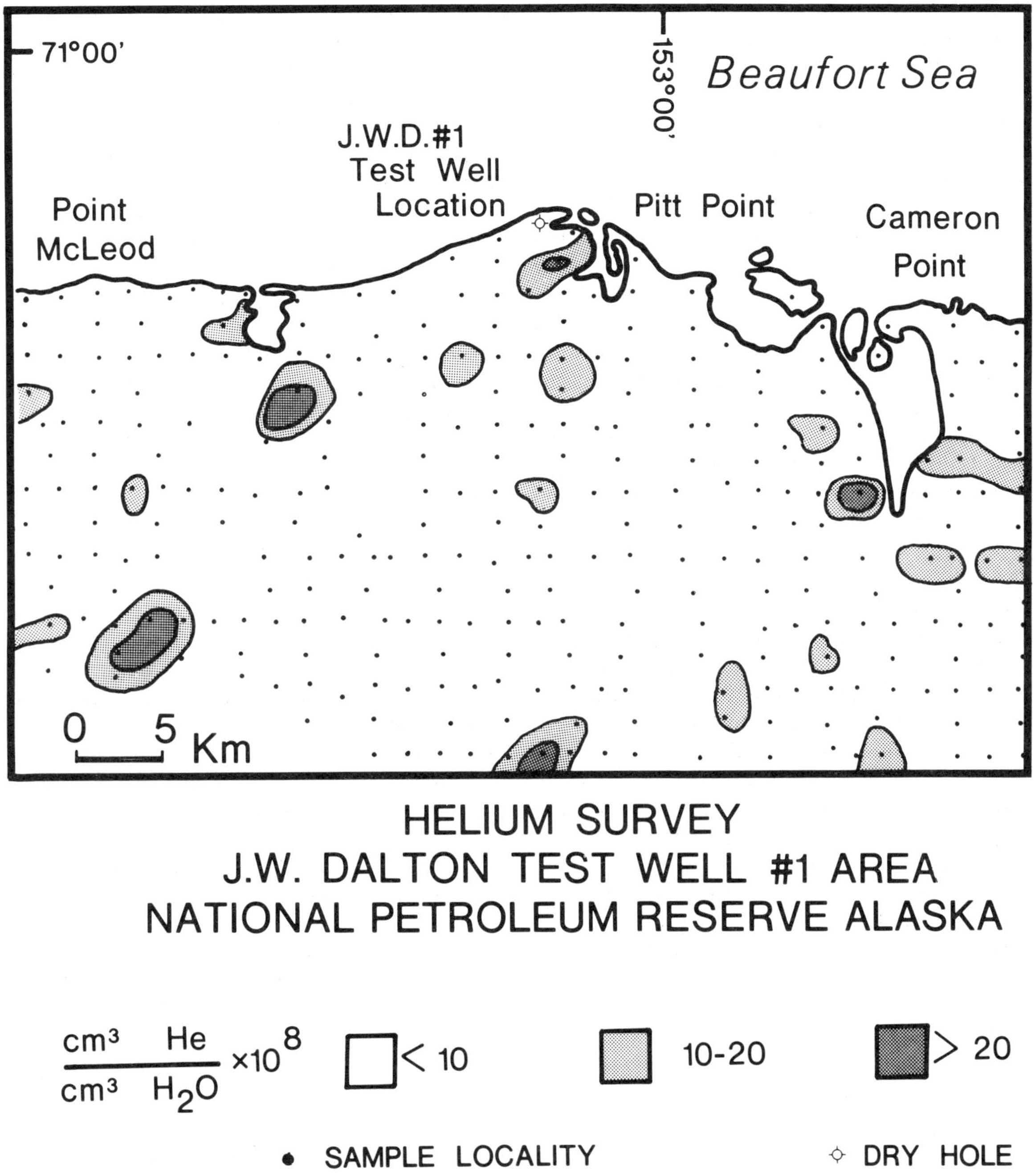

Fig. 7.

from which faults appeared to radiate in several directions" (Collins, 1961, p. 629). It is a highly faulted, structurally complex area with Jurassic and Cretaceous beds dipping at angles from 20° to nearly vertical and is thought by some to have been created by meteoric impact. The South Barrow East Area field is separated into two apparently distinct reservoirs. The two wells on the western side of the field produce natural gas that has a high helium content, ranging from 1% to 4%. The eastern wells produce gas that is much lower in helium content, ranging from 0.05% to 0.13%. This difference in helium content probably indicates a permeability barrier between the two reservoirs, both of which are in the Barrow Sand.

The picture produced by the helium survey is also very complex. About half of the samples from the northern and eastern portions of the survey area contained high (greater than 10×10^{-8} cm^3 helium/cm^3 water) helium concentrations. It is possible that this helium is related to the South Barrow gas fields or to as yet undiscovered gas reservoirs in the area, but no clear picture emerges from the distribution of these high-helium samples. A more closely spaced grid of samples around the South Barrow fields is necessary to determine if any such relationship does exist. It is also possible that the high helium observed is related mostly to the shallow depth to argillite basement here, which is encountered at about 750 m in the South Barrow field and drops off rapidly to the south.

A survey was also run around the J. W. Dalton site that was to be drilled near Camp Lonely on the National Petroleum Reserve. The concentrations of helium observed in the permafrost in this survey were much lower than those observed around Barrow, the average concentration being about half that in the Barrow area. Figure 7 displays the results of this survey. There is only one sample that contained a high concentration of helium near the test site, and no pattern is observed that appears to relate to leakage from a petroleum reservoir at this site. This survey is typical of surveys run in areas where no high-helium source is believed to exist. When the well was subsequently drilled, only a very minor show of oil was encountered, and that was in the form of a very viscous asphaltic substance.

The final survey to be described was run over the Simpson Peninsula on the National Petroleum Reserve in Alaska (Figure 8). Five oil seeps are located in the Simpson area and were a factor in establishing the Naval Petroleum Reserve No. 4, which is now called the National Petroleum Reserve, Alaska. Numerous shallow wells were drilled in the vicinity of the four seeps on the eastern side of the peninsula. Subsurface data from these wells, along with seismic data, have identified an unconformity within the Seabee Formation of the Upper Cretaceous Colville Group (Robinson, 1964). West of the unconformity is the deep Simpson Canyon, filled with as much as 350 m of Cretaceous (Seabee Formation) clay shale. East of the unconformity the wells encountered alternating sandstone and claystone in the Seabee Formation. These porous sandstones dip gently to the east and are truncated on the west by the canyon just west of the seeps. Wells drilled near the seeps encountered oil at about 100 m in the basal part of the Seabee and Ninuluk Formations (undifferentiated) or in the upper part of the Grandstand Formation. This oil may be in a small stratigraphic trap (or traps), and the seeps may represent leakage at the truncation zone.

The most obvious feature of the helium survey is an arcuate pattern of very high helium values, indicated in Figure 8 by the bold line. The northern, eastern, and western portions of this halo are generally located over what is believed to be the Simpson Canyon. This same area is strikingly delineated in a low-level aeromagnetic survey (T. J. Donovan et al., unpublished data) as an area characterized by numerous high-frequency peaks. These are believed to be related to near-surface diagenetic magnetite, which is formed through reduction of oxidized iron minerals in areas where seepage of hydrocarbons has occurred (Donovan et al., 1979).

These results are interpreted to represent possible seepage of hydrocarbons and helium from a petroleum reservoir within the halo, indicated by the solid line in Figure 8. The petroleum quite possibly could be in a stratigraphic trap formed where the Simpson Canyon truncates the southeasterly dipping strata. The high helium values and the inferred diagenetic magnetite would represent microseepage at this truncation zone.

Four deep wells (greater than 1,000 m) have been drilled within the survey area, none of which were within this prospective area. None of these wells encountered producible oil or gas.

The survey of the Simpson area was run in the spring prior to any significant thawing of the soil. In this survey the top 15 cm of frozen soil of each sample core (about 350 samples) was also collected and analyzed. No correlation was ob-

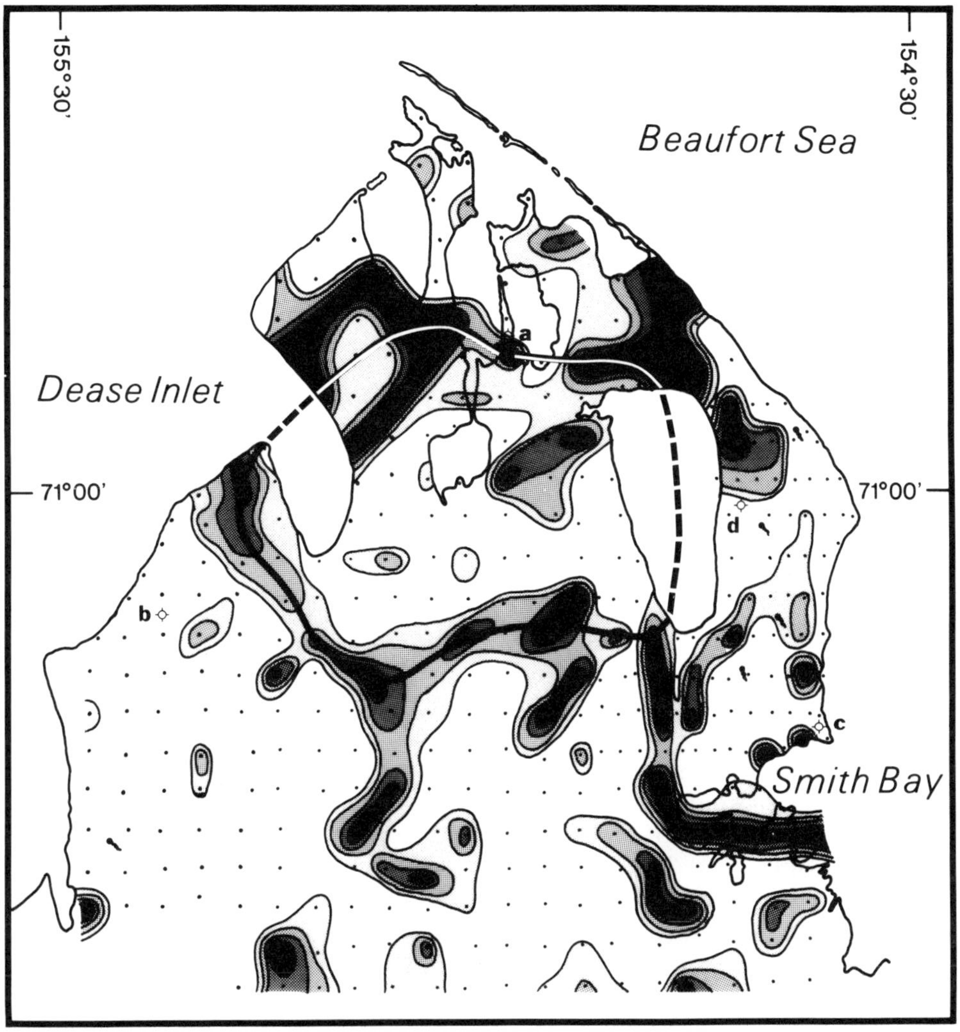

HELIUM SURVEY, CAPE SIMPSON AREA
NATIONAL PETROLEUM RESERVES ALASKA

0 1 2 3 4 5
KILOMETERS

◇ DRY HOLE
OIL SEEP
· SAMPLE LOCALITY

Helium concentrations shown as

$$\frac{CM^3\ He}{CM^3\ H_2O} \times 10^8$$

□ < 60

□ 60-70 □ 70-80 □ 80-90 ■ > 90

SIMPSON TEST WELLS

a NORTH SIMPSON **b** SIMPSON-1 **c** EAST SIMPSON #1 **d** EAST SIMPSON #2

HEAVY LINE INDICATES INTERPRETED ANOMALY

Fig. 8.

served between these shallow samples and those taken in the permafrost itself. The results from the shallow samples showed no discernible pattern of high helium values, but rather showed seemingly random pattern of highly variable helium contents. In the Barrow and Lonely areas, the survey was run in late summer at the time of maximum thaw of the active layer. At every fifth location (250 samples) a sample of the melted soil just above the permafrost was also collected and analyzed. The average depth to permafrost was about 0.4 m; these samples also showed no correlation with the deeper samples taken in the permafrost, nor did they display any discernible pattern of high helium values. These results indicate that in order for a helium survey to have meaningful results in the arctic, samples of the actual permafrost must be taken. Freezing and thawing of the soil in the active layer appears to produce a variability in helium concentrations that masks the anomaly resulting from leakage from a petroleum reservoir.

In conclusion, this paper has presented the results of seven helium serveys run over known or suspected petroleum accumulations. A total of fifteen such surveys have been completed, eleven over known oil or gas accumulations and four in prospective areas. Six of the surveys over known reservoirs revealed helium anomalies that would have been useful in locating or delineating the oil or gas accumulation. Three surveys showed definite, nonrandom helium anomalies, but it would have been difficult to determine the precise location of the reservoir they were related to. Two surveys over known oil fields showed no helium anomaly at all. Of the four surveys run in prospective areas, two showed no helium anomaly and were subsequently drilled as dry holes. Two gave definite, nonrandom helium anomalies, but have not as yet been drilled to determine if the helium source is an oil or gas reservoir. These studies indicate that helium surveys can be a valuable aid in a petroleum exploration effort when used in conjunction with conventional geophysical and geological studies. Because helium surveys are relatively inexpensive and environmentally nondestructive they may have special applicability in the fragile arctic environment.

ACKNOWLEDGMENTS

I would like to thank U.S. Geological Survey personnel M. C. Dalziel and V. C. Dean for their assistance in the collection of samples, and K. I. Cunningham for his assistance in sample collection and in drafting and revising this manuscript.

REFERENCES

Brenan, R. L., Peterson, B. L., and Smith, H. J., 1975, The origin of Red Wing Creek structure: McKenzie County, North Dakota: Wyoming Geological Association, Earth Science Bulletin, v. 8, no. 3 (Sept. 1975), 41 p.

Buis, O. J., and Evenson, R. E., 1977, Drilling under way on big Guatemala structure: Oil and Gas Journal, v. 75, p. 96-97 (May 23, 1977).

Bulashevich, Yu. P., and Bashorin, V. N., 1973, On the detection of faults along the Sverdlovsk DSS profile from high concentrations of helium in underground water: Physics of the Solid Earth, Academy of Sciences, USSR-Izvestiya, March, p. 185-189.

Clarke, W. B., Top, Z., Beavan, A. P., and Gandhi, S. S., 1977, Dissolved helium in lakes: Uranium prospecting in the Precambrian terrian of central Labrador: Economic Geology, v. 72, p. 233-242.

Collins, F. R., 1961, Core tests and test wells, Barrow area, Alaska: U.S. Geological Survey Professional Paper 305-K, p. 569-644.

Debnam, A. H., 1969, Geochemical prospecting for petroleum and natural gas in Canada: Geological Survey of Canada Bulletin, v. 177, p. 1-26.

Dikun, A. V., Korobeynik, V. M., and Yanitskiy, I. N., 1976, Some features of development of helium surveying: International Geology Review, v. 18, no. 1, p. 98-100.

Donovan, T. J., 1974, Petroleum microseepage at Cement, Oklahoma—evidence and mechanisms: American Association of Petroleum Geologists Bulletin, v. 58, p. 429-446.

———, and Dalziel, M. C., 1977, Late diagenetic indicators of buried oil and gas: U.S. Geological Survey Open-File Report 77-817, 44 p.

———, Forgey, R. L., and Roberts, A. A., 1979, Aeromagnetic detection of diagenetic magnetite over oil fields: American Association of Petroleum Geologists Bulletin, v. 63, no. 2, p. 245-248.

———, Noble, R. L., Friedman, Irving, and Gleason, J. P., 1975, A possible petroleum-related geochemical anomaly in surface rocks, Boulder and Weld Counties, Colorado: U.S. Geological Survey Open-File Report 75-47, 11 p.

Dyck, Willy, 1976, The use of helium for mineral exploration: Journal of Geochemical Exploration, v. 5, p. 3-20.

Eremeev, A. N., Sokolov, A. A., Solovov, A. P., and Yanitskii, I. N., 1972, Application of helium surveying to structural mapping and ore deposit forecasting: Fourth

International Geochemical Exploration Symposium, London, Proceedings, p. 183-192.

Friedman, Irving, and Denton, E. H., 1975, A portable helium sniffer: U.S. Geological Survey Open-File Report 75-532, 6 p.

Horvitz, Leo, 1954, Near surface hydrocarbons and petroleum accumulations at depth: Mining Engineering, v. 6, p. 1205-1209.

——— 1969, Hydrocarbon geochemical prospecting after 3 years, *in* Unconventional methods in exploration for petroleum and natural gas: Dallas, Southern Methodist University, p. 205-218.

——— 1972, Vegetation and geochemical prospecting for petroleum: American Association of Petroleum Geologists Bulletin, v. 56, no. 5, p. 925-940.

Hunt, J. M., 1979, Seeps and surface prospecting, *in* Petroleum geochemistry and geology: San Francisco, W. H. Freeman and Co., p. 425-433.

Mazor, Emanuel, 1977, Geothermal tracing with atmospheric and radiogenic gases: Geothermics, v. 5, p. 21-36.

———, and Fournier, R. O., 1973, More on noble gases in Yellowstone National Park hot waters: Geochimica et Cosmochimica Acta, v. 37, p. 515-525.

———, and Verhagen, B. TH., 1976, Hot Springs of Rhodesia: their noble gases, isotopic and chemical composition: Journal of Hydrology, v. 28, p. 29-43.

Moore, B. J., 1976, Analyses of natural gases: U.S. Bureau of Mines Computer Printout 1-76.

Oil and Gas Journal, 1977, International briefs, Exploration: v. 75, no. 38, p. 74.

Ovchinnikov, L. N., Sokolov, V. A., Fridman, A. I., and Yanitskii, I. N., 1972, Gaseous geochemical methods in structural mapping and prospecting for ore deposits: Fourth International Geochemical Exploration Symposium, London, Proceedings, p. 177-182.

Plyusnin, G. S., Volkova, N. V., Godumskiy, G. P., Smirnov, V. N., Papolitov, E. I., and Brandt, S. B., 1972, Trends in the distribution of argon and helium in contact regions and fault zones: Geokhimiya, no. 10, p. 1189-1196 (in Russian, English abstract).

Pogorski, L. A., 1980, Helium emanometry in exploring for hydrocarbons, I: This volume.

Reimer, G. M., 1976, Design and assembly of a portable helium detector for evaluation as a uranium exploration instrument: U.S. Geological Survey Open-File Report 76-398, 18 p.

———, and Adkisson, C. W., 1977, Reconnaissance survey of the helium content of soil gas in Black Hawk, Eldorado Springs, Evergreen, Golden, Morrison, Ralston Buttes, and Squaw Pass Quadrangles, Colorado: U.S. Geological Survey Open-File Report 77-464, 11 p.

———, and Denton, E. H., 1978, Improved inlet system for the U.S. Geological Survey helium sniffer: U.S. Geological Survey Open-File Report 78-588, 4 p.

———, Denton, E. H., Friedman, Irving, and Otton, J. K., 1979, Recent developments in uranium exploration using the U.S. Geological Survey mobile helium detector: Journal of Geochemical Exploration, v. 11, p. 1-12.

Roberts, A. A., 1975, Helium surveys over known geothermal resource areas in the Imperial Valley, California: U.S. Geological Survey Open-File Report 75-427, 6 p.

———, Friedman, Irving, Donovan, T. J., and Denton, E. H., 1975, Helium survey, a possible technique for locating geothermal reservoirs: Geophysical Research Letters, v. 2, p. 209-210.

Robinson, F. M., 1964, Core tests, Simpson area, Alaska: U.S. Geological Survey Professional Paper 305-L, p. 645-730.

Smith, G. H., and Ellis, M. M., 1963, Chromatographic analysis of gases from soils and vegetation, related to geochemical prospecting for petroleum: American Association of Petroleum Geologists Bulletin, v. 47, p. 1897-1903.

Torgerson, Thomas, and Clarke, W. B., 1978, Excess helium-4 in Teggan Lake: Possibilities for a uranium ore body: Science, v. 199, 769-771.

EXPLORATION VIA MAGNETOTELLURICS

Wayne J. Peeples

SOUTHERN METHODIST UNIVERSITY

ABSTRACT

The Magnetotelluric (MT) method is an electromagnetic geophysical technique used to determine variations of resistivity both laterally and horizontally in the earth. Since the resistivity of rock materials depends on mineral content, texture, porosity, and the amount and resistivity of pore-filling materials (water, oils, gas, etc.), knowledge of the resistivity structure can be useful in many types of geophysical exploration problems.

The MT method measures naturally occurring magnetic and electric (telluric) fields at the surface of the earth as a function of time. These time-domain measurements are converted by Fourier analysis to obtain (primarily) the apparent resistivity (r_a) as a function of frequency. Subsidiary quantities can also be calculated to indicate the dimensionality of the exploration area, quality of recorded data, strike direction, etc. The frequencies used range from 10^4 to 10^{-3} Hertz. Since lower frequency electromagnetic fields penetrate farther into the earth than higher frequency fields (the skin depth rule) the apparent resistivity curves can be viewed as an integrated resistivity versus pseudo depth profile of the earth.

The apparent resistivity curves are converted to true resistivity-depth curves by a number of interpretational schemes which use the solution of Maxwell's equations for incident plane waves on earth models where resistivity (1) is a function of depth only, e.g., layered models, (2) is allowed to vary in depth and one lateral dimension (a two-dimension model), and (3) varies with depth and both horizontal directions (a three-dimensional model). The pitfalls associated with each of these interpretation schemes are demonstrated.

Until fairly recently there have been no quantitative measures of the vertical and lateral resolution and precision of the MT method. The advent of Backus-Gilbert inversion techniques has now led to the first examples of resolution and precision for this particular geophysical technique. Resolution-precision analysis enables the exploration manager to determine whether an interesting subsurface structure is resolvable with MT data and to determine station density and required frequency range. These techniques thus promote more effectively designed exploration programs.

Several new developments in MT have appeared in the last several years: sophisticated instrumentation, cryogenic magnetometers (low noise levels), remote telluric references (reduced cost), remote magnetic references (to remove bias in MT data processing), algorithms to determine automatically the parameters of two dimensional models for experimental data (improve interpretation), and the simultaneous use of data from several geophysical experiments (including MT) to determine models for interpretation.

An analysis is made of the electrical transients method of exploration which indicates that the effect being measured is an IP effect. Further study of this technique would enable this tool to be added to the growing number of tools which detect hydrocarbon accumulations.

EXPLORATION VIA MAGNETOTELLURICS

Basic Concepts

In a sequence of papers in the early 1950s, Tikhonov (1950), Kato and Kikuchi (1950), and Rikitake (1950, 1951) observed the existence of a correlation between the variations of the naturally occurring geomagnetic field and the naturally occurring electric (telluric) field at the surface of the earth and investigated methods to use this correlation to explore the subsurface geoelectric profile to great depth. It was not until Cagnaird (1953) that the method attracted serious attention, when he presented a method of interpreting the magnetotelluric (MT) measurements that could be applied to geophysical exploration. Cagnaird's method assumed that the earth was horizontally stratified with each layer being electrically homogeneous and isotropic, that the source fields were plane electromagnetic waves impinging on the surface of the earth, and solved Maxwell's equations for this boundary-value problem.

The time variations of the electric and magnetic field at a site are measured over a wide frequency band and analyzed to obtain apparent resistivity as a function of frequency. Interpretation then consists of matching the observed apparent resistivity vs. frequency curves with curves predicted by simple models. Depth control in MT occurs because lower frequencies penetrate farther into the earth than do higher frequencies. Thus, by processing the magnetic and electric measurements,

Cagnaird's scheme purports to allow us to determine the electrical resistivity and thickness of each layer. By observing how layering changed with spatial location we are thus in a position to determine the subsurface structure in a particular target area. Results from the interpretation of MT measurements at a site can be compared to a highly smoothed induction log, and we have obtained these results without drilling a well. Our picture of the subsurface could then be turned into either a rock type or even a porosity/salinity distribution (since resistivity is primarily affected by these two physical parameters) if the resolution of MT is great enough.

The inherent capabilities of the MT technique have been perceived for almost thirty years, but the success of MT depends on a complex interaction of several disciplines, and the technique has received the necessary attention only recently. The unconventional technique of MT is being developed for the following reasons: (1) There exist very few geophysical experiments that allow us to estimate specific physical properties at depth solely by observing natural phenomena at the surface of the earth. Thus, any technique offering additional clues to earth structure is worth using. (2) The difficulties that have been encountered in nearly all phases of MT experiments have caused a lag in the development and subsequently the usefulness of this technique, but they are being resolved at a favorable rate. We would predict, therefore, that this presently unconventional technique will gain in respectability and become a conventional technique in the near future.

For our purposes here we will divide the typical MT experiment from the measurement of natural electromagnetic fluctuations at the earth's surface to a meaningful geophysical interpretation into several steps, and then examine the research difficulties and recent advances in each step to try to anticipate the potential contributions of the MT method in future petroleum exploration.

Maxwell's laws predict that as electric and magnetic source disturbances propagate through a region with uniform resistivity, the fields decay (energy is converted to heat via ohmic losses) in an exponential manner that depends on the resistivity and frequency of the wave. The skin depth, or distance that the fields travel before they are reduced to 37% of their original value is given by

skin depth in km $= \delta \approx \frac{1}{2}\sqrt{\rho/f}$ where ρ is resistivity in the ohm meters of the rock and f is the frequency of the wave disturbance (1)

This decaying magnetic disturbance in the earth is actually a diffusion rather than a wave propagation phenomenon. When this diffusing magnetic disturbance encounters a rock with different resistivity, part of the incident energy is reflected back toward the surface and partly transmitted into the second medium. This same process is repeated again and again as the propagating magnetic disturbance encounters new regions of differing resistivity within the earth. Thus if we measure the electric (E) and the magnetic (H) signals at the surface of the earth we are actually measuring a complex superposition of incident and reflected disturbances.

If for the moment we consider the earth to be homogeneous and isotropic, then the mathematical description of a plane electromagnetic wave electric (E) and an orthogonal magnetic (H) field would be

$$E_x(z) = E_{x_0} \exp\left[-i\omega t + (i-1)z/\delta\right]$$

$$H_y(z) = H_{y_0} \exp\left[-i\omega t + (i-1)z/\delta\right] \qquad (2)$$

It can be shown that

$$E_{x_0} = \frac{(1-i)\mu\omega\delta}{2} H_{y_0} \qquad (3)$$

$$\text{therefore impedance} = \frac{E_{x_0}}{H_{y_0}} = (1-i)\left(\frac{\mu\omega\delta}{2}\right)^{1/2} \qquad (4)$$

where the zero subscript indicates the values at the surface of the earth

δ is the skin depth

μ is the magnetic permeability $= 4 \times 10^{-7}$

i is $\sqrt{-1}$

ω is two pi $\times$ frequency of the wave

t is time

and z is depth at which we evaluate the field.

Solving for resistivity we obtain (in MKS units)

$$\rho(\omega) = \frac{i}{\mu\omega} \left[\frac{E_{x_0}(\omega)}{H_{y_0}(\omega)} \right]^2 \quad (5)$$

The resistivity of a homogeneous medium, ρ, can thus be calculated from the measured amplitudes of E_x and H_y at the surface of the earth.

When the resistivity is calculated from the above formula using the amplitudes $E_x(w)$ and $H_y(w)$ measured at the surface of a layered earth, the quantity calculated is called the apparent resistivity, denoted $\rho_a(w)$, and represents a weighted and smoothed average of subsurface resistivities.

Clearly, for a uniform earth $\rho_a(w)$ is identical to $\rho(w)$. The skin depth (Equation 1) along with the description of the propagating wave (Equation 2) indicate that low frequency waves propagate farther into the earth than high frequency waves for a given amount of attenuation and will consequently sample material deeper in the earth. Therefore, a plot of $\rho_a(w)$ against frequency indicates near-surface resistivity averages for high frequencies and deeper resistivity averages for the lower frequencies.

In discussing the possible types of geoelectric structure the electromagnetic wave might encounter in actual survey conditions, it is convenient to define one-, two-, and three-dimensional structures.

A one-dimensional (1D) structure occurs when resistivity is only a function of depth, e.g., a layered medium. In a two-dimensional (2D) structure, the resistivity varies along one horizontal direction and depth. The other horizontal direction is called the strike direction. The three-dimensional (3D) structure is encountered when resistivity in the earth varies along both horizontal directions and depth and thus has no strike direction.

The interpretation of MT can be made by using either one-, two- or three-dimensional models.

If the target area is one-dimensional and isotropic, then we need measure only one electric field component and an orthogonal magnetic field component and interpret the results as a layered or a continuous medium. This procedure is often referred to as scalar magnetotellurics.

If the target area is two-dimensional, then we must measure two orthogonal components of the telluric field and three components of the surface magnetic field. By carefully processing (see below) the MT data, we discover that ρ_a at a particular frequency varies with measuring direction. Then E_x is partly the result of H_y but is also the result of currents induced by H_x which have been deflected by the two-dimensional structure. In this case, we write that the relationship between E and H is written as:

$$E_x = Z_{xx}H_x + Z_{xy}H_y$$
$$E_y = Z_{yx}H_x + Z_{yy}H_y \quad (6)$$

where Z_{yx} (an impedance) gives the portion of E_y due to H_x and Z_{yy} gives the portion of E_y due to H_y. For a two-dimensional earth there is a direction (the strike direction) in which $Z_{xx} = Z_{yy} = 0$. We then calculate the apparent resistivities for both strike direction and orthogonal to strike as:

$$\rho_{a_{xy}}(\omega) = \frac{i}{\mu w}(Zxy)^2$$

$$\text{and} \quad \rho_{a_{yx}}(\omega) = \frac{i}{\mu w}(Zyx)^2 \quad (7)$$

Apparent resistivities calculated in the above manner will be denoted tensor apparent resistivities, and the data can be processed to determine the strike direction no matter what the measuring directions are at the surface of the earth.

If the target area is three-dimensional, again two horizontal components of field and three components of H (two horizontal and one vertical) are measured. These components are measured at a number of locations in and around the target area, and we then try to find a three-dimensional model which matches the observed ρ_a's versus frequency at each station

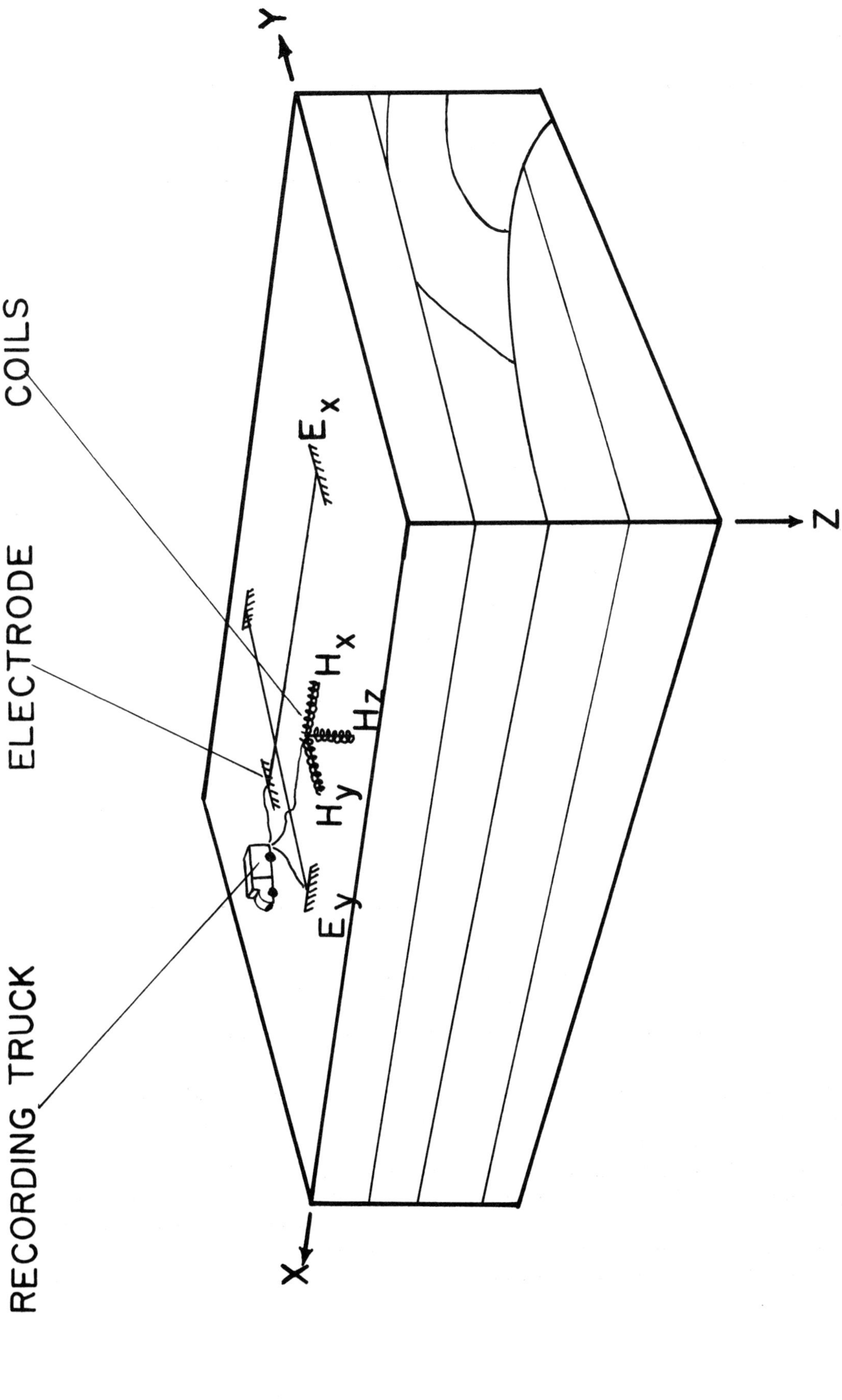

Fig. 1. The basic field setup for a magnetotelluric experiment. Electric field measured by determining the potential difference between two points, while magnetic field is determined by the voltage induced in a coil.

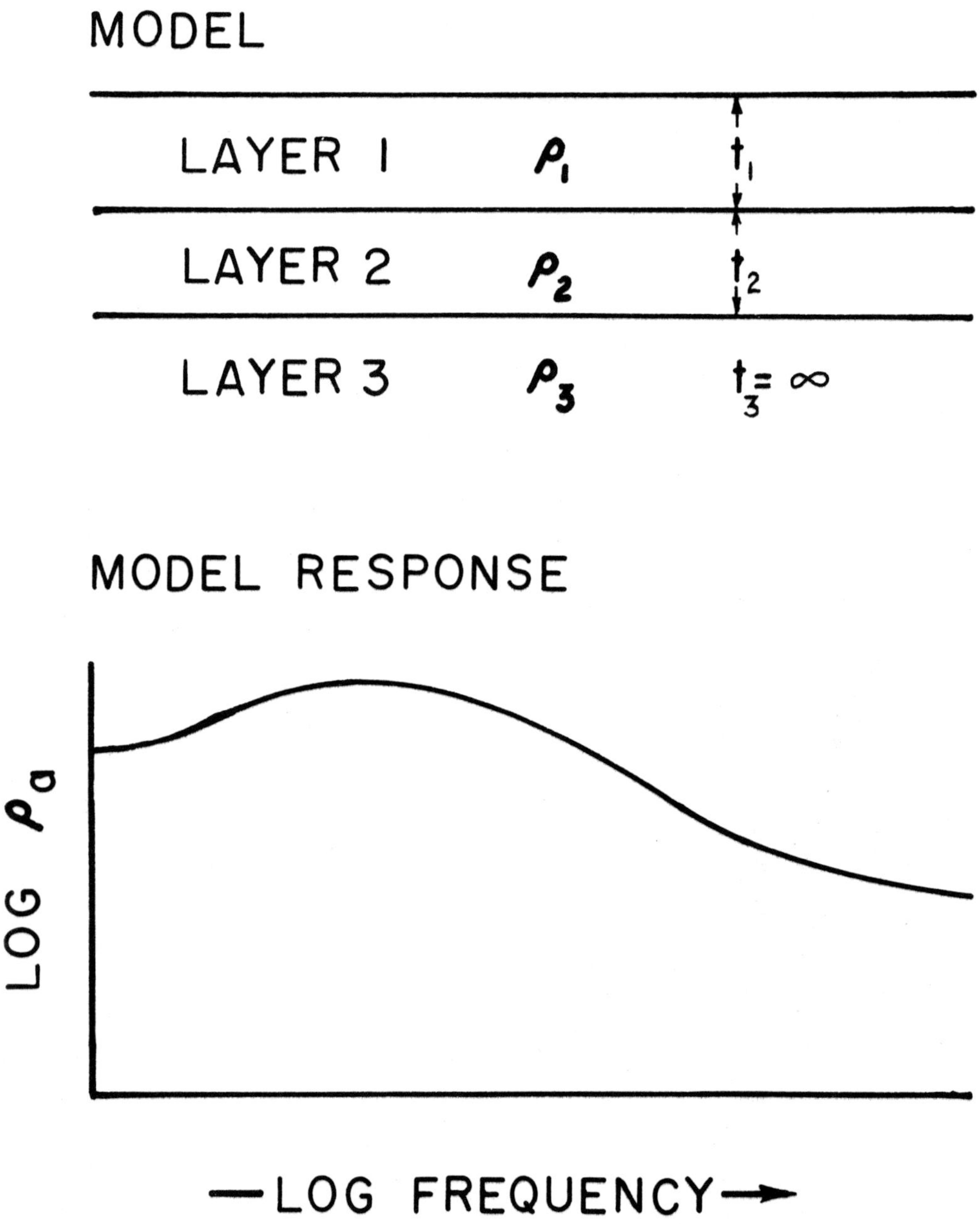

Fig. 2. A one-dimensional MT model (3 layers) and the apparent resistivity as a function of frequency of the model. $\rho_1 < \rho_2$, $\rho_3 < \rho_2$, $\rho_2 > \rho_1$.

location. When two electric and three magnetic components are measured and processed, the resulting relationships are interpreted via the inpedance tensor (Equation 6) and are called tensor apparent resistivities, as in the two-dimensional case.

The basic field setup is shown in Figure 1. The surface electric field is measured by electrodes and the magnetic field components are measured via coils (or lately a SQUID magnetometer).

In nearly all the MT interpretations in the literature, MT results were interpreted via one-dimensional modes. Seldom, however, are data collected at a location where the apparent resistivities were independent of aximuth and

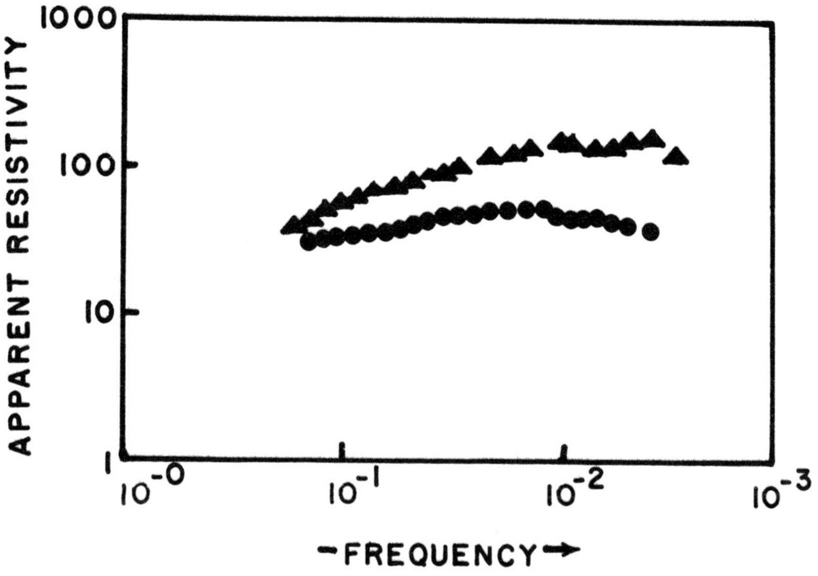

Fig. 3. Tensor apparent resistivities (parallel and orthogonal to strike) for a single station.

the two apparent resistivities are anisotropic—an indication that the proper model should be *at least* two-dimensional.

Data Processing and Presentation

The purpose of data processing in MT is to determine reliable values of impedances, apparent resistivities, and strike directions as a function of frequency from field records and to present these results in a form convenient for interpretation.

The processing state usually follows the following steps: (1) manual edition of records to remove bad record sections; (2) Fourier transformation of each record (Ex, Ey, Hx, Hy, and Hz) to go from time to frequency domain—this can be looked upon as a narrow band filtering operation to determine the field amplitudes at different frequencies; (3) calculation of impedance elements; and (4) determination of principal axes direction (strike direction) and the apparent resistivities parallel and orthogonal to the strike direction. The above operations are well understood and the reader is referred to Vozoff (1972) and Sims (1969) for complete details.

The earth-response functions, ERF (apparent resistivity, strike, angle, etc.), are usually presented on a station-by-station basis, as shown in Figure 2. The surface layer is assumed to be low resistivity and is indicated on the high frequency end of the ρ_a versus frequency curve. The intermediate layer is high resistivity and is indicated at intermediate frequencies; a resistive basement is indicated at the low frequency end of the curve. The tensor apparent resistivities for a single station will not be identical but will be anisotropic, as shown in Figure 3.

To indicate how resistivity changes along a profile, we show an example from Peeples and Rankin (1973) in Figure 4, in which we contour the theoretical values of ρ_a, indicated below each station. The ρ_a's are written in a column in order of decreasing frequency (increasing depth of exploration).

Analysis of Observations (Interpretation)

Interpretation of the previously determined earth-response functions ideally should proceed in five stages: (1) obtaining an overview of ERF (both MT and other experiments, e.g., seismic, well logs, gravity, etc.) for the target area, and formulating interpretational attack procedure (2D, 3D, etc.); (2) determining layered models for each site; (3) two-dimensional modeling of results along profiles through the target area; (4) three-dimensional

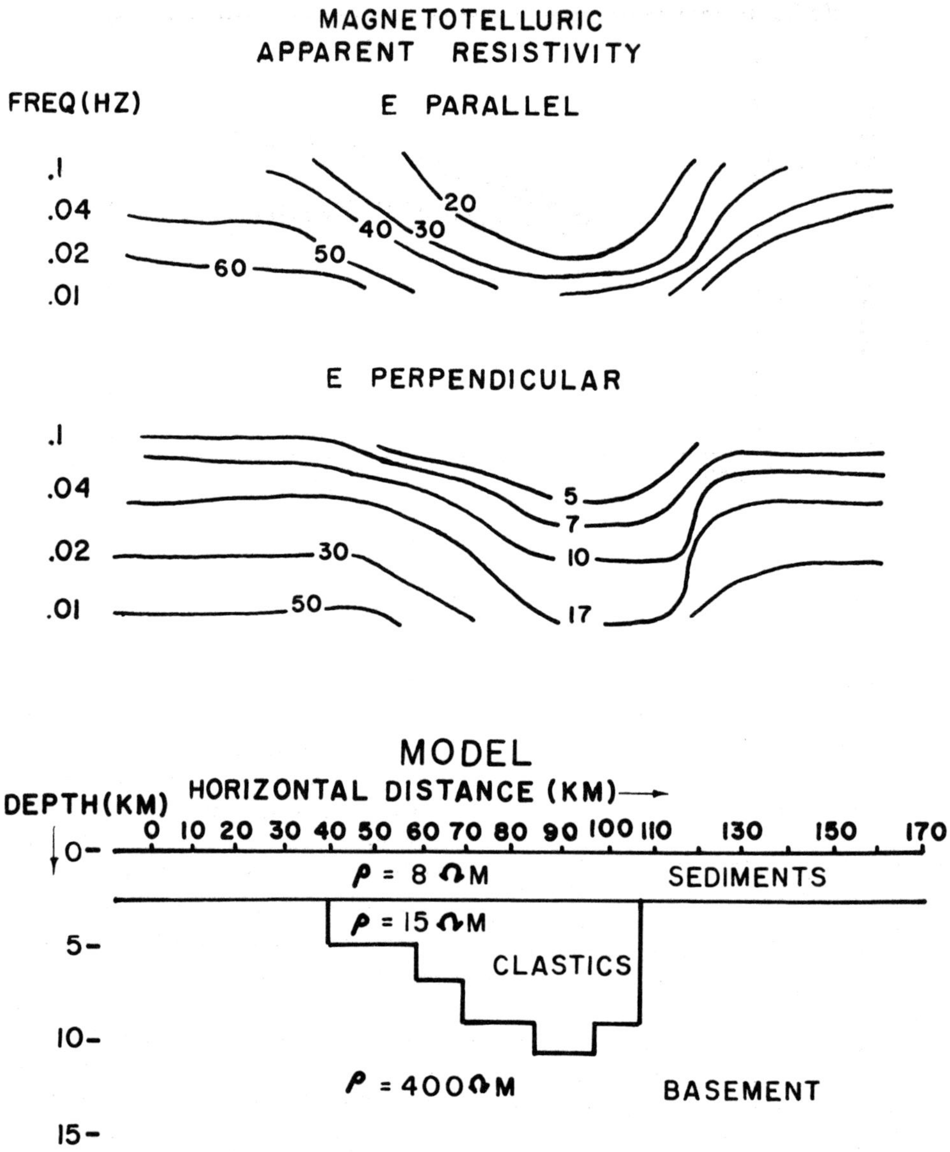

Fig. 4. Theoretical apparent resistivities for a two-dimensional sedimentary basin. E parallel and E perpendicular are the apparent resistivities for the electric field parallel and perpendicular to strike.

modeling of results in target area; and (5) assigning rock type to the 3D model parameters.

One-dimensional (layered models) forward problems are very simple, can be calculated exactly and inexpensively (e.g., Ward, Peeples, Ryu, 1974), and can even be solved on programmable pocket calculators. Inverse algorithms to determine $\rho = \rho$ (z) for the 1D problem are also inexpensive and can be done via programmable pocket calculator (Bostick, 1978). This explains the preponderance of 1D-type interpretations of MT data. Except under extremely limited conditions, however, ID interpretations are in error and can easily lead to totally erroneous interpretations of the data.

Two-dimensional modeling techniques are relatively more expensive and can be done by using a variety of techniques (e.g., finite elements—Rijo, 1978; integral equations—Hohmann, 1978; transmission surface—Swift, 1967; etc.), all of which are approximate solutions. These results for profiled MT data do

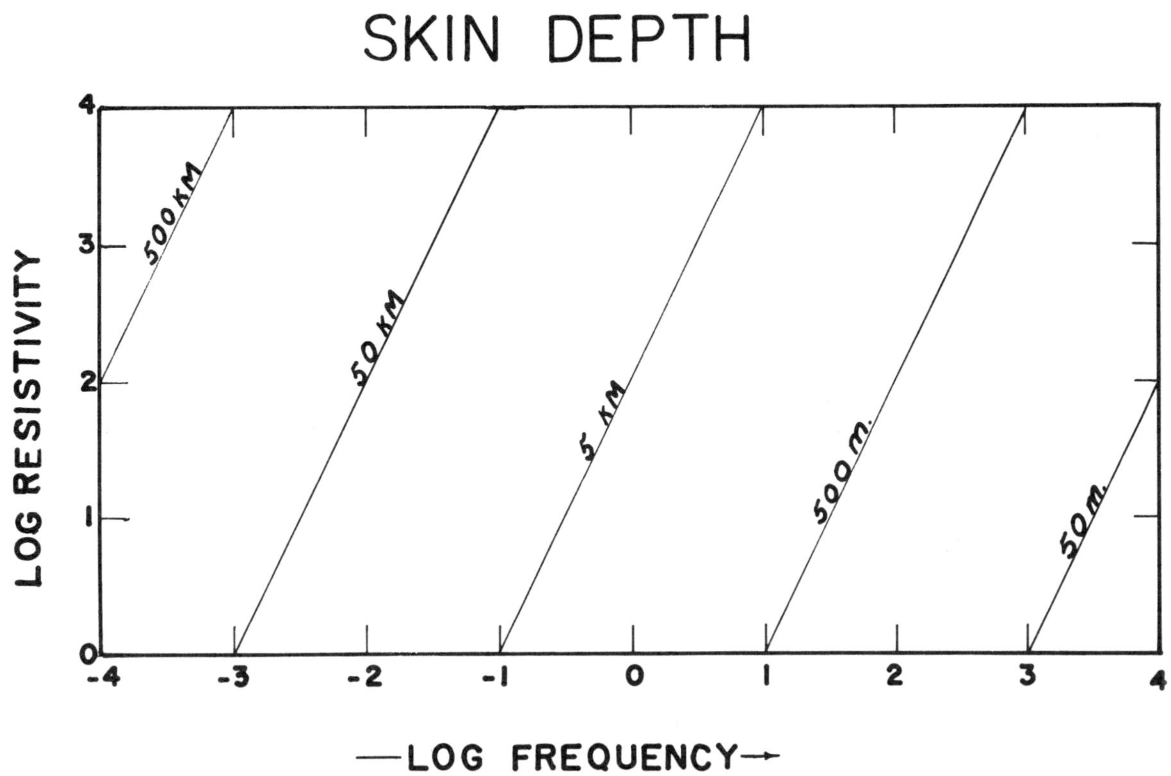

Fig. 5. Skin depths and wavelengths for various frequencies and resistivities.

begin to get at the heart of the problem, and in some areas are geologically and geophysically reasonable.

Three-dimensional models are *almost* intractable at present: cost is very much a dominant factor; e.g., Reddy and Phillips (1978) used 156 min. on a UNIVAC 1108 to calculate one frequency for a 3D model. Results to date have also been published using finite differences (Jones and Vozoff, 1978) and integral equations (Hohmann, 1978), but no cost estimates were given.

Areas Specifically Amenable to the MT Method with Present Technologies/Understanding

As indicated previously, one-dimensional target areas of interest seldom if ever exist. Two-dimensional areas (e.g., sedimentary basins) can be found and are indicated by parameters calculated during data processing (e.g., skew, Swift, 1967). Since two-dimensional models are not computationally expensive, this is the present state of the art. Two-dimensional inverse algorithms (automated techniques to determine model parameters) exist (Vozoff and Jupp, 1978) and can also be a distinct help in interpretation.

The Resolution of Magnetotellurics

Until approximately twelve years ago, the average geophysical interpreter would have found an acceptable model (one whose predicted data matched the observed data) and would have stopped there. We refer to this stage of the interpretation as curve matching. In 1970, Backus and Gilbert presented a mathematical technique, since labeled *inversion methods*, which was devised not only to find acceptable geophysical models but to investigate the nonuniqueness of these models. In other words, how well a given data set resolve the model parameters and the nonuniqueness of a geophysical solution to a particular problem. The methods of Backus-Gilbert to investigate precision-resolution

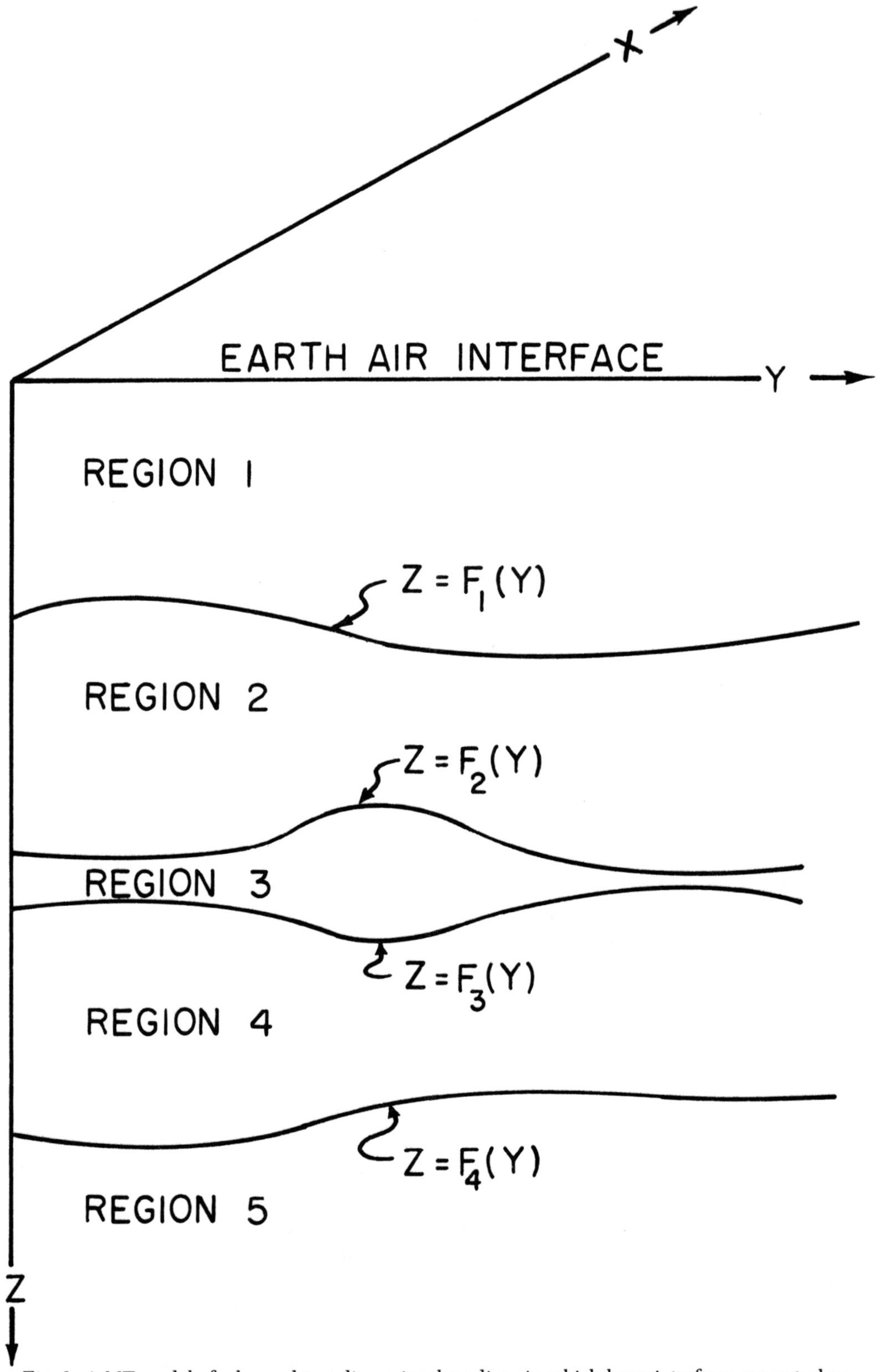

Fig. 6. A MT model of a layered two-dimensional medium in which layer interfaces are not plane parallel.

questions have been slowly incorporated into the field of MT. Statements regarding the equivalence of different resistivity structures exist (e.g., Porstendorfer, 1975, p. 29) but were difficult for us in an operational sense. Oldenburg (1979) has made the first quantitative published study of the resolution of the MT method. His model was one-dimensional where resistivity was a continuous function of depth only. Briefly, his analysis indicated that resolution at a specific depth will be good only when the EM wave passes through *and* is attenuated. Thus we can achieve reasonable resolution in narrow regions of low resistivity but not in regions of high resistivity. In other words, MT data can be interpreted to determine zones of low resistivity but *not* zones of high resistivity. Fine scale structure of a model can be determined if the scale of a feature is about as large as or larger than a wavelength. Since the wavelengths used in MT are fairly large, the scale of resolution of the MT method appears to be limited.

My recent work is an attempt to put more geology back into a model by considering a layered model with arbitrary nonplane parallel interfaces for a two-dimensional model, Figure 6. We have solved this problem and expect to publish these results and the inverse problem shortly. Although the resolution problem is not yet totally solved for complex models, it is tractable and will be understood soon.

Unsolved Problems

At this point, we are in a position to survey MT method and determine the problem areas:

1. Field instrumentation: During the last ten years, MT has gone to solid state digital recording devices, as have a number of other geophysical experiments. The problems with bulky field gear are being minimized, and cryogenic magnetometers (SQUID) for increased resolution along with low power, portable instrumentation are available. In fact, instrumentation is now better than our standard interpretational capabilities.

2. Data processing: In the last fifteen years, this area has been researched in great detail, and several viable processing techniques are in vogue. Some experimenters even do initial data processing in nearly real time in the field. Multistation data collection and processing (Gamble, Wombau, and Clark, 1978) show promise of producing even more reliable and accurate MT earth-response functions.

3. Data interpretation and analysis: This is the present primary area of concern. 2D inverse and 3D forward problems are presently being investigated and should be operationally useful in the next one to three years.

The three-dimensional response of MT models is still not completely understood. The work of Hohmann (1978) indicates that the 3D problem can be solved and does yield the information necessary for interpretation in regions of complex geology, but the modeling problem must be worked until reasonable cost/benefit ratios can be obtained.

Of even more use operationally perhaps are the evolving techniques of data interpretation for using data simultaneously from several experiments (e.g., M.T. gravity) to determine geological structure. These techniques are presently in their infancy but should become very useful in the next decade. When viewed as a whole, MT is seen as an evolving exploration technique whose real impact should be felt in the next decade.

The Electrical Transients Method

Another unconventional electromagnetic technique which has been touted as a direct detection experiment for petroleum exploration is the method known as "electrical transients." This method, based on a series of early technical articles (Evjen, 1943, 1945, 1948; Lewis, 1948), uses a Schlumberger array with the transmitter or current electrodes 0.5 miles apart and the receiver or potential electrodes centered within the current electrodes and spaced at 500 ft, as shown in Figure 7. A biopolar pulse train of current is injected into the ground and the potential is measured after a short period of time, as shown in Figure 8. It is claimed (Azad, 1973, 1977, 1979) that the electromagnetic pulse, upon injection into the ground, propagates until a highly resistive region is encountered (i.e., a hydrocarbon ac-

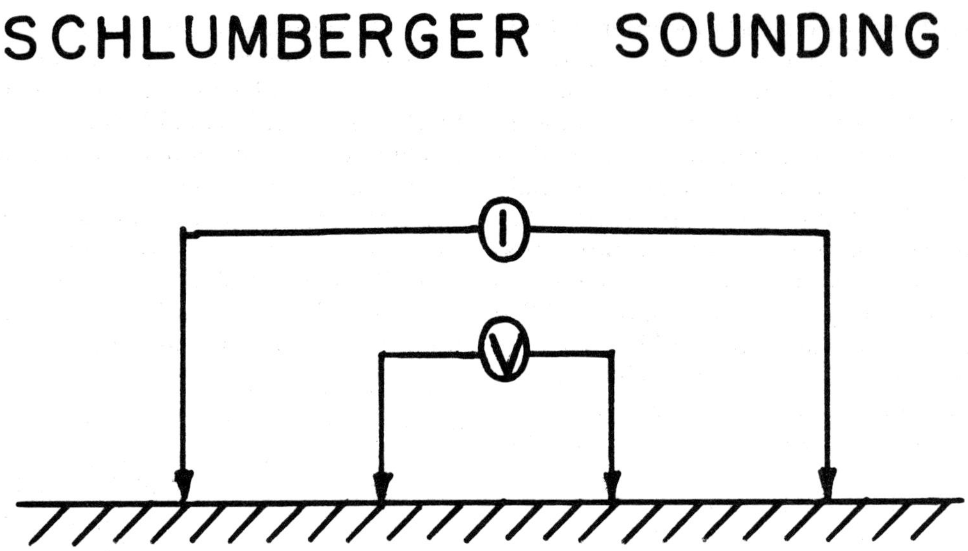

Fig. 7. The Schlumberger array used for the electrical transients method.

cumulation), at which point the pulse is reflected back to the receiver and measured as a potential difference between the receiver electrodes. I propose to examine this claim and show that although what is measured is a real effect, it is more likely to be a combination of mutual coupling and a near-surface induced polarization effect.

First let us examine the idea of obtaining electromagnetic reflections from a buried resistive region. Let us assume for simplicity that the electric field is a plane monochromatic wave propagating in the z direction and polarized in the y direction. Then at any depth z

$$E_y(z) = E_o \exp \{-iwt + i-1)z/\delta\}$$

Thus attenuation is again exponential, as in MT, and depth of effective exploration is governed by the skin depth.

For terrestrial geologic materials (see Figure 5)

$$\delta \approx \tfrac{1}{2} \sqrt{\rho/f}$$

and the wavelength is $\lambda = 2\Pi\delta$ and the phase velocity, V_p is given by $V_p = \omega\delta$.

If we take a typical sedimentary environment, $\rho = 10$ ohm m, we can construct the following table of values:

FREQUENCY	SKIN DEPTH (m)	V_p(m/sec)	λ (m)
.01	1.6×10^4	10^3	10^5
.10	5×10^3	3×10^3	3×10^4
1.0	1.6×10^3	10^4	10^4
10.	500	3×10^4	3×10^3
100.	160	10^5	10^3

If we then calculate the attenuation/unit wavelength, we see that the signal attenuates 55 db/wavelength. Receiver systems used in most electrical transients measurements have an (optimistic) dynamic range of approximately 100 db. Therefore, the total travel length (again optimistically) is two wavelengths. The maximum detectable transmitter target distance at depth is thus about one wavelength. This puts definite limitations on the effective depth of exploration of these systems, which is probably quite shallow.

If the electromagnetic reflections are the phenomenon causing the potential drop across the receiver electrodes, the target must be less than one wavelength deep. If we consider the timing gate of the receiver, the response would be missed because of lack of time resolution. The reflections probably do exist, but they are a second order effect and would be overwhelmed by other physical effects.

Induced Polarization (IP) Effects

Induced polarization is a geophysical exploration tool which measures the slow decay of

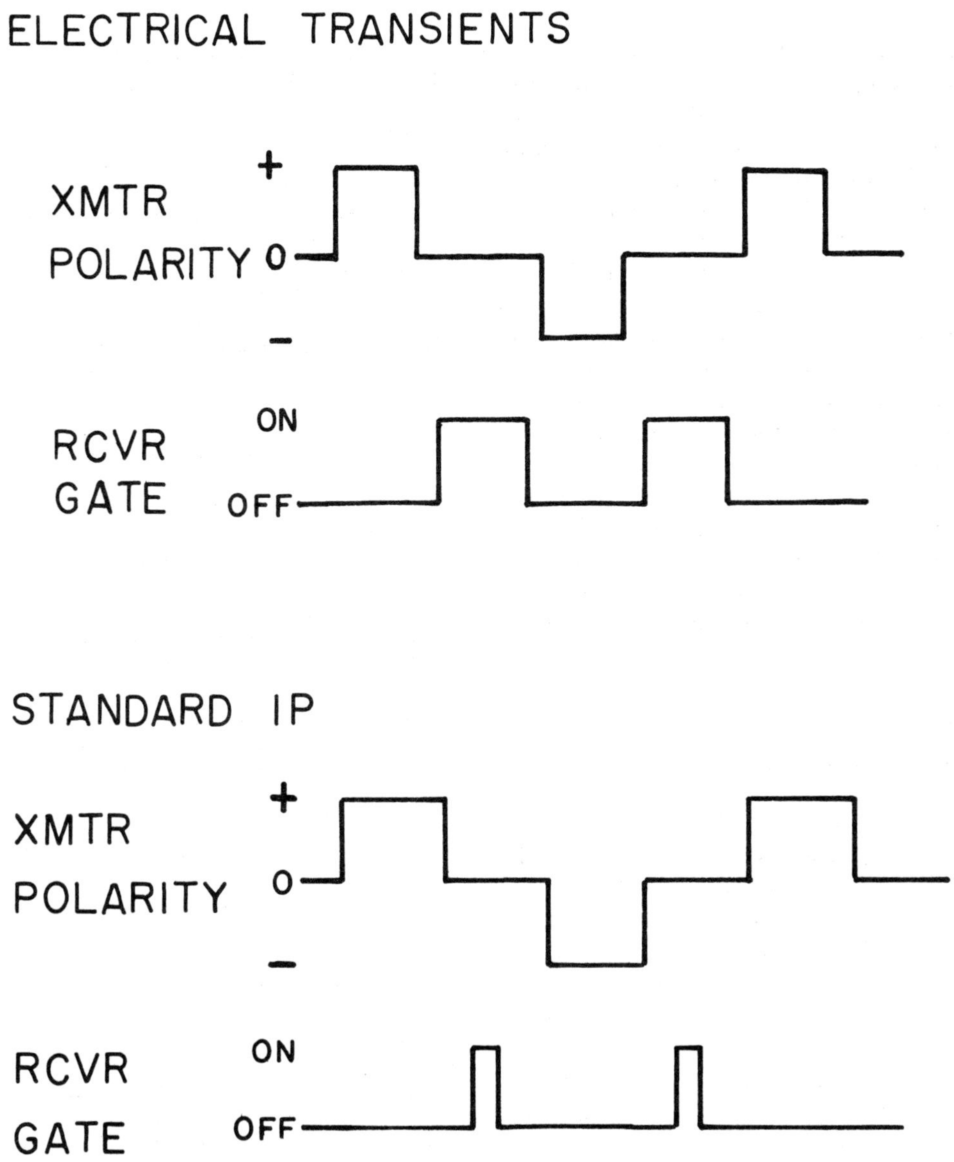

Fig. 8. Bipolar pulse train timing sequence used in electrical transients experiments and a typical IP timing sequence.

voltage in the ground after the excitation current has ceased (Sumner, 1976). This is caused by an electrochemical phenomenon in which an excess of accumulated ions are focused at rock interfaces by an impressed current. When the impressed current ceases, the Coulomb forces between ions that have accumulated at interfaces cause the ions to return to their equilibrium positions. This relaxation causes charge flow that can be measured and is the decaying polarization potential. The exact electrochemical reactions, ion diffusion, and mobilities are not well known in field environments (Sumner, op. cit.).

IP responses have been determined for graphite, sulfide mineralization, clay alteration, alluvium, limestone, etc., and even from buried pipes (Nelson, 1977). Figure 8 also shows one of the standard timing schemes used in IP measurements, and we see that it is very close to the electrical transients timing.

Mutual Coupling

Mutual coupling is caused when the current

flowing through wire induces a voltage in another nearby wire. This mutual coupling causes considerable difficulties with IP measurements as a result of the injection of undesirable noise into the measurement.

In view of the above analysis and information, we are led to hypothesize that the real phenomenon being measured during the electrical transients experiments is not an electromagnetic reflection from a resistive body (the petroleum accumulation), but a combination of mutual coupling and IP response. One possible explanation for the apparent repeatability of electrical transients anomalies in known areas of hydrocarbon deposits would be that hydrocarbons or hydrogen sulphide percolates upward from an oil accumulation at depth, interacts with ground water, and creates an electrochemical environment which is polarizable.

Summary

It would thus appear that the field measurements of the electrical transients methods are not actually detecting reflected electromagnetic transient signals. A more reasonable hypothesis is that a near-surface IP effect is actually being measured. A worthwhile research direction for the direct detection of hydrocarbon accumulations would be a combination of geochemistry and transient measurements by investigators more familiar with both phenomena and the techniques involved.

Magnetotellurics, another electromagnetic technique, is also at the development state in which it can become a reliable tool in identifying geologic structure settings as soon as the 3D modeling techniques become less computationally expensive.

REFERENCES

Azad, J., 1973, Direct oil prospecting with electrical transient reflections: Canadian Journal of Geophysics, v. 12, p. 1-11.

———, 1977, Direct oil prospecting uses electrical transient reflections: Oil and Gas Journal, Sept. 12, 1977, p. 83-94.

———, 1979, Mapping accumulations of oil and gas: Oil and Gas Journal, Jan. 15, 1979, p. 144-149.

Backus, G., and Gilbert, F., 1970, Uniqueness in the inversion of inaccurate gross earth data: Philosophical Transactions of the Royal Society, London, A315, p. 185-194.

Cagnaird, L., 1953, Basic theory of the magnetotelluric method in geophysical prospecting: Geophysics, v. 18, p. 605.

Evjen, H. M., 1943, Utility of the electric methods in geophysical exploration: Geophysics, v. 8, p. 146-156.

———, 1945, The principle of variation in path with an application to structural mapping: Geophysics, v. 10, p. 221-228.

———, 1948, Theory and practice of low-frequency electromagnetic exploration: Geophysics, v. 13, p. 584-594.

Gamble, T. D., Goubau, W. M., and Clark, J., 1979, Magnetotellurics with a remote magnetic reference: Geophysics, v. 44, p. 53-68.

Hermance, John F., 1973, Processing of Magnetotelluric data: Physics of the Earth and Plantary Interiors, v. 7, p. 349-364.

Hohmann, G. W., and Ting, S. C., 1978, Three dimensional magnetotelluric modeling: Topical Report for the Department of Energy, Division of Geothermal Energy, Contract No. EG-78-C-07-1701.

Jones, F. W., and Vozoff, K., 1978, The calculation of magnetotelluric quantites for three dimensional conductivity inhomogeneities: Geophysics, v. 43, p. 1167-1175.

Jupp, D. L.B., and Vozoff, K., 1977, Two dimensional magnetotelluric inversion: Geophysical Journal of the Royal Astronomical Society, v. 50, p. 333-352.

Kato, Y., and Kikuchi, T., 1950, On the phase difference of earth currents induced by the changes of the earth's magnetic field: Science Rpts. of Tohoku University, Pts. 1 and 2, ser. 5, Geophysics, v. 2, p. 139-145.

Lewis, W., Bradley, 1948, Electrical surveys of some shallow salt domes: Geophysics, v. 13, p. 595-599.

Madden, T., and Nelson, P., 1964, A defense of Cagnaird's magnetotelluric method: Geophysical Lab Report, MIT Project NR-371-401.

Nelson, P. H., 1977, Induced polarization effect from grounded structures: Geophysics, v. 42, p. 1241.

Oldenburg, D. W., 1979, One-dimensional inversion of natural source magnetotelluric observations: Geophysics, v. 44, p. 1218-1244.

Peeples, W. J., and Rankin, D., 1973, A magnetotelluric study in the western Canadian sedimentary basin: Pure and Applied Geophysics, v. 102, No. 1, p. 134-147.

Porstendorfer, G., 1975, Principles of magnetotelluric prospecting: Geoexploration Monographs, ser. 1, no. 5.

Reddy, I. K., Rankin, D., and Phillips, R. J., 1977, Three dimensional modelling in magnetotelluric and magnetic variational sounding: Geophysical Journal of the Royal Astronomical Society, v. 51, p. 313-325.

Rijo, L., 1977, Modeling of electric and electromagnetic data: Ph.D. thesis, University of Utah, Salt Lake City, 84112.

Rikitake, T., 1950, Electromagnetic induction within the earth and its relation to the electrical state of the earth's interior: Bulletin of the Earthquake Research Institute, Tokyo, v. 28, p. 45-98.

———, 1951, Changes in earth currents and their relation to the electrical state of the earth's crust: Bulletin of the Earthquake Research Institute, Tokyo, v. 29, p. 270-275.

Sims, W. E., 1969, Methods of magnetotelluric analysis: Ph.D. thesis, University of Texas, Austin.

Sumner, J. S., 1976, Principles of induced polarization for geophysical exploration: New York, Elsevier Publishing Company.

Swift, C. M., Jr., 1967, A magnetotelluric investigation of an electrical conductivity anomaly in the southwestern United States: Ph.D. thesis, Massachusetts Institute of Technology, Cambridge, Massachusetts.

Tikhonov, A. V., 1950, Determination of the electrical characteristics of the deep strata of the earth's crust: Doklady Akademiia Nauk., v. 73, p. 295.

Vozoff, Keeva, 1972, The magnetotelluric method in the exploration of sedimentary basins: Geophysics, v. 37, p. 98-141.

Ward, S. H., Peeples, W. J., and Ryu, J., 1974, Analysis of geoelectromagnetic data, *in* Bolt, B., Adler, B., Fernback, S., and Rotenberg, M., eds., Methods in Computational Physics, v. 13, New York Academic Press.

TRANSIENT ELECTROMAGNETIC SOUNDING IN THE NEAR ZONE

Alexander Kaufman

COLORADO SCHOOL OF MINES

A relatively new method of electroprospecting which was developed in the USSR is called transient sounding in the near zone (ZSBZ). There are few modifications of ZSBZ; as shown in Figure 1a, in all systems a separation between a transmitter and a receiver can be less than a depth of investigation.

The main feature of this method may be briefly described as follows:

Behavior of current in a transmitter is shown in Figure 1b. During "time on" the current in the loop (T) is constant, and therefore a magnetic field does not change with time. For this reason induced currents are not in a conducting medium. When the current in a transmitter is turned off, induced currents appear almost immediately. First of all, they arise near the open surface. This is the early stage of the transient field. In this stage, measuring a magnetic field in a receiver, caused by induced currents, we obtain information about resistivity of an upper layer. With increasing time, currents penetrate into a conducting medium.

Distribution of currents with time is shown in Figure 2. Curves of current density with increasing time show that relative influence of currents in upper parts of medium becomes less and, correspondingly, a depth of investigation increases, regardless of a separation between a transmitter and a receiver. At the late stage, unlike the early stage, the magnetic field, measured on the open surface, is defined by conductivity of deep layers only. In other words, the upper part of medium becomes transparent. This fundamental fact was confirmed by theoretical investigations, vast numerical calculations of apparent resistivity curves in horizonally layered medium, and, of course, by practical results in various geoelectrical conditions.

Thus regardless of the distance from the transmitter and a receiver, one can realize investigation of a geoelectrical section; and the transient soundings in the near zone constitute the first method of structural electroprospecting, when the distance between the transmitter and the receiver can be less than the depth of investigation.

The method has the following merits:

1. The field, measured on the open surface, is usually more sensitive to a change of parameters of geoelectrical sections with respect to the vertical soundings, magnetotelluric soundings, and the frequency soundings. For example, in the case of high resistive basement the magnetic field in a certain range of time turns out to be proportional to S^3, where S is a longitudinal conductance of a sedimentary cover ($S = \frac{H}{\rho}$). Thus relatively small change of S leads to essential change of a field.

2. Because of a small separation between a transmitter and a receiver, it is possible to obtain more detailed information about a change of resistivity in a horizontal direction, and to reduce the influence of various types of heterogeneities which usually essentially distort magnetotelluric curves and apparent resistivity curves of the vertical soundings.

During the last ten years the transient soundings in the near zone have been successively applied to solve various problems of petroleum and mining geology. In petroleum geology the method can be used for solving three main problems: (1) mapping the surface of relatively high resistive basement, (2) determing the main features of a geoelectrical section of a sedimentary cover, (3) outlining oil (gas)-water contact under certain saturation of mineral water in a producing horizon.

Following is one example of the successful application of this method for the determination of oil-water contact in one of the oil de-

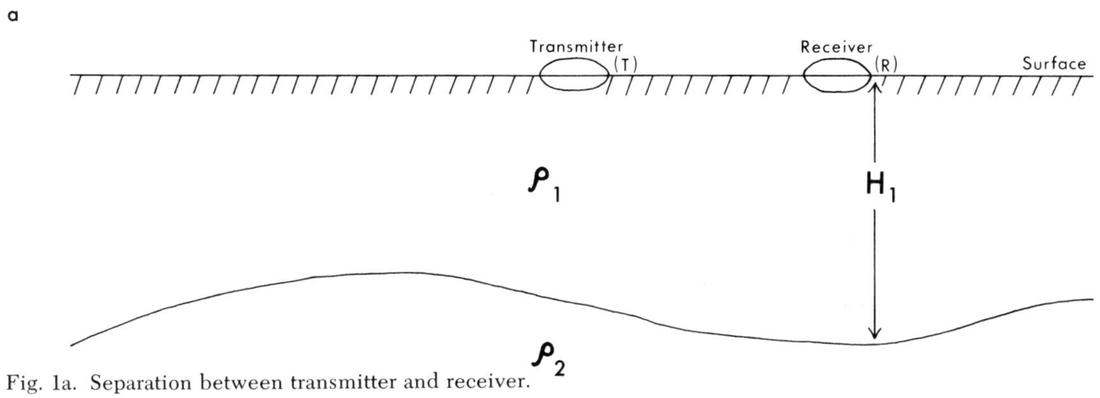

Fig. 1a. Separation between transmitter and receiver.

Fig. 1b. Behavior of current in a transmitter.

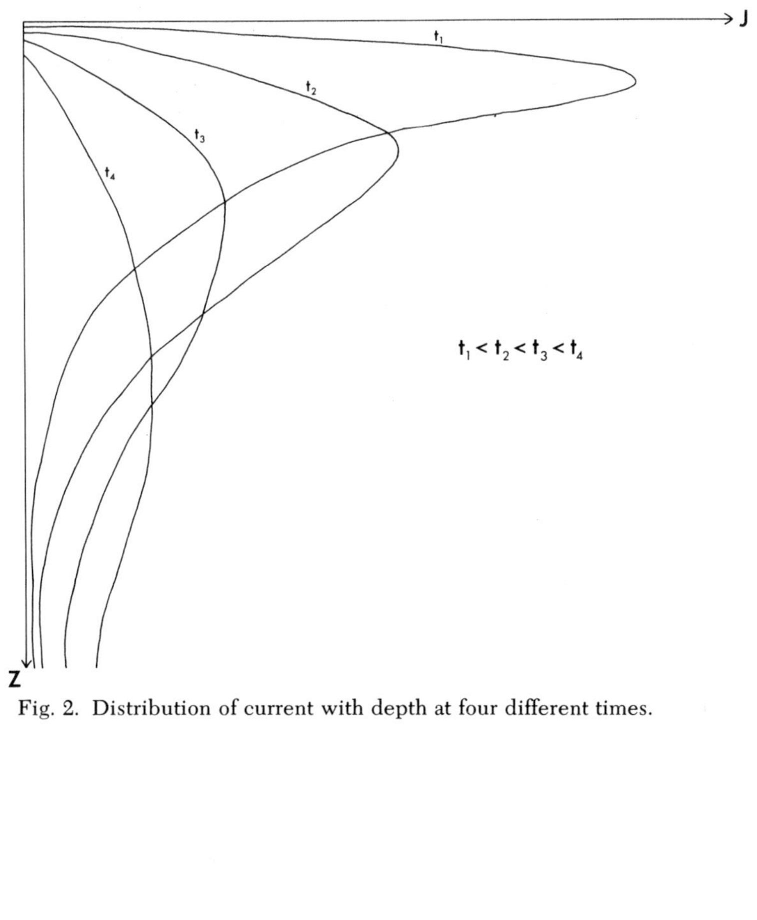

Fig. 2. Distribution of current with depth at four different times.

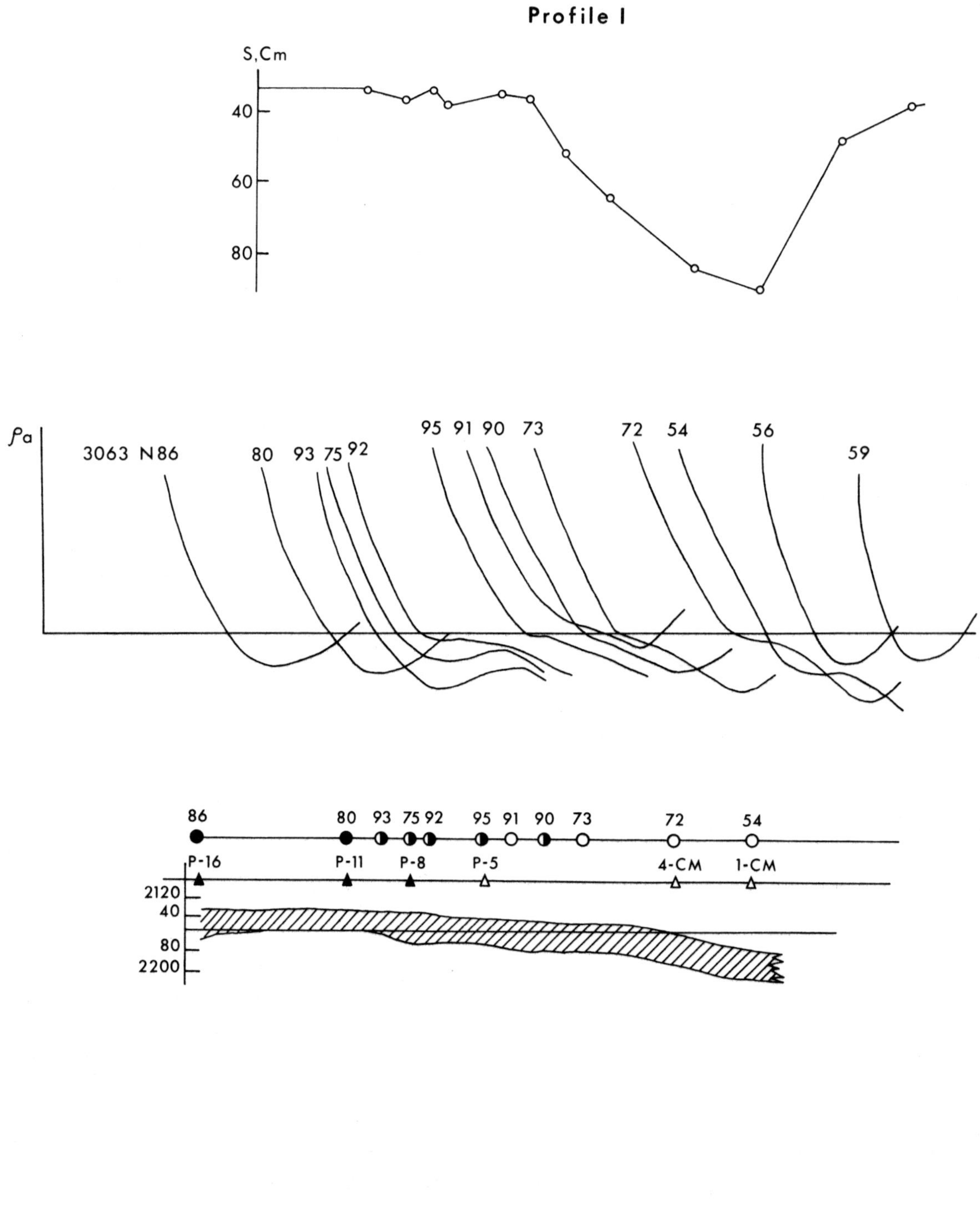
Profile I
S,Cm
40
60
80
ρa
3063 N86
80 93 75 92
95 91 90 73
72 54
56
59
86
80 93 75 92
95 91 90 73
72
54
P-16
P-11
P-8
P-5
4-CM
1-CM
2120
40
80
2200

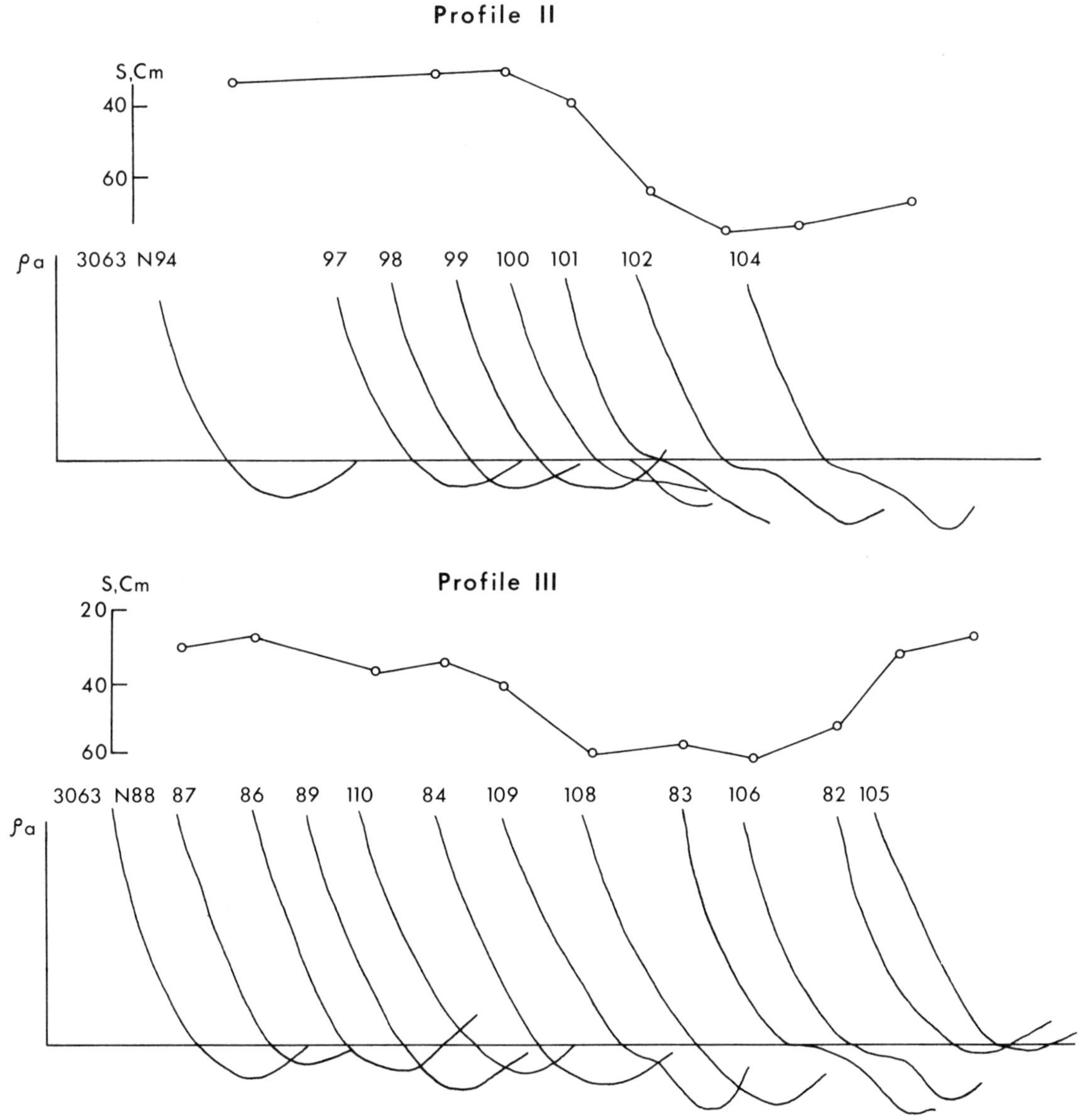

Fig. 3. (Profiles I, II, III). Apparent resistivity curves of the transient soundings.

posits in East Siberia. Parameters of geoelectrical sections are shown in Table 1.

TABLE 1

GEOELECTRICAL HORIZONS	RESISTIVITY δ	THICKNESS H
1	200	200
2	40	520
3	200	1500
4	25	315
5	$>10^3$	basement

The producing horizon (sands and thin layers of argillites), is located on the depth of about 2100 m and a thickness of 20 m. Curves of the apparent resistivity show that a transitional zone from nonconducting oil-gas part of the horizon to the part of the horizon saturated by conducting water clearly manifests itself.

Since most of the oil and gas deposits are located in areas where halogen formations are spread and mineralization of water in salt and under salt formations can reach 200-300 and sometimes 400-500$^{g}/\Lambda$, the transient soundings in the near zone can be very effective for the determination of productive zones.

SIGNIFICANT ADVANCES IN MAGNETOELECTRIC EXPLORATION

Sylvain J. Pirson

UNIVERSITY OF TEXAS, EMERITUS

ABSTRACT

The SP curve shale-base-line drift in thick shale sections is a neglected feature of electric logs, oftentimes regarded as a nuisance by the logging engineer, and something which must be eliminated from the final product. This is done without any indication that the log has been tampered with. Fortunately, in some geologic basins the SP curve drift is slight and the SP curves do not run off the recording paper. It is then possible to record quantitatively the amount of drift (such as mv/500 ft) and to correlate it with the occurrence of hydrocarbons at depth. When viewing the electric log from the bottom upward, an *oil gradient* is a drift to the right (positive) and a *water gradient* is a drift to the left (negative). A number of examples will be shown together with some examples of oil and gas fields that were predicted (and later discovered) by judicious interpolation and extrapolation from known wells.

Unfortunately, there are not enough wells, especially dry wildcats with satisfactory electric logs, to perform exploration in rank wildcat regions.

It is an axiom that the SP shale-base-line drift is a measure of the electrotelluric (ET) current flux that flows vertically within the sediments traversed by the borehole, although some extraordinary environmental sedimentary conditions might create such drift.

Since the SP curve-base-line drifts are the result of vertical ET currents, such currents create magnetic field perturbations by virtue of Ampere and Bio-Savart's laws. Such perturbations are combined into the magnetic field intensities that are measured in magnetic surveys either at the surface of the ground or above ground, such as in airborne surveys. Since the magnetic perturbations created by ET currents have *no potential*, it is therefore possible to filter them out by means of repeated line integrations of the total earth magnetic field along closed paths, according to Stokes and Gauss theorems of physics. Such filtering and integration techniques have been highly developed during the past six years.

Magnetoelectric exploration is a semidirect method of delineating oil and gas accumulations in the earth, certain mineral deposits, and geothermal energy by mapping the total earth magnetic field (or its horizontal component, or its vertical gradient) and relating variations in such field(s) to the electrotelluric currents that are generated at the geochemical interfaces (redox potential contrasts or geothermal contrasts) associated with such deposits in the earth.

Five small oil and gas fields have been discovered by ME surveys. Various statistical evaluations applied to ME data indicate its predictive capability for oil and gas pools to be about 80%.

Prediction of the outcome of exploratory drilling in the Baltimore Canyon Trough (offshore east coast) has been thus far 90% correct: two discoveries and ten dry holes, one of which is the Texaco confirmation well in Tract 598 which proved nonproductive and established the limit of production at 2 ma/acre. Two other dry holes, the Conoco Tract 590 and the Exxon Tract 684, are not statistically significant, the first being too shallow (12,000 ft) and the other being an edge well. The discoveries predicted are the Texaco (594) 12 MMCFD and the Tenneco-Aminoil (642), 3700 B/D and 12 MMCFD.

The significant advances made in magnetoelectric exploration pertain to the application of well-known methods of magnetic downward continuation into the earth to levels that are in close proximity to (but preferably above) the sources producing the electrotelluric currents. At such depths, the magnetic field so continued satisfies the first Maxwell equation (without displacement current because of steady state), and closed line integrals of the continued magnetic field in the horizontal plane do not vanish, but rather the values of such integrals are a measure of the electrotelluric current flux density within the area enclosed by the lines of integration. The depths of the desirable downward projection are determined from the subsurface geology of the probably producing reservoirs, as inferred from electric logs and from other well logs. The downward projection process involves the expansion of the measured magnetic field into Taylor series comprising the first and second vertical derivatives of the surface magnetic field. The derivatives are determined by well-known surface integrating procedures carried out over the surface magnetic measurements. The patented line integral procedure is then applied to the analytically continued magnetic field values as if they had been measured in the subsurface. The results are expressed in milliamperes per acre for the vertical magnetoelectric current flux flowing inside the earth; these values are related quantitatively to the hydrocarbon in place. When promising anomalies in the vertical electrotelluric current flux so mapped are found, the values of the flux densities are converted into hydrocarbon in place per acre and over the whole petroliferous areas so delineated. An appropriate recovery factor is then applied in order to estimate the monetary value of the discovered reservoirs.

Certain verification techniques are applied during a magnetoelectric survey. If wildcat wells are present in the area of investigation, and especially if such wells have been logged electrically and exhibit reliable SP curves, the values of the electrotelluric flux densities of various levels must check reasonably well with the values calculated at the same depths from magnetoelectric surveys, whether positive or negative. A number of such verifications have already been made: Lund, Granger, Garfield, Creedmoor, all in Central Texas.

An example of application of these techniques is made to an oil and gas prospect in the Ouachita Overthrust Belt in Central Texas which resulted in the delineation of channel-fill gas sand accumulations in the Travis Peak formation. This prospect is verified not only by SP oil gradient in a nearby wildcat well, but also by a surface self-potential sink caused by near-surface ET currents converging into the redox chimney overlying the prospect.

PART ONE

To the log analysts, one of the most neglected curves in electric logs is the SP curve. From the infancy of logging this curve has been tampered with by the logging engineers who impress on it a variable and continuous emf by literally "cranking-it-in" by hand during the recording process in an attempt to "rectify" it, i.e., make it look straight from bottom to top. This is an attempt to keep the SP curve within the bounds of the recording paper. At times this is done by an abrupt "SP manual shift," which is then obvious on the record. It is when a gradual compensation is applied in order to eliminate the SP drift that a lot of valuable information is lost concerning the migration and the accumulation of hydrocarbons.

Fortunately, logging engineers are like the rest of us—they are lazy—and in regions where there is no substantial SP natural drift, no rectification potential has been applied to the SP curve. Such a region is the Denver-Julesburg basin, where most electric logs exhibit an uncompensated SP curve drift above the D-J sands. Most of the early log environmental observations were made in this basin. In other areas, and especially in very deep wells, the SP drifts (+ or −) were partially cranked out in the parts of the logs where the SP curves might go off scale. Perhaps 90% of the electric logs have had their SP curves partially to fully tampered with while logging. For the purpose of the study that follows, it is thus necessary to ascertain the geologic sections that have not been tampered with through a long process of correlation and of acquaintance with logging practice in the various areas of interest.

Logging engineers are not the only persons that rectify SP curves; when logs are reproduced by draftsmen they also like to eliminate drifts from SP curves; one such case will be shown in which an "artistic rectification" was performed.

Let us consider some electric logs, especially in the intervals (generally shale sections) wherein it is believed no tampering was performed.

Figure 1 illustrates a group of three logs from Montana (Bell Creek field area) where the SP curve shale base lines exhibit strong drifts toward the negative, when viewing the log from bottom to top. This drift will be called a "negative" or "water gradient," as it has been found statistically associated with shale sections overlying water-bearing reservoirs: sands, shaly-sands, carbonates, etc., even fractured shales and hard rocks. The sands in Figure 1 at 4,400 ft to 4,500 ft are all water-bearing and nonproductive of hydrocarbons.

Figure 2 is an electric log from a well in the Larremore field, Caldwell County, Texas, that still produces oil from the up-to-the-Coast-faulted Edwards limestone at 1,250 ft. The electric log exhibits a positive gradient, i.e., drifting toward the positive throughout the full Taylor shale section and extending all the way up to the surface of the ground. While only one log is shown thus far, a similar positive gradient, called "*oil gradient,*" has been found to be statistically associated with shale sections overlying hydrocarbon producing reservoirs.

These SP shale-base-line drifts called "*water-gradient*" and "*oil gradient,*" respectively, have an obvious genetic relationship to the migration and accumulation of hydrocarbons, to the exclusion of many possible but inadequate explanations:

a. Change in shale lithology. This would be apparent in the resistivity of the section.
b. Change in shale depositional environment. This would also be reflected in shale resistivity but more specifically on a *Redox* log, if one had been taken. Unfortunately, there is only one such log in the whole country, and it shows that SP drift and redox variations do not correlate.
c. Streaming potential and mud filtrate in shales is not a likely phenomenon that would explain SP drift. This would be caused by an upward water flow under the impelling force of compaction and of evaporation. The electrokinetic coupling, i.e., ratio of the emf generated to the pressure differential, is given by $\frac{RES}{4\eta\mu}$ wherein the variables are not easily evaluated:

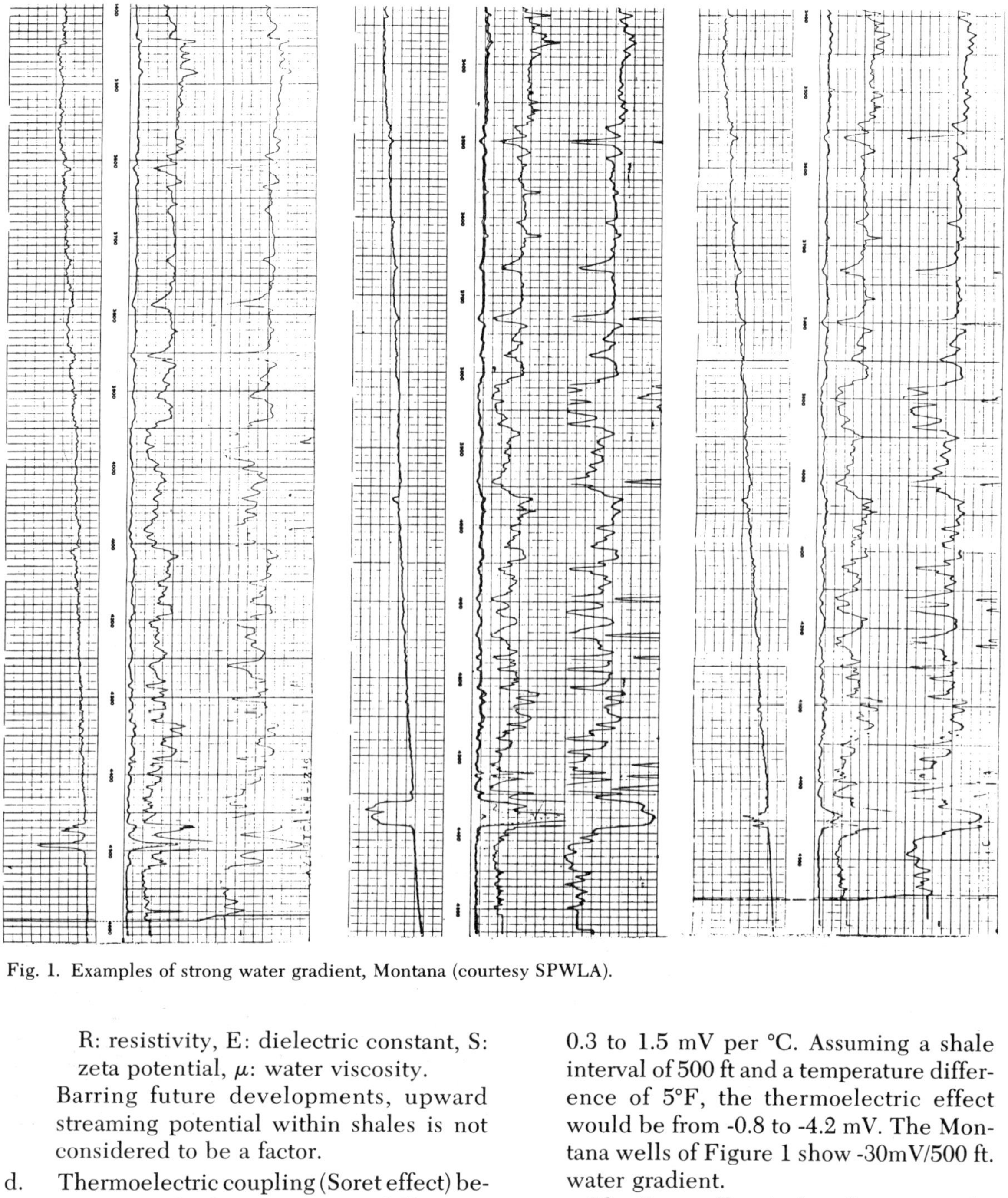

Fig. 1. Examples of strong water gradient, Montana (courtesy SPWLA).

R: resistivity, E: dielectric constant, S: zeta potential, μ: water viscosity.

Barring future developments, upward streaming potential within shales is not considered to be a factor.

d. Thermoelectric coupling (Soret effect) between applied temperature difference and the generated emf. For a variety of sandstones, limestones, and shaly rocks, the thermoelectric effect is in the right direction to explain the *water gradients*, but it is rather small compared to observed SP gradients. The thermoelectric coupling coefficients of the rocks vary from 0.3 to 1.5 mV per °C. Assuming a shale interval of 500 ft and a temperature difference of 5°F, the thermoelectric effect would be from -0.8 to -4.2 mV. The Montana wells of Figure 1 show -30mV/500 ft. water gradient.

The Soret effect is therefore not negligible in all cases, and if used as a correction in the SP gradient, it will enhance the value of the oil gradient.

Other unlikely physical and chemical possibilities that could affect an SP gradient in a well bore are:

e. Cosmic currents escaping vertically into

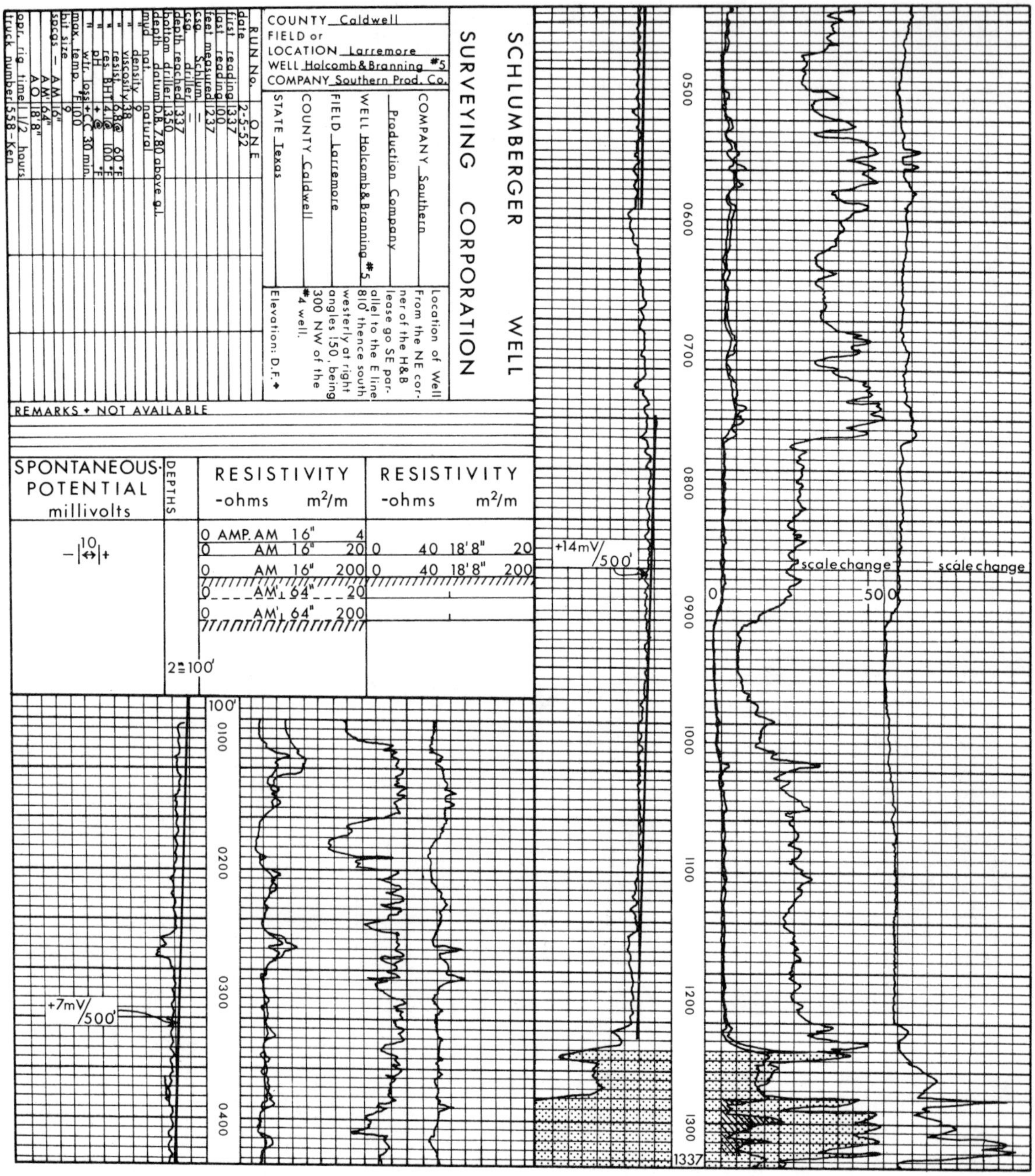

Fig. 2. Electric log from the Larremore Oil Field, Caldwell Co., Texas, with SP oil gradient.

the atmosphere and originating in magnetic storms.

f. Bimetallism refers to unwanted potential produced by galvanic action between dissimilar metals exposed to the mud. As a rule these metals either do not exist or are insulated. If a bimetallic pair exist, the SP is affected mainly at resistive beds contacts.

g. Salinity gradient in shallow formations. An example is given in Figure 3. A salinity gradient in various porous formations containing formation water will indicate smaller deflections from an SP base line as

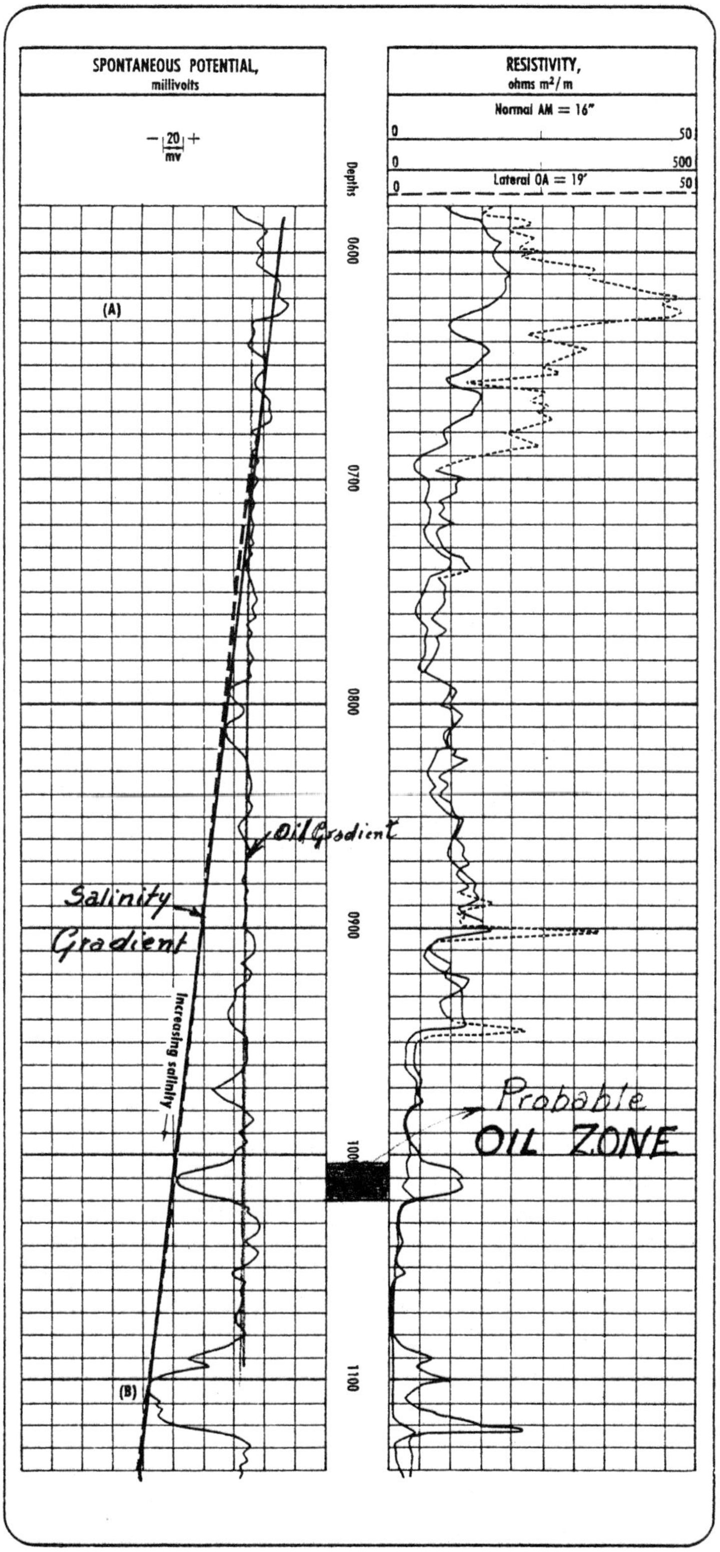

Fig. 3. Example of salinity gradient from SP deflections and of an oil gradient from the shale base-line drift (courtesy *Oil and Gas Journal*).

the water-salinity decreases, and it may even become negative for salinities less than the mud. This is not the SP gradient we are interested in but rather the *shale base-line drift*. In the case of Figure 3 there is an oil gradient, perhaps indicating the presence of oil, in the interval 1005-20 ft. This oil accumulation is suspected because of a well-rounded SP deflection, the short normal measurement in the invaded zone, and the critical thickness behavior of the 19 ft lateral curve. It is not believed that the salinity of shale is variable (increasing with depth) and we are faced with a true oil gradient.

h. The electrode potential (i.e. of lead) given by the Nernst equation $E_P^0 = \frac{RT}{nF} \ln \frac{OP}{ESP}$ where n is the lead valence (2), F is the Faraday (96540 Coulombs), OP is the lead osmotic pressure in atmosphere in the mud and ESP in the electrolytic solution pressure in atm. of lead into the mud. For a homogenized mud, the last two are constant. R is the gas constant, an invariable. The only variable is T, the mud temperature around the electrode. In 500 ft, Δt is 5°F and E_P^0 varies 0.5 mV.

None of the above physical phenomena are strong enough to explain the water and oil gradients. We call on electrotelluric currents.

Electrotelluric Currents

Let us examine the current intensities required to generate the SP gradients observed in Figures 1 and 2. For the purpose of standardization let us consider a volume of shale of one acre (43,560 ft²) in area and 500 ft long, its resistivity being 5Ωm. The resistance of this earth volume is $5 \times 3.28 \frac{500}{43,560} = 0.19\Omega$. Assume that this rock is drilled by a well bore 8 in. in diameter and filled with a mud of one Ωxm resistivity. The resistance of the 500 ft mud column is 4,700 Ω. It is along this resistant column that the SP variations are measured, such as SP gradients in shale. It is as if they were measured by means of a high impedance millivoltmeter, and the mV readings are those of a voltage drop existing in the cylindrical shale volume. The conditions are approximately those that prevail in the Montana wells (Figure 1, Well #2). Over a 500 ft length, the SP water gradient is -45 mV. Applying Ohm's law (Ohmic coupling: $\frac{e}{I} = r$) the vertical current flowing upward in the 500 ft cylinder, one acre in cross section, is $\frac{.045}{.19} = 236.8$ ma.

Looking now at the Larremore well of Figure 2, where the shale resistivity is 2.0 Ωm and the mud resistivity is 4.0 Ωm at 100°F, we observe the SP oil gradient to be in two parts. The lower part is +15mV/500 ft and corresponds to 197 ma/acre, whereas the upper is 7mV/500 ft and corresponds to 98.5 ma/acre. Obviously, the downward flowing current over the Larremore accumulation is flowing from two separate "fuel-cells" in the earth, with a separation in the current flux at the level of the Pecan Gap formation at 600 to 760 ft.

The above calculations indicate that the main current fluxes responsible for the water and oil SP gradients are the result of vertical electrotelluric currents flowing in opposite directions to create the water gradients (upward) and the oil gradient (downward).

I have observed a multitude of such cases over the past seventeen years, since the inception of what was called then "Projective Well Log Interpretation" (Pirson, 7a).

Similar calculations have been made in the past to bring further evidence to the concept of electrotelluric currents being associated with the accumulations of oil and gas in the earth.

Additional Examples of Oil and Water SP Gradients

Figure 4 exhibits three logs in the Denver-Julesburg basin in the vicinity of the Latigo field, Arapahoe County, Colorado. The well to the left is producing from the J sand and shows a well-defined oil gradient. The well to the right is water-bearing in the J sand and shows a well-defined water gradient. This set of logs was redrafted for publication, notably for patent #3,943,436, and shows an example of "artistic rectification" of the water gradient. The center well exhibits SP gradients that may be the result of an overenthusiastic recording engineer who overcompensated a drifting SP and the necessary recall in order to maintain the SP curve on the recording paper. There is no way of knowing. Hence, the various SP drifts are

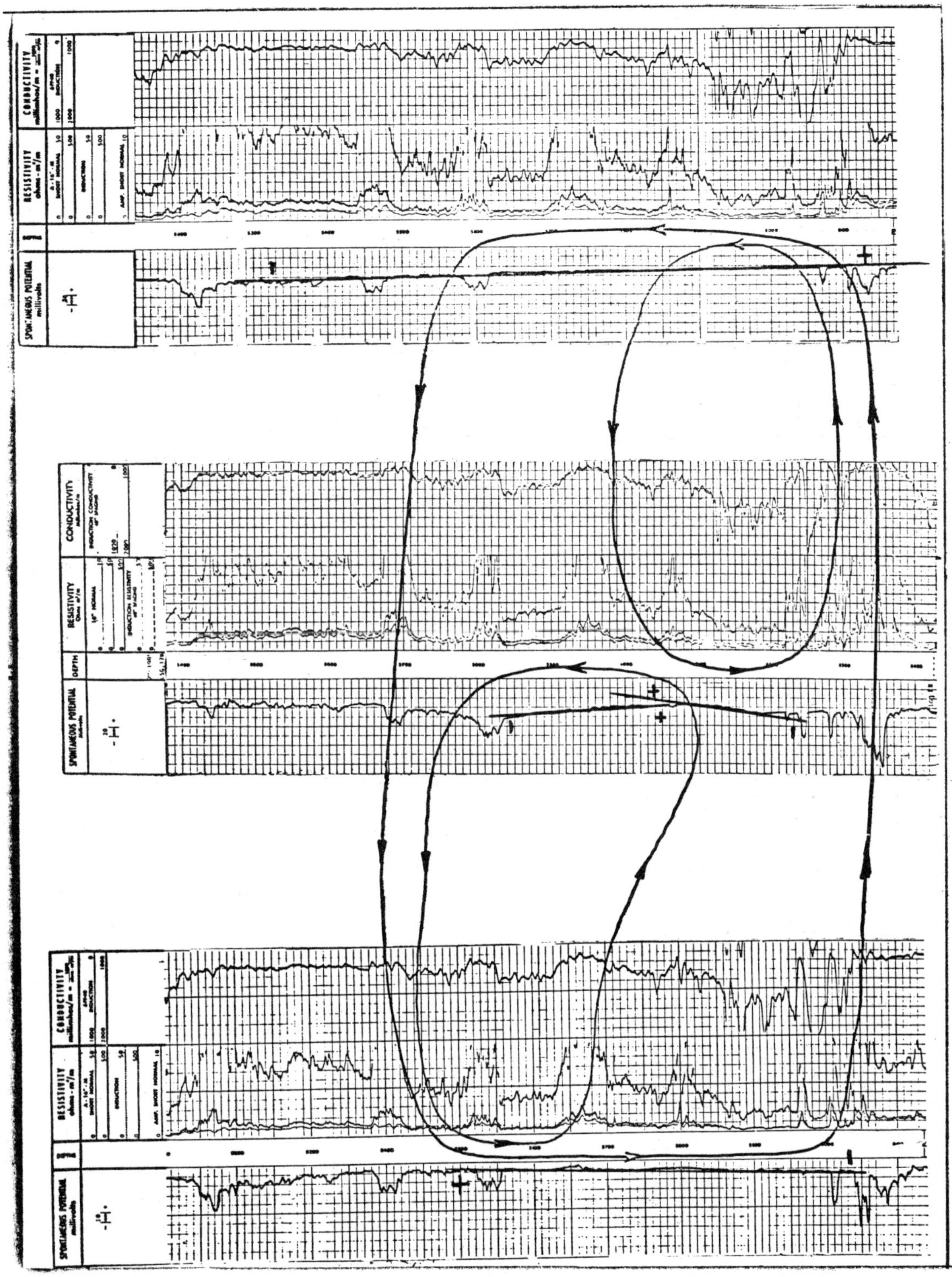

Fig. 4. Evidence of ET effects on the SP curves.

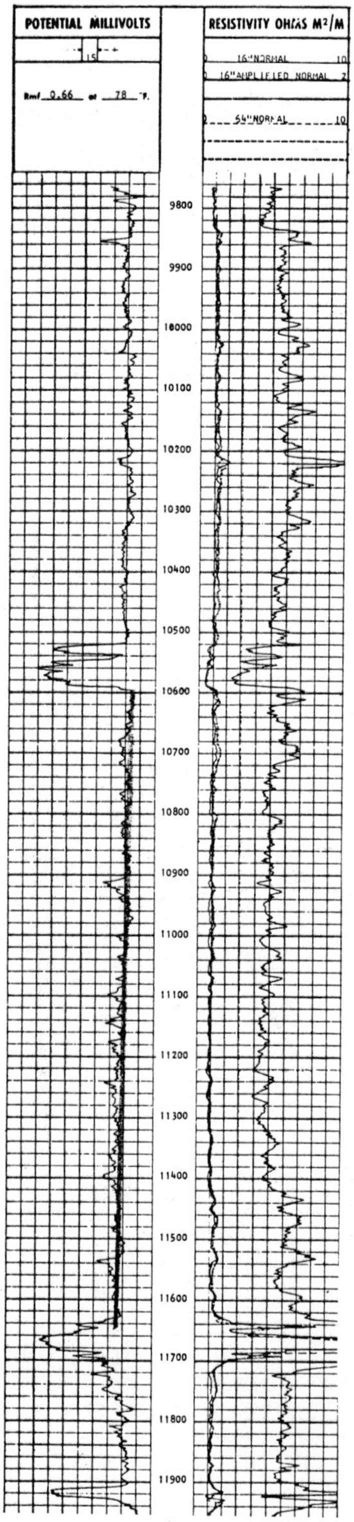

Fig. 5. Electric log from a St. Mary Parish, Louisiana, producing oil well with SP oil gradient.

interpreted as is, by the graphing of major and minor fuel-cell-current distribution as shown by the electrotelluric-current flow lines.

Figure 5 is a log in St. Mary's Parish, Louisiana, which produces oil at 11,630-11,700 ft and shows an oil gradient up to 10,600 ft. This is in the geopressured zone. Accordingly, magnetoelectric exploration techniques are applicable to abnormally pressured zones. Figure 6 is a log from Landry's Parish, Lousiana, with production from a fractured shale section at 8,750-65 ft, which it is not possible to evaluate by conventional log analysis techniques. The oil saturation was discovered from the cutting samples and the productivity from drill stem tests.

Figure 7 is a log from Venezuela (Anzoategui), which shows a strong oil gradient over a very thick reservoir (top: 8,300 ft).

Figure 8 is a log of the Peoria field, Colorado, with SP oil gradient over the producing J sand.

Figure 9 shows the previous prediction made by means of eight logs from dry holes in the vicinity of what was to become the Peoria field. The SP gradients in mV/500 ft of shale section are plotted and contoured using considerable extrapolation and interpolation. An oil-favorable area was outlined in which eventually the three fields of Peoria, No. Peoria, and Latigo were developed. This was done five years before the discovery of Peoria.

Figure 10 illustrates how the SP gradient mapping technique was used in the South Florida Embayment in 1963 in order to outline the promising petroliferous trends. As shown in Figure 10b, eight oil fields were discovered within those petroliferous trends at later dates.

Other oil fields predicted by SP gradient mapping are the ABM and McGill, Pecos County, Texas, and the Womack Hill field, Alabama.

As an exploration tool, the mapping of SP gradients from electric logs is at best a reconnaissance tool for the purpose of establishing the petroliferous nature of certain areas. It requires, of course, that a number of wildcat wells be already drilled, and therein lies the limitation of the method. However, the validity of the SP gradient technique as an exploration tool led to the discovery of other valid approaches:

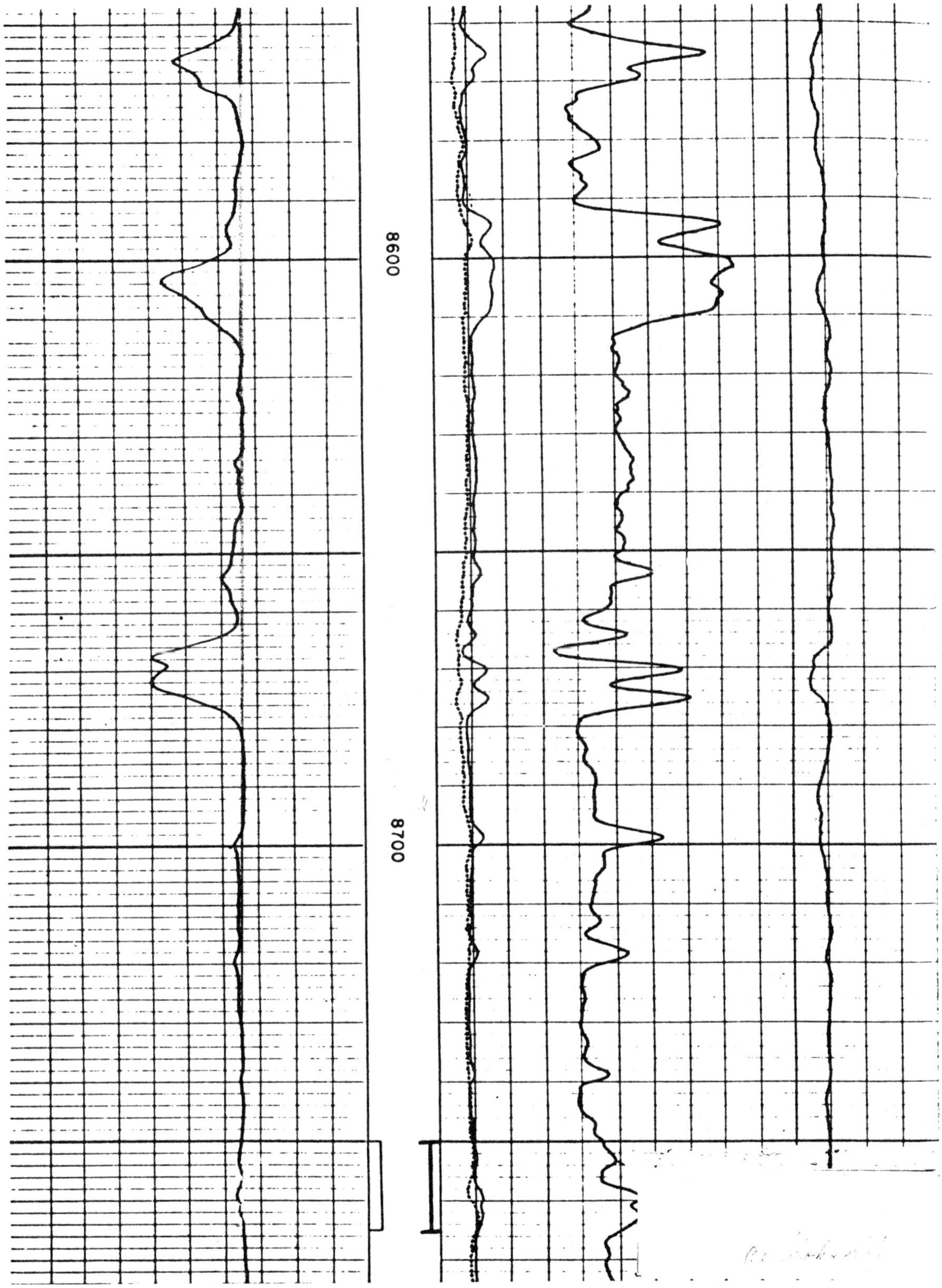

Fig. 6. Electric log from a St. Landry Parish, Louisiana, producing oil well with SP oil gradient.

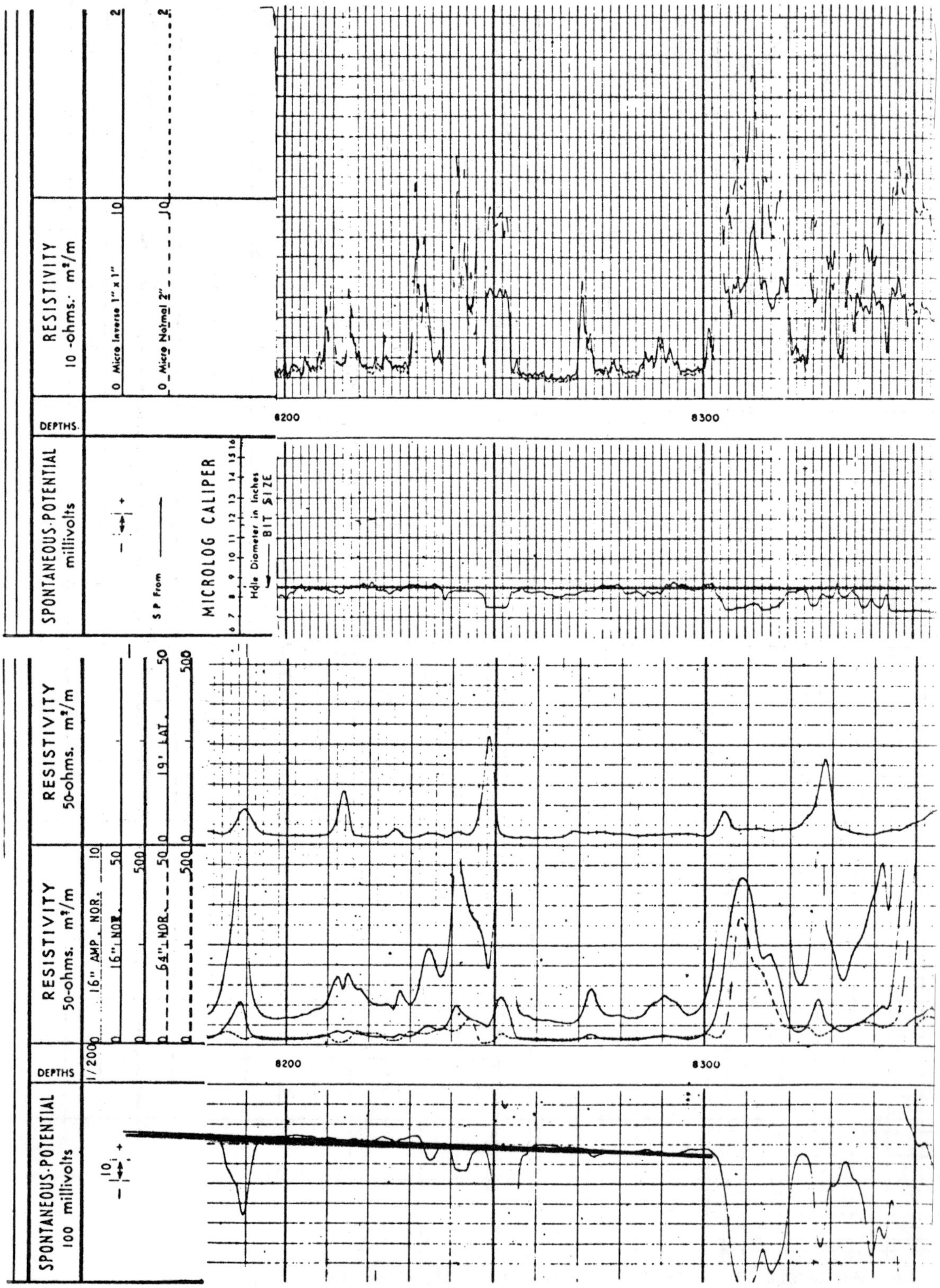

Fig. 7. Electric log from an Anzoategui, Venezuela, producing oil well with SP oil gradient.

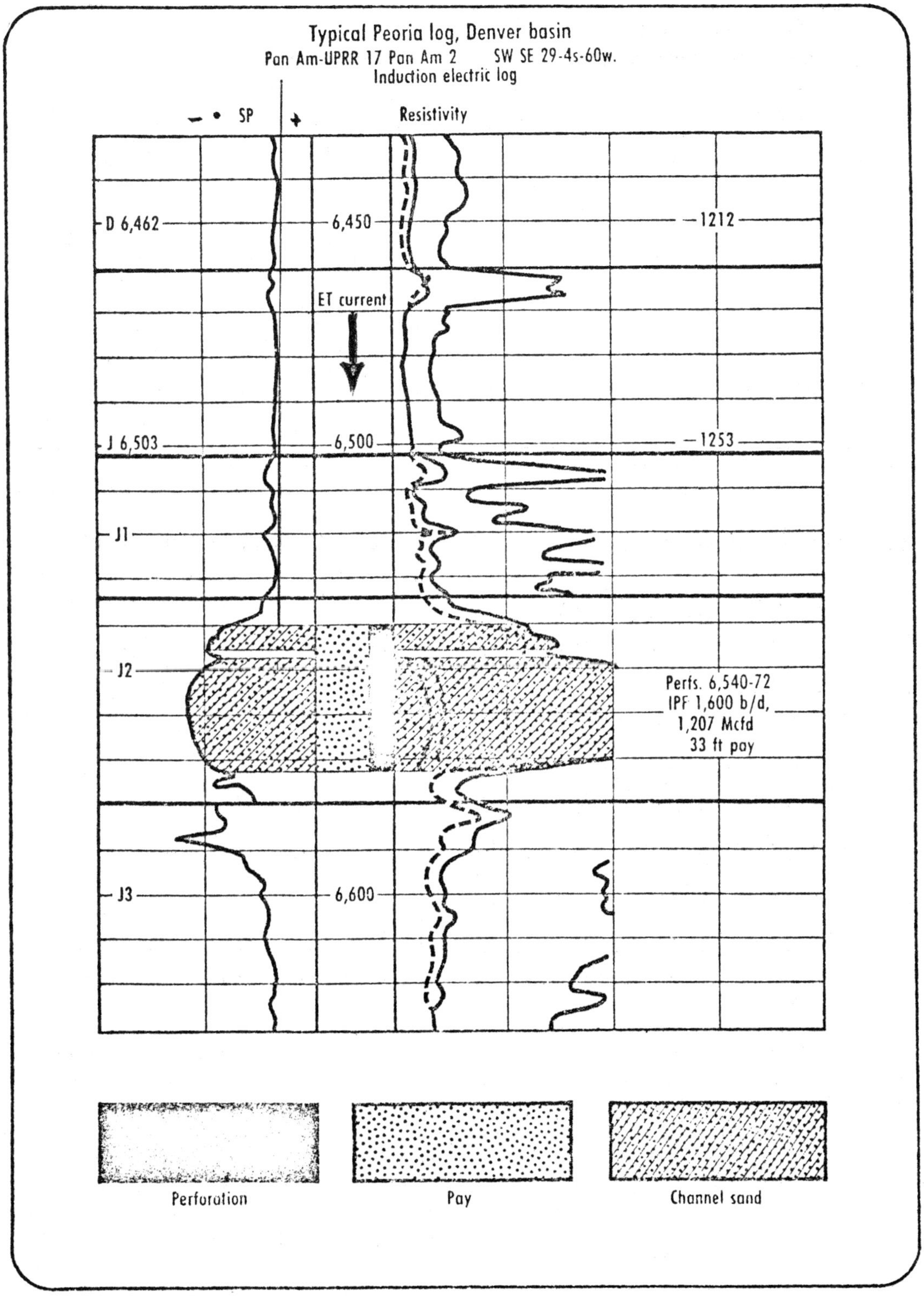

Fig. 8. Example of "oil gradient" on SP curve (courtesy *Oil and Gas Journal*).

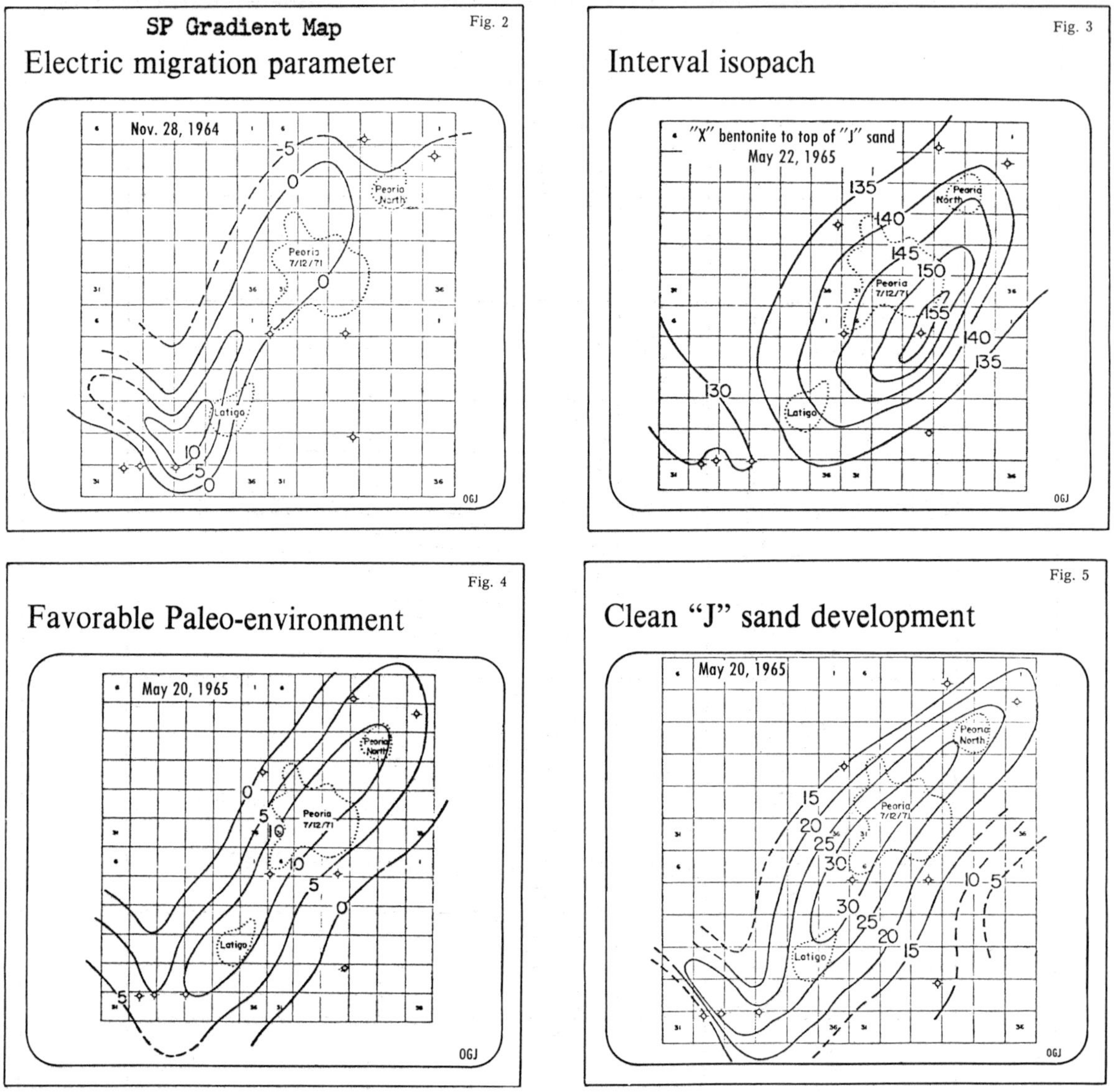

Fig. 9. SP gradient map and other log derived maps in the region of the future Peoria field, Arapahoe Co., Colorado (courtesy *Oil and Gas Journal*).

the redox potential surveys, the electrotelluric or SP survey, and the magnetoelectric surveys. Only the latter will be discussed further. But first a likely mechanism must be established for the generation of electrotelluric currents from the sites of accumulation of oil and gas.

PART TWO

The Hydrocarbon Redox Fuel Cell

Our knowledge of hydrocarbon origin, generation, migration, and accumulation is mostly speculative. Perhaps the most probable mechanism is that it is a water-borne process that uses as raw material organic compounds contained in source beds.

According to modern concepts of the formation of geochemical anomalies associated with oil and gas fields, such anomalies are viewed as *fossil remnants of past upward hydraulic leakage* from the bottom of sedimentary basins during their formation, compaction, uplifting, and erosion. This hydraulic leakage took place

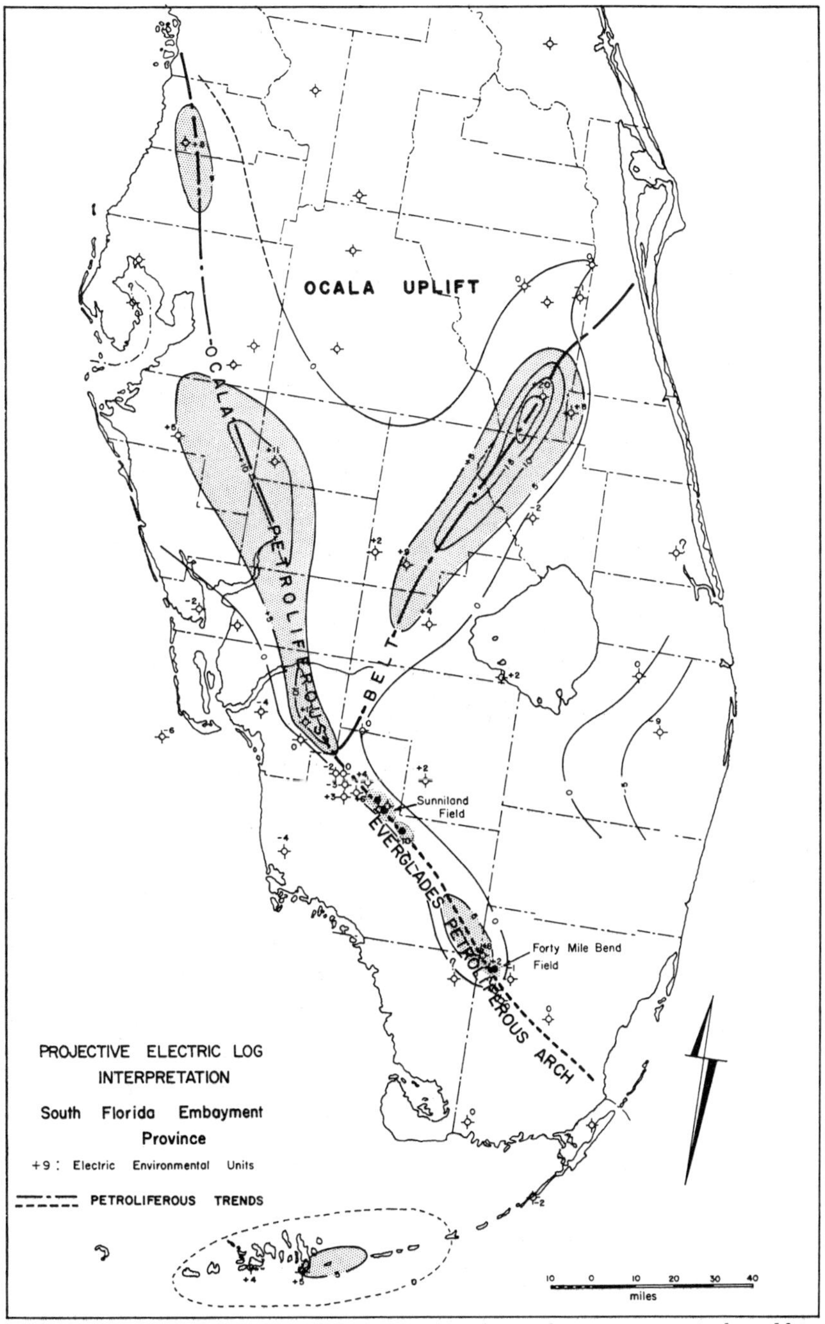

Fig. 10a. Favorable South Florida trends indicated by projective electric log interpretation, based largely on Sunniland field and forty wildcats drilled outside of Sunniland. Reprinted from *World Oil*. Copyright 1963, Gulf Publishing Co.

mostly over the period of active sedimentation and compaction. Hydrocarbons of all kinds, salts derived from rocks, and trace chemical elements usually associated with oil and gas in the process of generation migrated upward with the waters in motion. Last but not least, change took place in environmental conditions ranging from oxidation to reduction.

Genesis of geochemical anomalies over and around oil and gas fields while they were being formed is best illustrated by Figure 11. Here a marine basin is viewed to the left of a shore

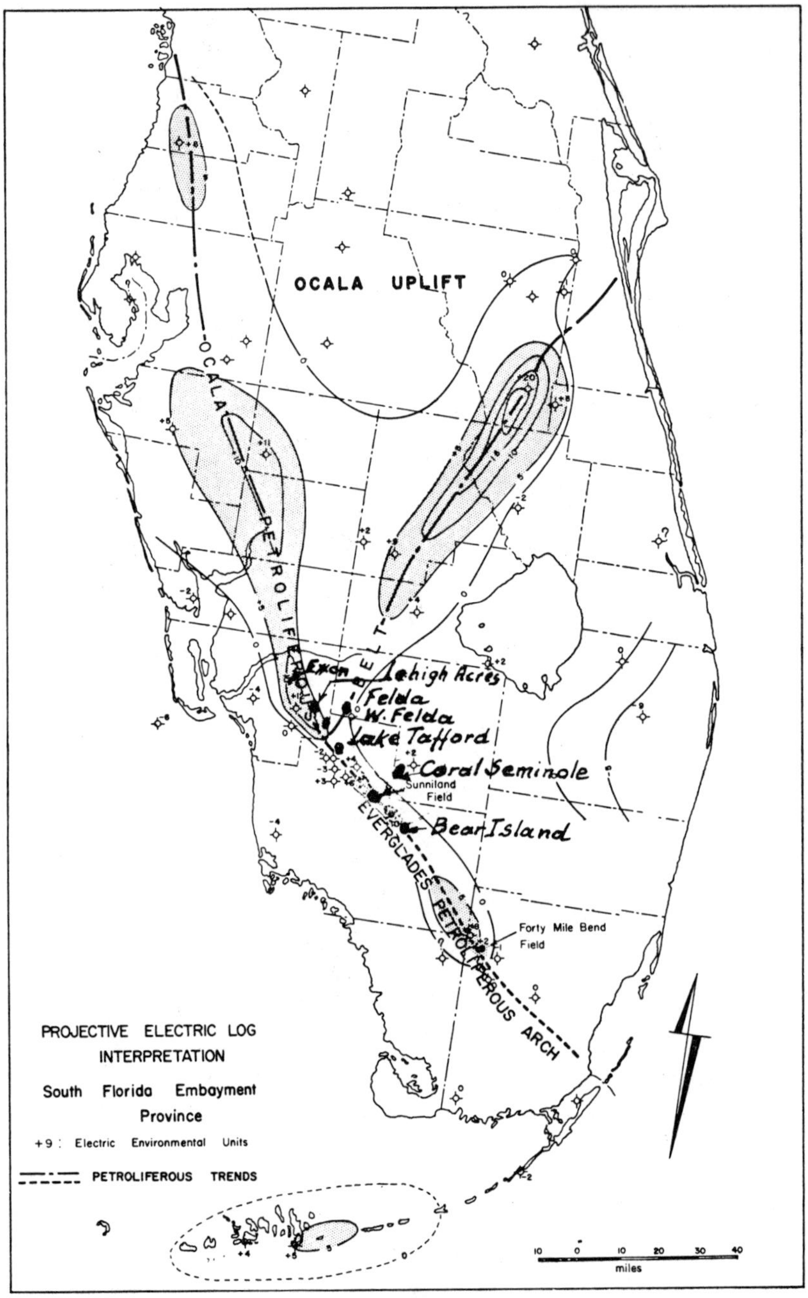

Fig. 10b. Oil fields discovered in South Florida located on SP gradient map published in 1963 (courtesy *World Oil*).

line. Detrital sediments are transported for deposition in the basin, subsiding under the burden of sediments. It is known from recent sediment examination that petroleumlike material is deposited together with sediments and ooze in sufficient disseminated volume to explain all the known oil fields of the world. All that is required to form an oil field is a process of migration, concentration, and accumulation in suitable geologic traps.

Yet more recent views consider organic matter in source beds (shales, shaly rocks, bituminous rocks, etc.) as having been subjected to a process of maturation by heat, pressure, and

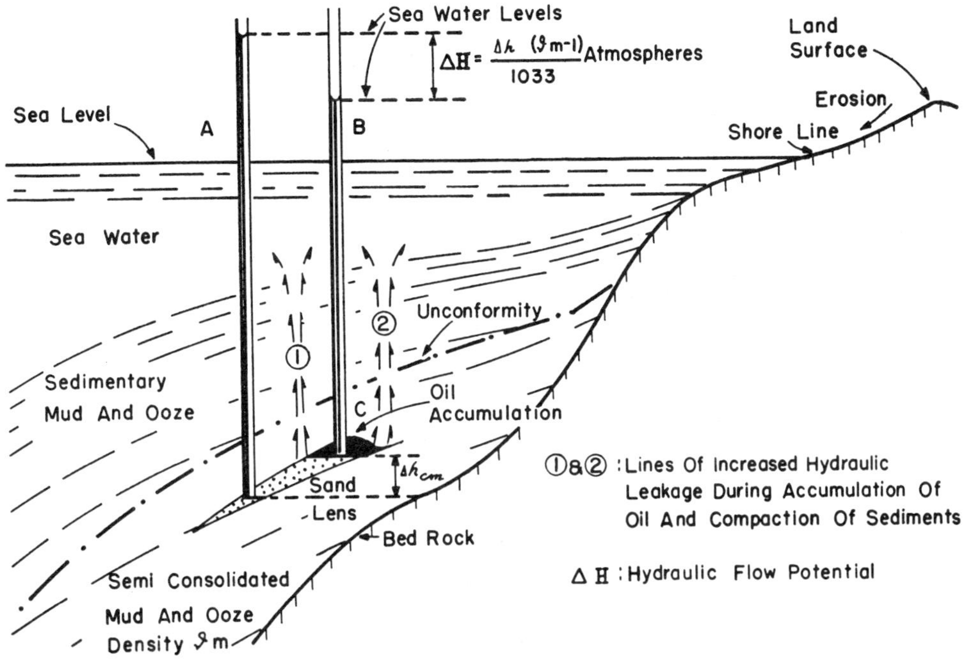

Fig. 11. Illustrates formation of a stratigraphic trap and attendant hydraulic relationships, including leakage around the edges of the oil deposit as it reaches a mature development stage. According to the theory expressed, it would not be necessary for fracturing provided by structural deformation to cause escape of hydrocarbons, as may be common in the case of structural traps. Hence, geochemical exploration methods should be effective in finding stratigraphic as well as structural deposits. Success with geochemical methods indicates that this is so. Reprinted from *World Oil,* April, 1960. Copyright 1960, Gulf Publishing Co.

time, and to have generated the disseminated petroleumlike material in the detrital sediments.

Consider the sedimentation trap (Figure 11) to be a buried sand bar or lens C made of relatively clean sand relative to the mud and ooze that enclosed it. Now visualize manometers A and B (i.e., well casing) embedded in the sediments but terminated at their bottom in the clean sand lens. In each manometer clear salt water rises to such a height as to be in equilibrium with both hydrostatic and geostatic (overburden) pressures at the respective bottom of each well. The water level will be higher at manometer A than in manometer B because of the greater thickness of ooze and mud around well A, and particularly in view of the unconsolidated nature of the sediments and of their incompetency.

According to Darcy's law of fluid motion in porous and permeable media, a hydraulic flow of reservoir water will take place from A to B within the sand lens. In so doing, this fluid contains countless tiny globules of oil. At the top of the lens, these globules cannot escape into the overlying ooze and mud because of its fine capillary size. The oil droplets are filtered out at the upper end of the lens because oil is the nonwetting phase and it would have to overcome a large capillary pressure gradient to reenter the mud and ooze which eventually will become shale.

Hence, oil and gas accumulates at the top of the sand lens, and this continues as long as the hydraulic flow is sustained. Permeability to water no longer exists in the accumulated oil zone, however, and hydraulic leakage is forced around the accumulation as shown in Figure 11. This oil accumulation process appears as a very realistic explanation not only of oil and gas accumulation in stratigraphic and structural traps, but also of mineralization and environmental effects that occur in unconsolidated sediments overlying developing oil and gas accumulations.

As formation water migrates around the pe-

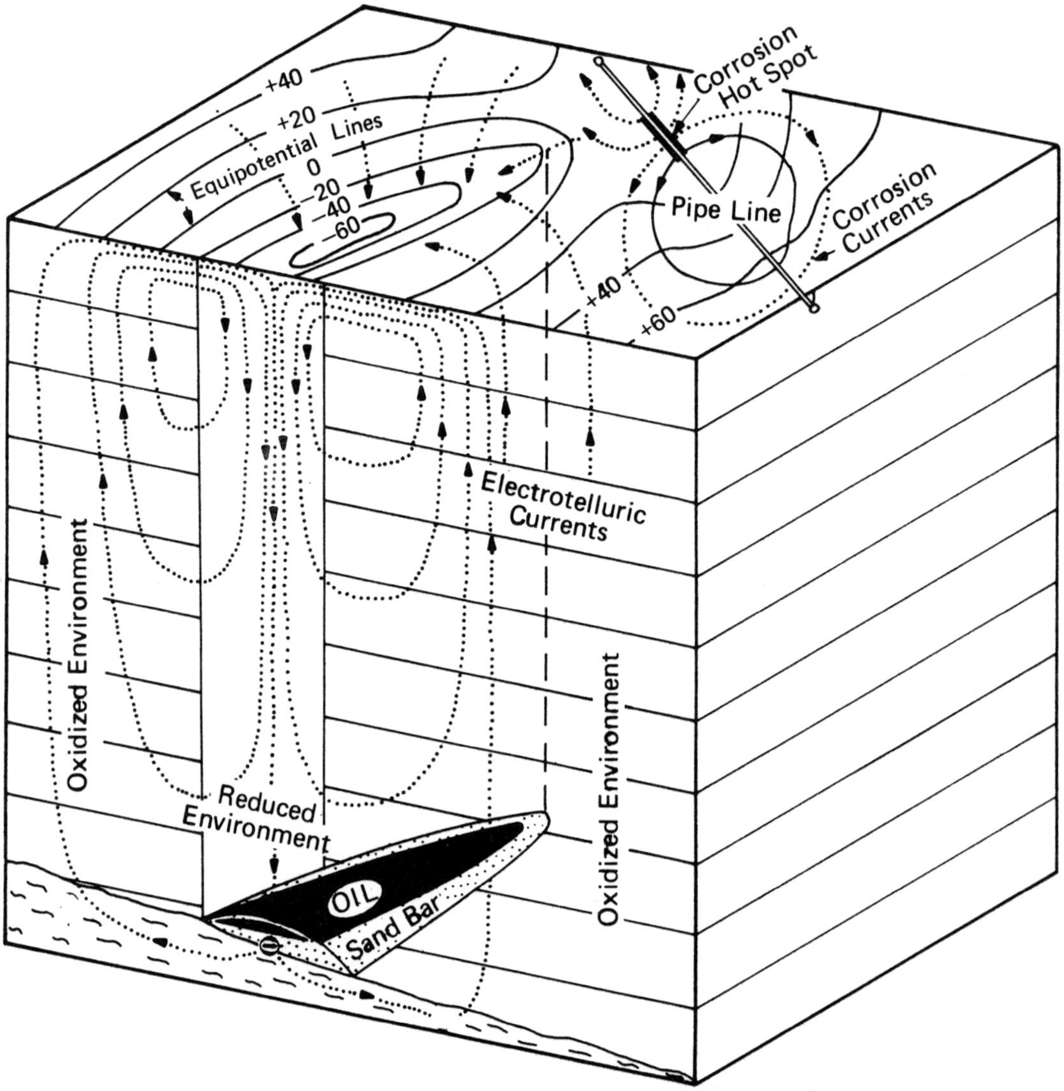

Fig. 12. Distribution of electrotelluric currents generated by an oil field before drilling and its associated self-potential sink at the surface of the ground as delineated by equipotential lines with possible disturbances by pipelines.

troleum deposit from greater to shallower depths, it cools and precipitates salts within the geologic section, *giving rise to a zonal distribution around the oil field.* Absorption of salts, ions, and hydrocarbons may also play a part in this mineralization effect.

The environmental effects of the above process must be emphasized. It has been shown by repeated experimentations (Pirson, 7h, i, j) that the injection of hydrocarbons in a shaly rock (sand or limestone) creates reduced conditions in such rocks, i.e., an excess of electrons. This phenomenon involves the cracking and polymerization of hydrocarbon molecules. Essentially the hydrocarbon accumulations may be viewed as negative poles which are within the bulk of massive sediments surrounding the oil pools and which have not been subjected to environmental modifications: i.e., they may be considered as oxidized (with an excess of positive charges) with respect to the reduced oil pools. We thus have the making of a *redox fuel-cell* or *electrochemical coupling* ($Eh = \frac{RT}{nF} \ln \left(\frac{(Red)}{(Ox)}\right)$) between the two environments. In a conductive medium such as the earth sediments, electric currents—the *Electrotelluric* (ET) currents—will flow in the external circuit from the oxidized region (Ox) to the reduced region (Red).

Figure 12 indicates how such ET currents are flowing in a stratigraphic type oil field. This figure gives a generalized idea of the physico-chemical modifications which the presence of oil and gas at depth creates in the overlying sediments. The fundamental mechanism of oil and gas migration and accumulation which is implied here, and which is also generally accepted by the petroleum geology profession, is that formation waters which were expelled from shales under the compacting effect of the weight of the rock overburden, and which contained highly disseminated hydrocarbon globules or hydrocarbons in solution because of the elevated formation water temperatures at depth, escaped vertically upward above the loci of oil and gas entrapment; i.e., structural or stratigraphic traps. Such expelled waters were filtered out of their suspended hydrocarbons; oil and gas pools were thus formed. In their vertical escape paths, the waters released the hydrocarbons still in solution in a gradual manner as temperature declined at shallow depths, and these hydrocarbons rendered the overburden rocks more reduced than when they were originally laid out, thus forming a "chimney" or "funnel" of reduced rocks which extended to the surface of the earth. When surface rocks were exposed to weathering agents, the reduced nature of the upper part of the chimney may have been partially oxidized. However, the escape of compaction waters and of hydrocarbons takes place continuously during geologic time and thus the "fuel cell" is being replenished continuously.

A permanent and durable physico-chemical contrast is thus created above oil and gas pools, which is manifested in many ways; but for the present purpose, this contrast is mostly observed in the near-surface rocks by their ability to generate electric potentials and currents as if from a giant "fuel cell," the source of potential being distributed on the external surface of the subsurface chimney.

This "fuel cell" generates current flow lines closed on themselves, a few of which are represented on Figure 12. Normally, the direction of the earth currents, or electrotelluric currents, so generated is as shown over this stratigraphic trap. Such currents and their distribution in the earth last indefinitely during geologic time, or as long as the "fuel cell" is being resupplied with "fuel" in the form of escaping hydrocarbons in solution or in suspension in the compaction expelled waters, or by molecular gas diffusion from the accumulation of oil and gas.

The mechanism of oil and gas migration and accumulation, represented in Figure 12, and the development of a cylindrical geochemical anomaly over the oil accumulation as represented have been verified by well-documented articles.

Figure 13 illustrates the various SP gradients that would be observed in various wells drilled in and around an oil field, in a stratigraphic trap such as represented in Figure 12. At well #2 oil was discovered in the reservoir sand trap, and its SP curve gradient is toward the positive when traveling upward. Wells #2, #3, and #4 are dry holes, and the upward SP gradients are reversed. In addition, Figure 13 illustrates the radioactive modifications of a marker shale bed as its potassium and uranium (also radium) ions are removed from above the oil fields by the action of the outward flowing electrotelluric currents through the process of electrophoresis. The reduction of the radioactivity of shales directly overlaying oil fields has been observed over at least thirty known oil and gas fields from radioactivity (gamma ray intensity) well logs. This observation has also been used successfully in predicting favorable petroliferous areas wherein the N. Heluma (Ellenburger), the ABM, the McGill, and the XBC (Devonian) fields were subsequently discovered in Upton County, Texas.

Detecting Subsurface ET Currents

Figure 14 illustrates three basic methods by which ET currents may be detected in the ground:

A. By measuring the *redox potential* (or Eh) of rocks in situ by means of a probe that extends below the weathering zone of surface rocks.

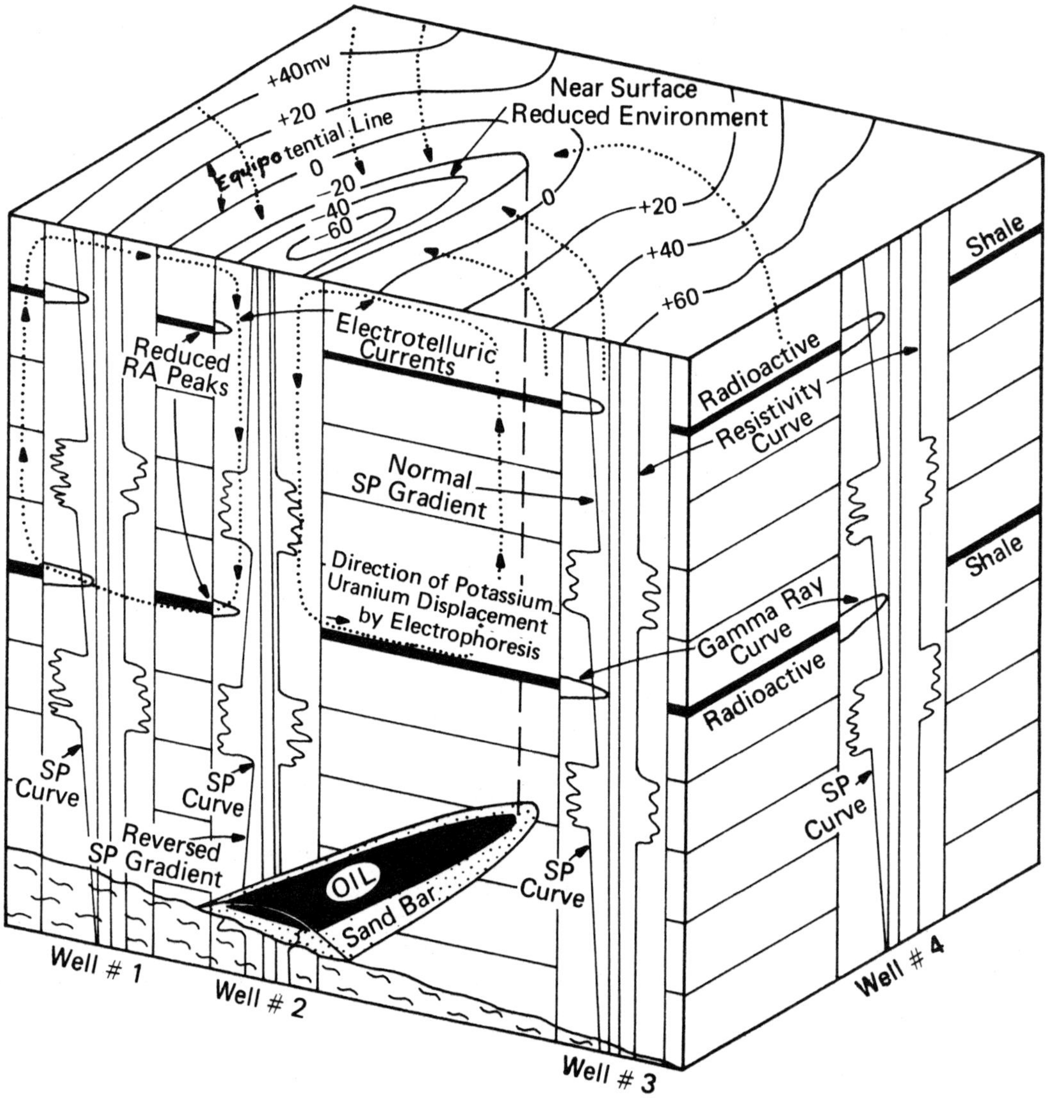

Fig. 13. Effect of electrotelluric currents on the SP curves of the wells drilled in and around an accumulation of oil and on the radioactivity of shale beds.

The thickness of this zone may be from a few inches to ten to twenty feet. Readings are often variable with season, rainfall, atmospheric pressure, etc. A redox potential sink is observed over an oil and gas field under favorable surface conditions.

B. By measuring the emf or *potential drop* (ET) along profiles at the surface of the ground between nonpolarizable electrodes embedded a few inches to a foot or two in the surface soil. When these emf are cumulated, a potential sink is observed over an oil and gas field because of the ohmic potential created by the horizontal ET current flux in the near-surface rocks, and that converges toward the apex of the structure or stratigraphic accumulation. The results may be presented in two manners:

1. As an *equipotential map* when the misclosures around closed profiles are distributed.
2. As an *ET map* when the observed values are misclosures around small closed profiles. Then the magnitude of the vertical ET current flux may be calculated in a

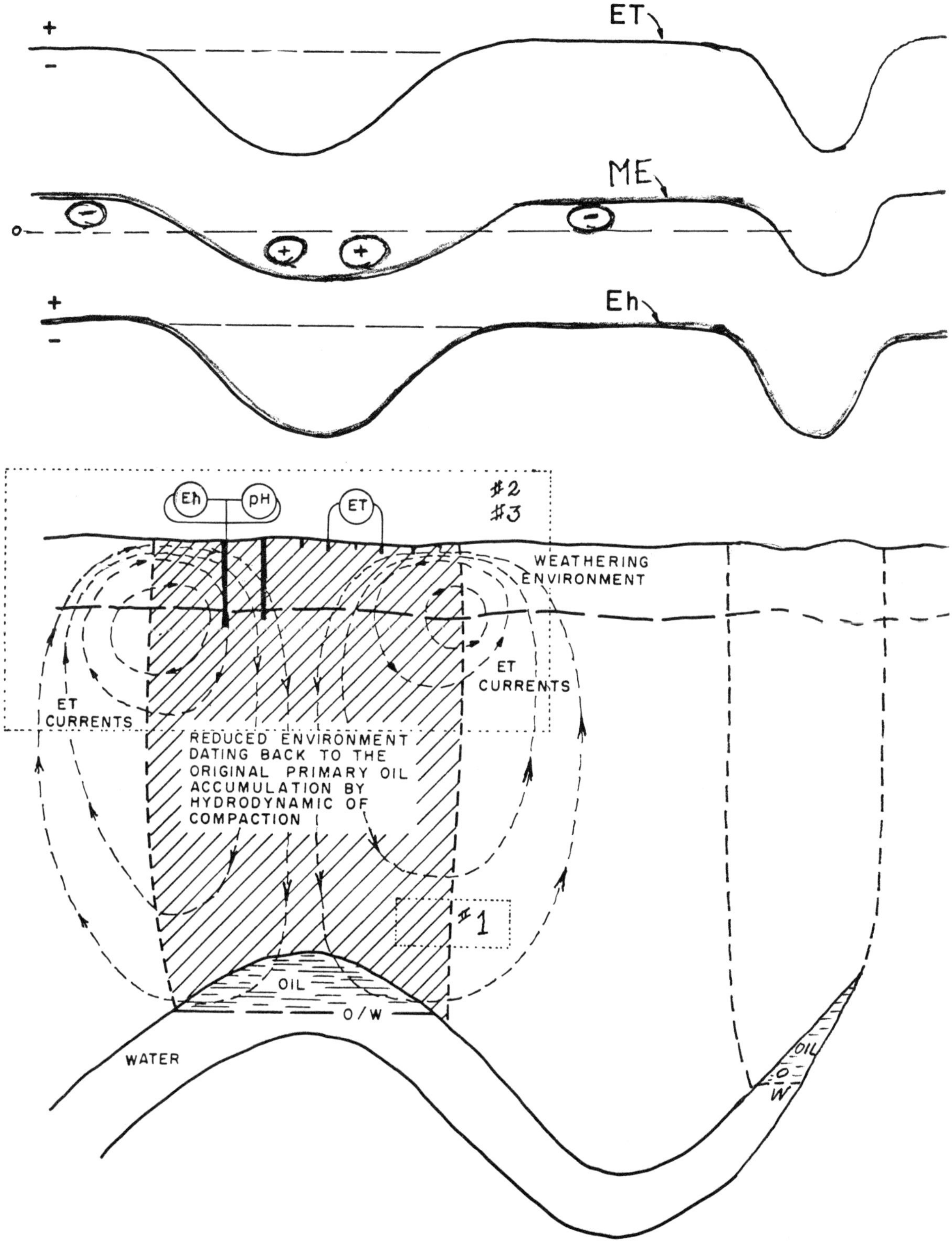

Fig. 14. Three techniques by which subsurface electrotelluric currents may be detected: electrotelluric surveys, magnetoelectric surveys, redox surveys.

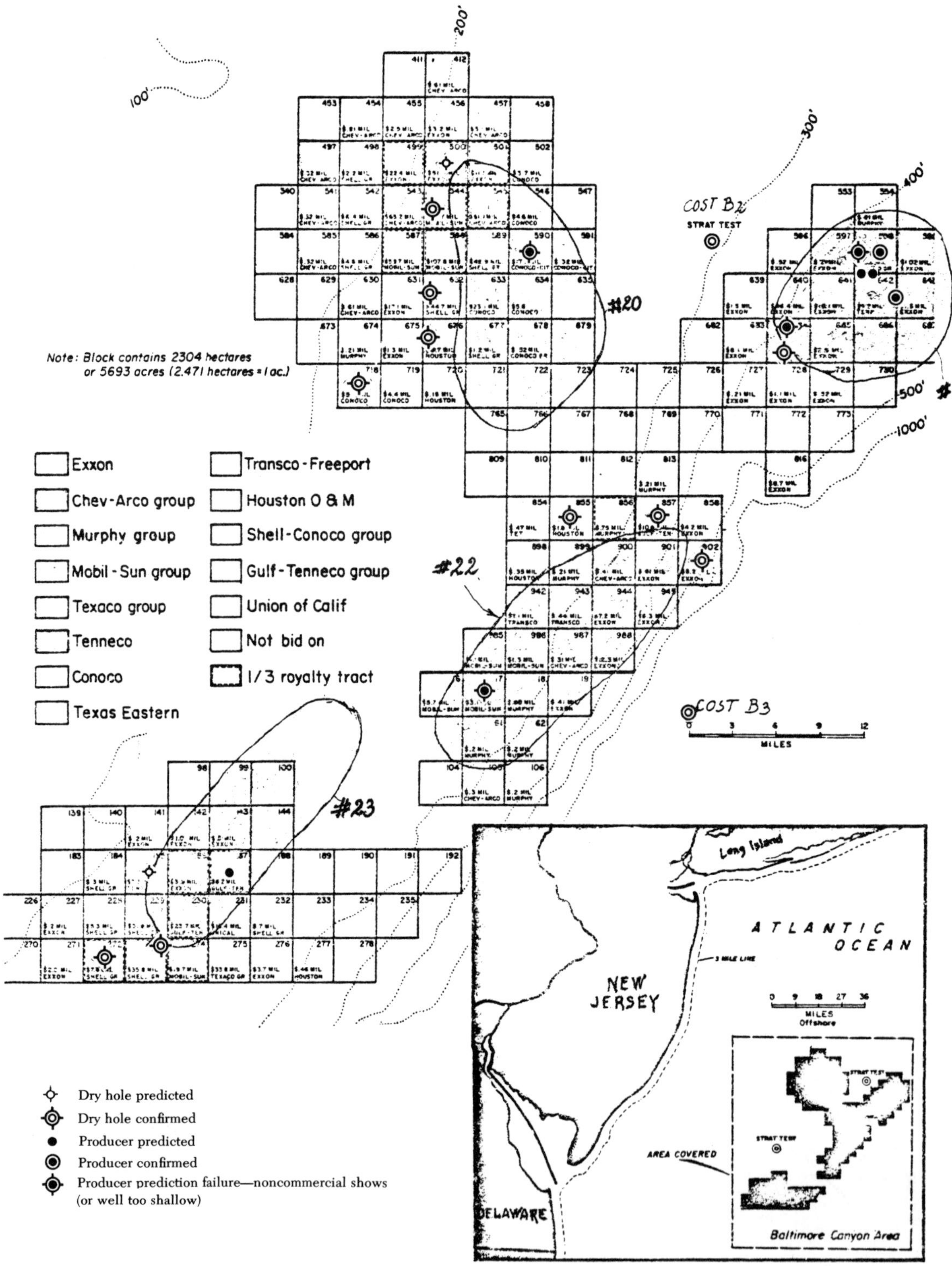

Fig. 15. ME predictions of wildcats in the Baltimore Canyon by the circle integration method. Base map courtesy *Offshore News*.

manner entirely similar to what is done in ME surveys. However, ET surveys using large numbers of closed profiles are slow, expensive, difficult to make, even impossible over water. ET surveys have been superseded by ME surveys. Much work of this type was carried out by the writer in Oklahoma with the discovery of the N.W. Norge field, the N.W. Oackdale field, and the S. W. Vici field. An unsolved difficulty of this type of survey has to do with the electrode junction potential, which must be eliminated at each location in order to obtain satisfactory results.

C. By measuring the magnetoelectric coupling that exists between the subsurface ET currents and the perturbations in the earth magnetic field that may be ascribed to it. Such an effect had already been detected by Zashekinskey over sulfide ore bodies that create intense magnetic anomalies (Metodika Tekhnika, 5th conf. 1963). These types of measurements are the object of Part Three.

PART THREE

Magnetoelectric Exploration

The principles upon which magnetoelectric (ME) exploration is based have already been elaborated on in voluminous literature and will not be repeated here. The following is a summary of statistical proof:

1. A *field statistical verification* of the validity of ME exploration was first made in 1972 by means of a computer interpretation of a 1,500 square-mile area in which a total of 240 wells had been drilled, 143 being oil and gas producers and 97 dry holes. This gives a success ratio of 59.6% by conventional exploration methods. Within the favorable area delineated by ME exploration, 140 wells had been drilled of which 105 were oil and gas wells, for a success ratio of 75%. ME exploration was then responsible for a 15.4% improvement in discovery efficiency. This study included both wildcat and development wells.

2. A *second field statistical verification* was made in the Delaware basin, West Texas and New Mexico. Using the USGS Aeromagnetic Map GP661 and the Midland Map Company "Permian Basin Wildcat Map," which is issued every month, fifty wildcat wells were selected during the period from 1973 to 1977 that fell within ME prospective areas, as determined from the USGS Map 861 by the visual scanning interpretation procedure (explained elsewhere). Out of the twenty-seven oil and gas producers predicted, eighteen became actual gas producers and six oil producers, for a *producer* predictive confidence ratio of 88.9%. Of the twenty-three predicted, nineteen were actually dry and four were producers, for a *dry hole* predictive confidence ratio of 82.6%.

3. Presently a statistical prediction analysis in progress involves the drilling activity in the Baltimore Canyon Trough. In 1977, a complete evaluation of the oil and gas potential of the province was made and presented at the Offshore Technology Conference (Paper OTC 2837, May 2-5, 1977). The wildcat map pertinent to the present activity is shown in Figure 15, where four ME prospects (N° 20, 21, 22, and 23) are indicated by reference to the leased tracts. Of the sixteen wells presently drilled and completed, ten were predicted dry and were dry, four were predicted producers and *two* produced: Texaco #1, Tract 598, 14 MMCFD; and Tenneco-Aminoil #1, Tract 642, 3700 BOD and 12 MMCFD. Other predictions are in progress as shown in Table I. The Exxon well #684 and the Conoco well #590 are not considered statistically significant.

Discoveries

Five small oil and gas fields have been entirely discovered by means of ME Surveys: (1) The OHB Co. Ltd.—O'Brien-Harkins # C1 and C2, Goliad Co., Texas; (2) Amity Oil Co.—Crawford #1A, Young Co., Texas; (3) Seventy-Six Oil Co.—B. W. King #1, Young Co., Texas; (4) The OHB Co. Ltd.—O'Brien-Harkins #D3, Bee Co., Texas; (5) a small gas field in Tennessee—3 producers, one dry hole.

Recent Advances in ME Exploration

Significant advances made in magnetoelectric exploration pertain to the application of well-known methods of magnetic downward continuation into the earth to levels that are in

TABLE 1

Exploration Predictions Track Record by the Magnetoelectric Method in the Baltimore Canyon Trough, Atlantic Offshore
September 1, 1979

PAST PREDICTIONS

10 dry holes predicted: outcome—all were dry. Prediction ratio: 100%

Mobil	Tract # 544
Houston O&M	# 676
Shell	# 632
Shell	# 273
Gulf	# 857
Houston O&M	# 855
Shell	# 272
Gulf	# 718
Exxon	# 684
Exxon	# 902

4 producers predicted: outcome—2 producers, 2 dry holes. Prediction ratio: 50%

Mobil	# 17	dry
Texaco # 1	# 598	discovery: 14 MMCFD
Texaco # 2	# 598	dry
Tenneco-Aminol	# 642	discovery: 3700 BOD, 12 MMCFGD

Success ratio by geology-geophysics 14.3%
Success ratio by ME exploration 85.7% (6 times greater than by conventional exploration)

Not statistically significant wells

Exxon	# 684	edge well, too close to call
Conoco	# 590	predicted discovery, well not deep enough, 12,000 ft

FUTURE PREDICTIONS

Predictions on currently drilling wells or locations

Exxon	# 500	probably dry
Tenneco	# 187	marginal discovery
Tenneco	# 185	probably dry

close proximity to (but preferably above) the sources producing the electrotelluric currents. At such depths, the magnetic field so continued satisfies the first Maxwell equation (without displacement current because of steady state), and closed line-integrals of the continued magnetic field in the horizontal plane do not vanish, but rather the values of such integrals are a measure of the electrotelluric current flux density within the area enclosed by the lines of integration. The depths of the desirable downward projection are determined from the subsurface geology of the probably producing reservoirs, as inferred from electric logs and other well logs.

The downward projection process involves the expansion of the measured magnetic field into Taylor series comprising the first and second vertical derivatives of the surface magnetic field. The derivatives are determined by well-known surface integration procedures carried out over the surface magnetic measurements. The patented line integral procedure is then applied to the analytically continued magnetic field values as if they had been measured in the subsurface. The results are expressed in mil-

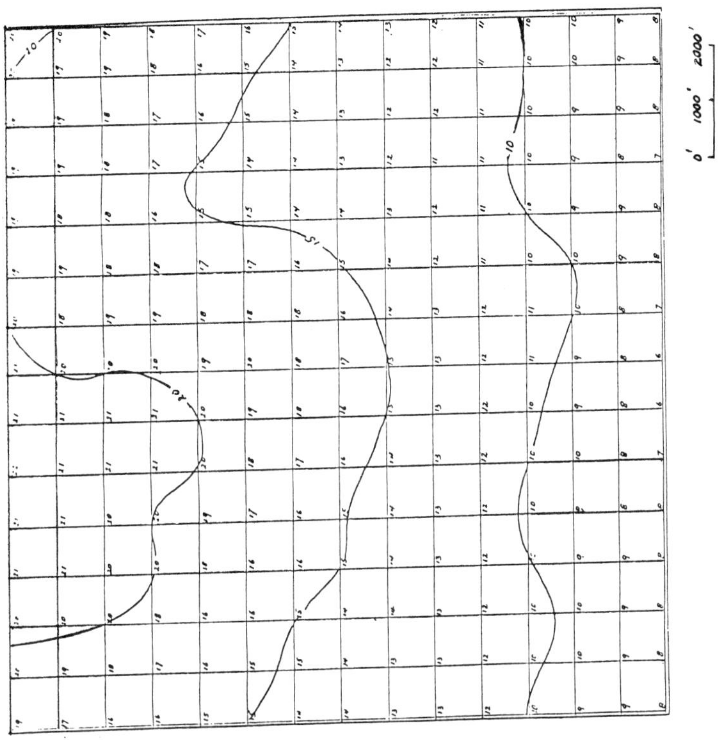

Fig. 17. Magnetoelectric survey. Surface integration: ma/acre.

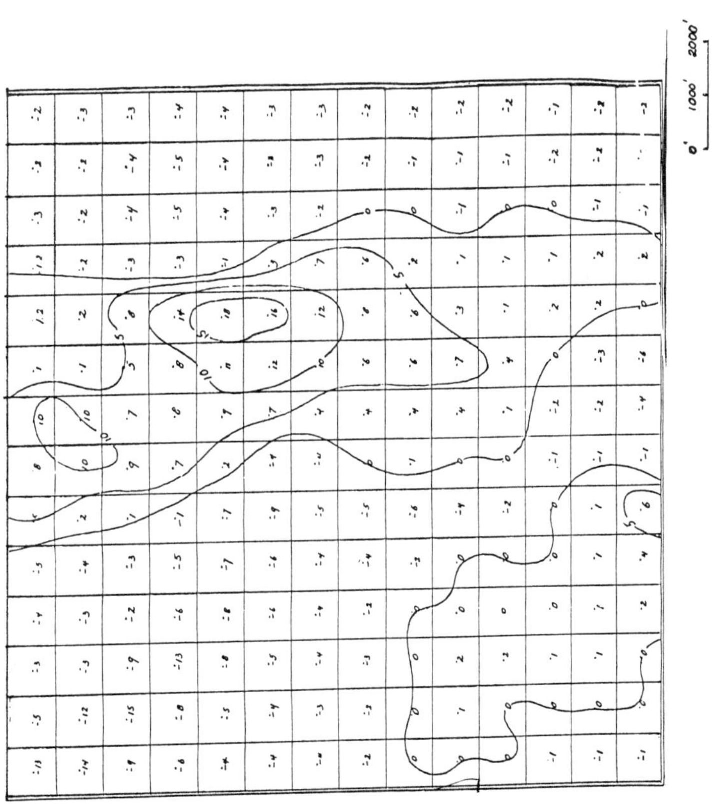

Fig. 16. Magnetoelectric survey. Variations in total earth magnetic field (δ).

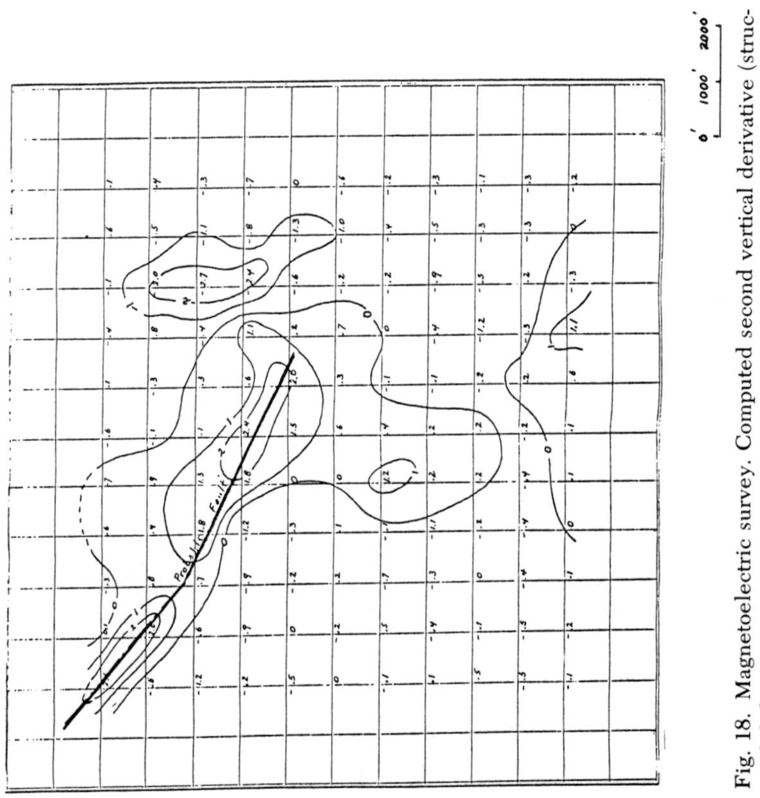

Fig. 18. Magnetoelectric survey. Computed second vertical derivative (structural deformation or curvature).

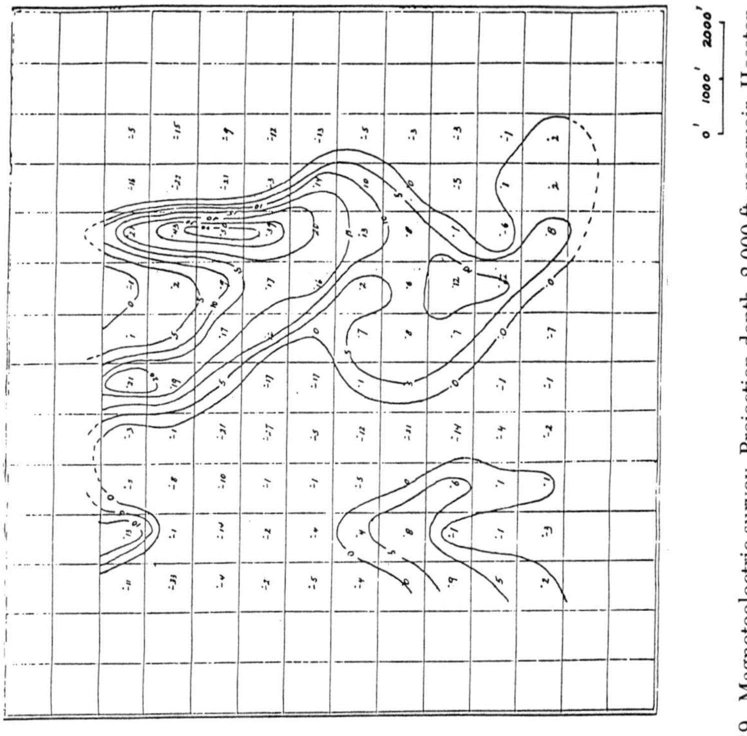

Fig. 19. Magnetoelectric survey. Projection depth, 2,000 ft; reservoir, Hosston (Travis Peak); probable productive area, 750 acres; hydrocarbon in place, 20 million cu ft/acre; total, 15 billion cu ft.

liamperes per acre for the vertical magnetoelectric current flux flowing inside the earth, which values are related to the hydrocarbon in place. When promising anomalies in the vertical electrotelluric current flux so mapped are found, the values of the flux densities are converted into hydrocarbon in place per acre and over the whole petroliferous areas so delineated. An appropriate recovery factor is then applied in order to estimate the monetary value of the discovered reservoirs.

Example of Application

A prospect was found in the Ouachita Overthrust Belt of Texas by means of maps of the total earth magnetic field and of its vertical gradient. A first, interpretation was carried out by means of the circle integration procedure and an ME prospect was thereby discovered. In order to verify the validity of the prospect further, a ground magnetic survey was carried out using a Geometrics Proton Precession portable magnetometer, Model 816. The map so obtained was digitized on a 1,000-by-1,000-ft-square grid oriented in the magnetic north; it is shown on Figure 16. The data were punched on IBM cards for reading into a FORTRAN V program for the Univac 1108 available at Tracor, Inc., Austin, Texas. This program is written to calculate and print in grid form the values of the line integral of surface data around each grid, to calculate the vertical gradient and the second vertical derivative, to multiply each value so obtained by the appropriate factor in order to calculate the continued value of the total magnetic field at specified depths, and to perform additional line integrations at such depths and thereby map the ET flux density at such depths in milliamperes per acre.

Figure 16 depicts the variations in the total earth magnetic field over the prospective area; no feature of great significance is immediately apparent.

Figure 17 depicts the variations in the vertical ET currents, as calculated by a surface line integration around closed paths around each grid: 1,000 by 1,000 ft in size. The values are shown in ma/acre and they depict the current intensities after a considerable part of the vertical flux has already been deviated horizontally by the nonconductive atmosphere. Nevertheless, a significant NNW-SSE trend is observed which indicates the presence of a geochemical source of current at depth.

Figure 18 is the computed second vertical derivative of the surface data, a function which is used in the downward continuation of the data as well as in tectonic studies.

In Figure 19, the variations in the vertical ET currents are calculated by the same integration procedure as in Figure 17, but at a depth of 2,000 ft. The continued values of the magnetic field to that depth, which is immediately above the Hosston (Travis Peak) formation, are given as if magnetic data had been obtained at that depth. Considering the area enclosed by the 5 ma/acre contour, 750 acres will be found gas productive, containing 15 billion cu ft of gas in place. Assuming a recovery of 75% and $2.00 per 1,000 cu ft, a value for the gas production of $22,500,000 is determined.

Confirmation of the petroliferous nature of this prospect was obtained by means of an electrotelluric, or self-potential, survey of the area by Oscar Dudley. This survey shows the electromotive potential sink in millivolts over the area created by the horizontal radially converging ET currents toward the vertical plume overlying the gas pool. A picture of the ET current potential sink is shown in Figure 20. The area of the "throat," or funnel, is approximately 200 acres, and an average of 15 ma/acre flows downward into it for a total of 3.0 amperes. The radial current sheet is perhaps 100 m thick, formed of rocks of 5Ωm resistivity (as observed on electric logs). A flow potential drop of 20mV takes place over a radial distance from 3,000 to 1,000 ft from the center of the sink. Applying the radial electric flow formula

$$I = 2 \frac{b}{R} \cdot \frac{\Delta e}{\vartheta n \; re/rw}$$

where:

I is the total centripetal radial current flow (in amperes which eventually becomes the total downward ET current flux)

b is the thickness in meters of the radial flow sheet, estimated at 100 m

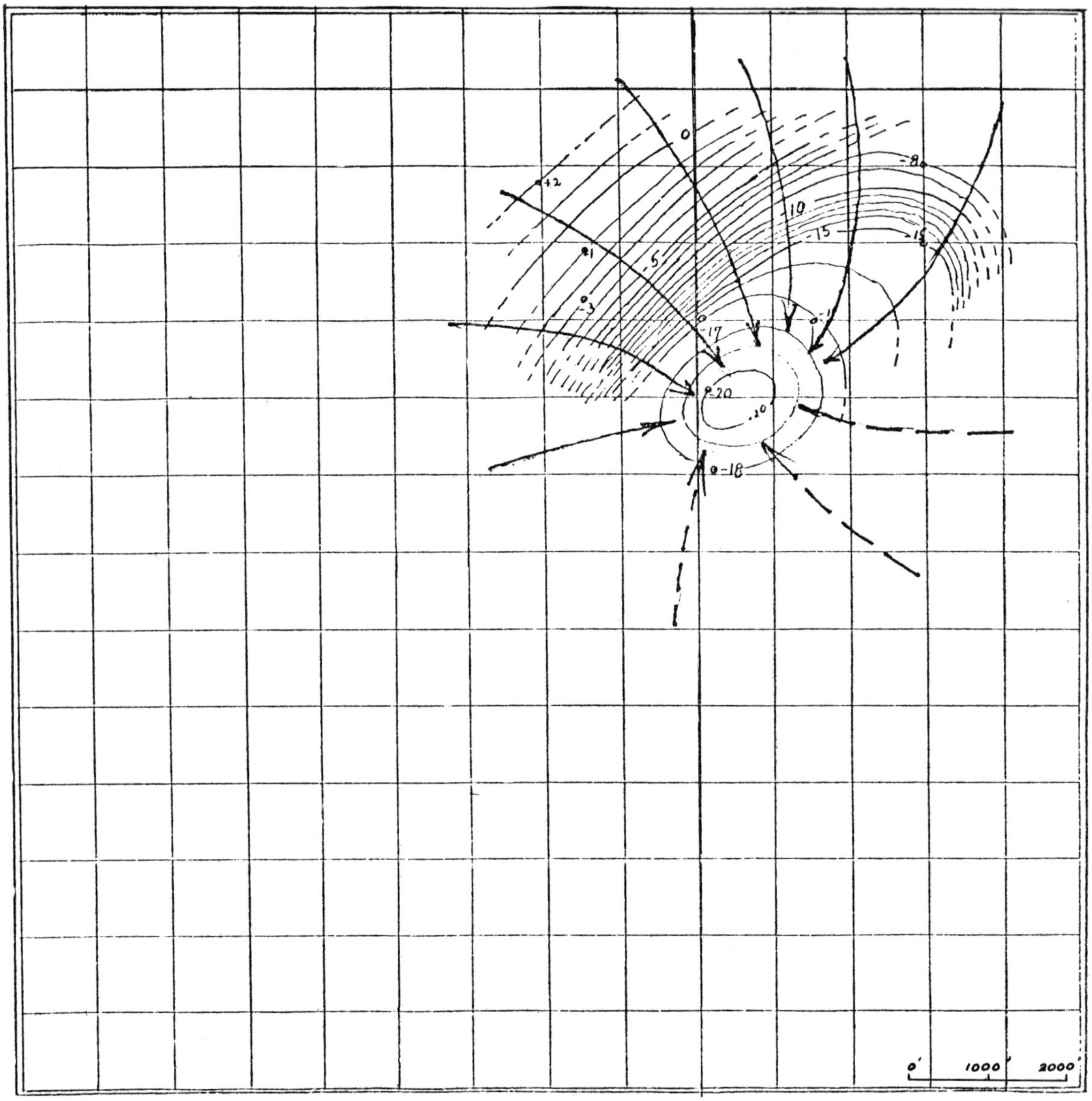

Fig. 20. Electrotelluric survey. Self potential (SP in mV). Arrows indicate ET current flow lines in the subsoil.

R is the near-surface rock resistivity: 5 ohm-meters

Δe is the electric potential drop in volts, between the two radii: re = 3,000 ft and rw = 1,000 ft

We find I = 2.28 amperes after substitution of the above values. This is a remarkable order of magnitude check of the gross approximation involved, especially with respect to the thickness of the horizontal current flow sheet just below the surface of the ground.

It is observed that the ET anomaly corresponds best with the surface integration ME anomaly (Figure 17). This is as it should be because both maps represent near-surface effects of the ET currents. By contrast, Figure 19 represents the ME anomaly at a depth of 2,000 ft, where it exhibits a dendritic pattern, as if the ME effect were produced by hydrocarbon accumulations in channel- or stream-filled sands, a hypothesis which is highly probable in the case of Travis Peak. In this area, during Travis

Peak time, the paleo-streams were transporting arkosic sediments from the North-West; i.e, from the Llano uplift toward the East Texas basin. The ME downward continuation method resolves the hydrocarbon accumulations into various individual channels, as observed in Figure 19. Accordingly, the type of interpretation under review allows us to pinpoint the emplacement of exploration wells with the maximum chances for successful discoveries.

Structural Studies

As a bonus, magnetoelectric studies also contribute to structural evaluation of the area, namely: rock deformation and defaulting. This relationship is based on the common expectation that sedimentary deformation is often associated with basement rock relief and deformation. Basement rocks are generally more magnetic than sediments and, therefore, it is possible to deduce geological subsurface information from magnetic measurements made at the surface of the earth, or by aeromagnetic surveys. Although there are exceptions, sediments are made up of nearly nonmagnetic minerals. Accordingly, the variations and convolutions of the isogam lines of magnetic surveys are related to the composition and the structural relief of the igneous (or metamorphic) basement rocks. If there is no anomalous basement magnetic polarization effect, there is a direct correspondence between the main part of the earth magnetic field (that which is derivable from a potential) and structural basement relief. Therefore, if one computes the curvature of the surface magnetic data, i.e., the second vertical derivative (Pirson, 1978, p. 207-211), one obtains a map depicting the structural trends of the basement rock. This is particularly significant when faulting occurs, as this is revealed by unmistakable lineations. The area under study is a case in point which reveals as a probable fault a NW-SE trending lineation (Figure 18) that passes through the prospective oil and gas accumulation.

It may also be concluded that the overlying brittle reservoir rocks (Austin Chalk and Edwards Lime) may be highly fractured along this trend. Drilling may be hampered by loss circulation and the presence of cavernous sections.

Conclusion

Additional evidence is here adduced concerning the validity of ME exploration; namely, that the patented line integral method (Pirson and Pirson, 1976) of exploration based on magnetic surveys (ground or airborne) can now be performed on magnetic earth field values as if they had been measured in the subsurface. This is accomplished by downward projection of the surface magnetic measurements by analytical continuation methods that are well known in the geophysical sciences. This should clarify the existing misunderstanding concerning the basic principles on which ME exploration is based.

The downward continuation method in ME exploration is equivalent to adding a "telescoping lens" to the "magnetic camera" that measured the earth magnetic field. This may be considered as a method of "magnetic focusing" that yields greater contrast and better definition with depth of investigation. The ME anomalies obtained by surface integration, although significant, are blurred and merged, whereas the downward projected ME anomalies are distinct and sharp, and they represent the separate individual anomaly producing oil and gas accumulations that are associated with individual structures or stratigraphic traps.

REFERENCES

Elkins, T. A., 1951, The second derivative method of gravity interpretation: Geophysics, v. 16, p. 29-50.

Evjen, H. M., 1936, The place of the vertical gradient in gravity interpretation: Geophysics, v. 1, p. 127-136.

Henderson, R. G., et al., 1960, A comprehensive system of automatic computation in magnetic and gravity interpretation: Geophysics, v. 25, p. 569-585.

Jakosky, J. J., 1940, Exploration geophysics: Times Mirror Press, Los Angeles, p. 84-85.

Peters, L. J., 1949, The direct approach to magnetic interpretation and its practical application: Geophysics, v. 14, p. 290-320.

Pirson, S. J., 1945, Quantitative interpretation of gravity meter surveys: Oil Weekly, April 16, 1945.

———, 1947, An evaluation of present day geophysical exploration for oil: Oil Weekly, March 31, 1947.

———, 1963-1964, Projective well log interpretation: World Oil, October and November, 1963; August, September, October, and November, 1964.

———, 1966, Oil finding by systematic well log analysis, Log Analyst, v. 6, no. 5, January-March, 1966, p. 4-17.

———, 1968, Geological, geophysical and chemical modifications of sediments in the environment of oil fields: Unconventional Methods in Exploration for Petroleum and Natural Gas, SMU Symposium, Dallas, p. 159-186.

———, 1969, Environmental logging and mapping in the search for minerals: Society of Professional Well Log Analysis, Annual Logging Symposium, Tenth, Transactions, Paper I, Houston, Texas, May 25-28, 1969.

———, and Negut, Aurelian, 1969, Preliminary model experiments in redox well logging (with applications to mineral exploration): Society of Petroleum Engineers Paper No. 2593, October, 1969. Also in Revue Roumaine de Géologie, Géophysique et Géographie, v. 16, no. 1, p. 145-167, 1972.

———, and Pirson, J. E., 1969, Exploration by surface electrical measurements: Oil and Gas Journal, October 27, 1969, p. 108-111.

———, 1971, New electric technique can locate gas and oil: World Oil, April 1971, p. 69-72, and May 1971, p. 72-74.

———, 1973, Advantages and limitations of direct oil-finding methods: World Oil, April 1973, p. 63-66, and May 1973, p. 73-76.

———, 1973, Direct oil and gas finding methods: an answer to the energy crisis in the United States?: Texas Business Review, v. 47, no. 7, July, 1973, p. 123-160.

———, 1975, Aid to oil hunters: ME exploration: Oil and Gas Journal, March 31, 1975, p. 121-127.

———, and Pirson, Jacques E., 1976, Line integral method of magneto-electric exploration: U.S. Patent 3,943,436, March 9, 1976.

———, 1976, Prediction of hydrocarbon in place by magneto-electrotelluric exploration: Oil and Gas Journal, May 31, 1976, p. 82-86.

———, 1976, Track record in ME exploration: Oil and Gas Journal, September 20, 1976, p. 241-249.

———, 1977, Discoverable hydrocarbon reserves from the Baltimore Canyon trough: Offshore Technology Conference (OTC 2837), Houston, May 2-5, 1977.

———, 1978, Hydrogeology I—hydrodynamics of compaction, *in* Geologic Well Log Analysis, 2d ed., Gulf Publishing Co., Houston.

———, 1978, Physical and chemical signals around oil and traps: American Association of Petroleum Geologists Research Symposium, Oklahoma City, May, 1978.

———, 1978, Computerized magneto-electric exploration—case history: Giddings, Texas: Oil and Gas Journal, June 19, 1978, p. 168-178.

DETERMINATION OF SUBSURFACE LITHOLOGY FROM SEISMIC OBSERVATIONS

Carl H. Savit

WESTERN GEOPHYSICAL CO.

INTRODUCTION

In 1922 seismic refraction was a new and unconventional method of exploration for oil. That technique was a simple extension and adaptation of the World War I method of sound-ranging to locate enemy artillery. By 1927, seismic refraction had become a conventional method that had successfully located most of Louisiana's shallow piercement salt domes. In that year, however, the newer, unconventional seismic reflection method was being introduced. Within a short time the simplest form of seismic reflection technique had become the conventional method of hydrocarbon exploration.

In the intervening years, a succession of improvements and extensions of seismic reflection technology have been introduced to the exploration industry. Each has been unconventional at the time of its introduction but most have been accepted and shortly thereafter have become "conventional."

When this paper was first proposed for Symposium II on Unconventional Methods in Exploration for Petroleum and Natural Gas, the application of the acoustic velocity transform to seismic data was distinctly unconventional. At this writing, that process is being widely applied and is well on the way to universal acceptance.

As with most advances in science and technology, the practical application of the acoustic velocity transform could only take place when all of the necessary antecedent and ancillary technologies had matured sufficiently. The exploration community also had to be receptive, basing their acceptance on the perception that the older, conventional technologies had reached a point of diminishing returns.

SEISMIC DELINEATION OF STRUCTURE

In its first forty years, the seismic reflection method had been devoted exclusively to the determination of the structural configuration of subsurface sedimentary rocks. The seismic reflection method was practiced by injecting a transient acoustic signal into the earth near the surface and measuring the time required for that transient to reach each of a number of sensitive detectors deployed along one or more lines radiating from the source. Comparison of arrival times of the acoustic signals for a large number of different source-receiver pairs yielded enough data to construct a cross-sectional representation of the interfaces between buried rock layers of different acoustic properties.

During those forty years, virtually all efforts at improving the system were devoted to refining the derived structural cross sections. A host of signal-generating, recording, processing, and conditioning techniques was developed to improve the signal-to-noise ratio so that it would be possible to detect the weaker reflections from interfaces between rocks of similar properties and from interfaces that were deeper in the earth. Some techniques improved the resolution of the method so that thinner layers and finer details could be seen. Still other methods were generated to determine, ever more accurately, the exact positions of subsurface features relative to surface coordinates.

Some kinds of "stratigraphic" traps could be found by knowing layer configurations in fine detail. Among such traps were pinchouts and bioherms (Seelis, 1959; Savit, 1969; Luttes et al., 1956). Only rarely, however, was the seismic method used to furnish information on lateral changes in the composition of rock layers.

All of these painstakingly developed processes and more were prerequisite to the eventual emergence of seismic reflection methods that could routinely yield information not only of shape or form, but of the type or kind of rocks below the surface.

GROSS MEASUREMENT OF ROCK PROPERTIES

Seismic reflection times and time differences had been used for many years to estimate the velocity of sound in subsurface formations. Such estimates were used primarily to translate the time-measurements into estimates of depth to specific reflecting interfaces. Velocity estimates derived from time information could be used to identify average rock types of thick layers or zones. Resolution was rarely, if ever, fine enough to distinguish producing zones on the basis of an anomalously low velocity that should be identified with anomalously high porosity.

For a long time, however, it had been known that the amplitude and form of received seismic signals contained information related to properties of the materials through which those signals had passed. The 1948 Master's thesis of J. Miller at Rice University derived in quantitative detail the relationship between the strength of a reflected seismic signal from a porous sandstone and the nature of the fluid (gas, oil, or water) contained in its pores (Miller, 1948). Miller demonstrated that the reflections from a gas- or oil-saturated sandstone should be stronger and of opposite polarity relative to reflections from the same formation when saturated with water.

Unfortunately, early seismic instruments were limited in dynamic range and could not reliably record variations in the strength of received seismic reflections. No significant effort was devoted to preservation of signal strength because the exploration industry was content with arrival times only. By the late 1950s, however, the introduction of magnetic-tape recording made possible the recovery of some relative signal strengths. As shown in a 1960 publication (Savit, 1960), tape-recorded information from oil fields in California and Alberta indicated that reflection strengths did, in fact, behave as predicted by Miller. Because the observed effects were only relative and were difficult to verify, general use of reflection-strength information had to await the introduction of digital recording instruments whose dynamic range was greater than that of any conceivable seismic exploration signals.

Even then the highly touted "bright-spot" method used only qualitative measures of seismic reflection strength to point the way to a number of hydrocarbon discoveries in the late 1960s and early 1970s (Savit, 1974).

QUANTITATIVE MEASURES OF A LITHOLOGIC PROPERTY

By 1978, techniques and instrumentation had matured to the point that it became possible to derive and present detailed estimates of acoustic velocity in the sedimentary section from the amplitude of seismic traces.

Mathematically it is quite simple to derive the acoustic impedance of the subsurface section if we assume that the seismic trace is entirely composed of a sequence of pulses, each representing normal-incidence, primary reflections from a simple succession of interfaces. In that case the amplitude of each individual pulse is proportional to the reflection coefficient at the corresponding interface. Since the reflection coefficient is simply related to the impedances on the two sides of the interface, a layer-by-layer calculation of impedances relative to the first layer used in the calculation is possible. For ease of application it is assumed that there are infinitely many layers. This assumption is permissible because the reflection coefficient of an "interface" between two identical layers is zero, and a zero reflection coefficient implies identity of acoustic impedance (Lindseth, 1979).

To convert the derived "log" of relative impedance to an acoustic-velocity log requires one more assumption and a calibration step. The assumption is that the contribution of density to impedance changes can be neglected. In nearly all cases it has been found that velocity changes from layer to layer are much greater than are the density changes.

Because the impedance values are relative, they are, at best, determined only up to a multiplicative constant. The further transformation to velocity is dependent on the density, which changes slowly with depth. The cumulative errors resulting from noise and from the inevitable errors of measurement combine with the velocity transformation problem to mandate calibration against an independent measurement of velocity. Such independent measurements are preferentially obtained from acoustic logs in nearby drilled holes. When no well information is obtainable, absolute velocity information can be derived from analysis of reflection times—the so-called normal-moveout-velocity analysis.

The processes and instruments needed to make the velocity-transform technique generally applicable are those required to validate the assumptions upon which the technique is based. Most important has been the complex of data-gathering methods, instruments, and computer processes that has yielded a severalfold improvement in signal-to-noise ratio. With this improvement it became possible to apply signal-conditioning steps to transform complex reflected signals into a reasonable approximation of a simple pulse. Other new processes sorted out and removed signals that had undergone more than one reflection, and still other processes collapsed diffraction patterns and correctly located the reflecting points in three-dimensional space.

Finally a computer-controlled plotter became available that facilitated the plotting of velocity values in the form of color-coded images on ordinary paper. Only a color-coding system can effectively represent a complex, multivalued function of two variables in a way that can be readily apprehended by a human being.

Final details included the devising of a computer algorithm to control the finished color representation of acoustic velocity in cross-section form. The method chosen succeeded in producing pictures that look like geologic cross sections and could be readily accepted and understood by the exploration community.

RESULTS

Within a short time of its introduction the cross-sectional representation of acoustic velocity in color was put to widespread use in making exploration decisions.

The examples presented here include one case from the Midland Basin of Texas that vividly demonstrates the value of acoustic velocity in locating porous zones in a limestone bed (Figure 1).* A dry hole in a structural high is associated with a tight formation corresponding to a high value of acoustic velocity, while the off-structure productive zone corresponds to an anomalously low velocity in the same bed.

A second illustration from the cover of the ODECO annual report for 1978 shows the perfect agreement between well logs and the color-velocity representation of a gas field in the Gulf of Mexico (Figure 2).

Velocity representations in color have successfully been used for reserves calculations. When a three-dimensional survey has been performed, it is possible to prepare velocity "slices" in any place through the surveyed block. In one case such slices have been used to make three motion pictures representing motion downward, northward, and eastward through the surveyed block. This particular survey was performed to supply the basic data for a unitization program of a major oil field (Figure 3).

PROGNOSIS

As new forms of seismic data, such as those generated by shear waves, become available, it can be expected that other unconventional exploration methods will be attempted. Some will prove generally useful and will quickly become conventional. The process will end only when the explorationist, whether geophysicist, geochemist, or diviner, can predict with 99% or better accuracy exactly what the drill will find at any specified depth at any selected locality.

REFERENCES

Lindseth, Roy O., 1979, Synthetic sonic logs—a process for stratigraphic interpretation: Geophysics, v. 44, no. 1, January 1979.

*See atlas for Figs. 1-3.

Luttes, L. L., Savit, Carl H., and Brustad, John T., 1956, The Glen Park field, a reef detailing problem: Geophysical Case Histories, v. 2, p. 169-178.

Miller, J. W., 1948, An investigation of the possible relation of reflection seismic record amplitudes to the presence of oil and gas in a simple structure: Master's thesis, Rice Institute, May 1948.

Savit, Carl H., 1960, Preliminary report: a stratigraphic seismogram: Geophysics, v. 25, no. 1, February 1960.

———, 1969, Use seismic data to find stratigraphic traps: Oil and Gas Journal, April 11, 1969.

———, 1974, Bright spot in the energy picture: Ocean Industry, February 1974, p. 60-65.

Seelis, Karl-Heinz, 1959, Das reflexionsseismiche Bild des Bentheimer Sandsteines in Raum von Meppen: Erdöl und Kohle, v. 12, p. 953-957.

CARBONATES AND ISOTOPE RATIOS FROM SURFACE ROCKS: A GEOCHEMICAL GUIDE TO UNDERLYING PETROLEUM ACCUMULATIONS

William Duchscherer, Jr.

GEOCHEMICAL SURVEYS, INC.

ABSTRACT

After early research showed that geochemical techniques of prospecting for hydrocarbons were only partly satisfactory for petroleum exploration, research was directed toward the effect of hydrocarbon leakage on alteration of the soil above petroleum accumulations.

Imperfect capping rock seals above petroleum accumulations allow large volumes of low-molecular-weight hydrocarbons to leak and diffuse to the surface. It was reasonable to assume that this leakage could chemically alter the soil. Not only could hydrocarbons be adsorbed and absorbed by the soil particles, they could be altered to CO_2 by oxidation and form a stable carbonate or compound with the soil. Obviously, many carbonates unrelated to hydrocarbon leakage are present in soils. Through research covering many years, a complex carbonate related to buried petroleum accumulations has been discovered in the colloidal fraction of soils and has been designated ΔC.

The technique for measuring ΔC requires mechanical separation of clay from soil material. The clay fractions are treated by differential thermal analyses, and the complex carbonate dissociates into CO_2 at a specific temperature range that is different from the temperature at which any known soil carbonate dissociates. The released CO_2 is measured in millivolts.

After statistical studies are made of the analysis values, the values are posted on a prospect map at the sample locations and are contoured.

As valid anomalies are revealed, selected carbon-isotope ratio values are obtained. A field example was performed over the Fox-Graham Field, Carter County, Oklahoma, where the ΔC halo pattern was seen to agree with the carbon-isotope ratio pattern which outlines the productive limits of the underlying petroleum accumulation. This is principally because the pore-filling cement is isotopically lighter with respect to carbon toward the interior of an anomaly overlying a petroleum accumulation. The gradient produces a distinct isotopic signature that agrees with the ΔC halo.

A geochemical model for hydrocarbon exploration incorporates the ΔC and carbon-isotope analyses as important techniques to be added to geochemical exploration.

INTRODUCTION

It is a matter of grave concern among exploration geochemists that in the United States geochemical methods have greatly lagged in development and acceptance, whereas in the U.S.S.R. there reportedly has been great progress in the use and application of numerous forms of geochemical exploration techniques.

It is hoped that symposiums such as this one will encourage the use of geochemical methods as direct hydrocarbon exploration tools. We who employ these tools and know of their potential do not view them as unconventional.

The main purpose of this paper is to present a direct oil-finding tool that Geochemical Surveys has employed for more than thirty-seven years with great success. The following will develop a geochemical model for hydrocarbon exploration that will support the basic concept that hydrocarbons migrate from subsurface petroleum accumulations to the surface and leave their imprint in the near-surface soils.

BACKGROUND ON GEOCHEMICAL PROSPECTING

Modern geochemical prospecting methods using gases collected from soils were first employed in Germany and Romania by Laubmeyer (1933) in 1929. Sokolov (1933 a, b) did similar experiments in the U.S.S.R., which were followed in 1938 by Rosaire (1938) and Horvitz (1939) and others in the United States. Rosaire and Horvitz analyzed adsorbed hydrocarbons in the soil itself, rather than gases collected from soil air.

Results of their studies suggested a direct relationship between these soil hydrocarbons and subsurface petroleum accumulations. As more data were obtained, this relationship became more apparent.

The first American company organized to employ geochemical methods to explore for oil and gas was the Subterrex Company, organized in 1937 by Dr. E. E. Rosaire. The company's primary purpose was to furnish field parties and laboratory facilities to oil companies interested in using geochemical methods in hydrocarbon exploration. In its first year of operation, Subterrex sold operating licenses under its patent rights to two Dallas firms—Magnolia Petroleum Company and Geophysical Service, Inc.

Geophysical Service concluded, after conducting a geochemical survey of half of Refugio County, Texas, that the geochemical methods then available were not very practical.

FOUNDING OF GEOCHEMICAL SURVEYS

Shortly thereafter, a meeting was held in the Houston offices of Humble between Dr. Ludwig Blau, Dr. E. E. Rosaire, Eugene McDermott, and his assistant, C. V. A. Pittman. McDermott had called the meeting to inform Blau and Rosaire that after the Refugio County geochemical survey, G.S.I. felt that the geochemical methods then being employed needed further research.

On the drive back to Dallas after the meeting, McDermott told Pittman that the gas leakage, which was without question taking place, must cause a valid soil alteration of some sort over real petroleum accumulations which should be measurable. If so, a cumulative factor, which would indicate over a broad area where maximum leakage had taken place, could be used by a geochemical survey to locate underlying hydrocarbon accumulations.

Upon returning to Dallas, McDermott met with Dr. J. C. Karcher and Everette DeGolyer and convinced them that a research and development program for $1,000,000 should be undertaken for one year to explore and verify his idea for finding the soil alteration thought to be present over oil and gas fields.

By 1940, the G.S.I. geochemical division, headed by E. C. Reagor and Jim Toomey, had investigated thirteen different geochemical and soil-analysis methods. This involved a laboratory and team of specialists operating twenty-four hours a day for one year and several G.S.I. field parties taking soil samples in all kinds of geologic environments.

During 1941, G.S.I. was trading with various oil companies in order to sell its oil properties developed during the depression years. Finally, in December, 1941, G.S.I. was sold to Stanolind with a condition that the Seismograph Division would revert to Eugene McDermott, Bates Peacock, Cecil Green, and Erik Jonsson. At the same time, G.S.I. thought it best to discontinue its geochemical division, which was then sold to Wm. R. Ransone, C. V. A. Pittman, E. C. Reagor, and Eugene McDermott, the last being a silent partner. The company thus formed was called Geochemical Surveys. During 1952, Dr. E. E. Rosaire joined the group and remained as a senior partner until his death in 1959.

As a direct result of the extensive research done by G.S.I., a technique was developed which measures a unique carbonate caused by oxidation of methane migrating from underlying petroleum accumulations. By this technique high carbon-dioxide values are measured from soil samples taken over areas thought to be potentially hydrocarbon producing when analyzed by differential thermal analysis.

Since 1942, substantial improvements have been made based on the original concept suggested by McDermott to Pittman in order to improve the effectiveness of the geochemical method in oil and gas exploration. The technique developed and now owned by Geochemical Surveys remained proprietary until McDermott's death in 1973.

The above has been mentioned to emphasize that geochemical oil and gas explorationists owe a debt of gratitude to all these men for keeping the American geochemical effort alive during the early years of development.

METHODS OF GEOCHEMICAL PROSPECTING

The search for hydrocarbon deposits by means of geochemical prospecting is directed toward localizing zones of maximum interest in areas thought to have oil and gas potential. Explorationists the world over have long searched for a direct oil-finding tool and as a result have utilized surface seepages and other surface indicators of petroleum as a guide for the location of exploratory drill sites.

In geochemical prospecting, petroleum accumulations are located in the subsurface by analyzing soil samples taken at or near the surface. Soil, in the sense used here, is the finely divided material that lies on a consolidated rock bed, at or near the earth's surface. Soils are formed by the weathering or disintegration of rock material and decomposition of organic matter. Soils, therefore, may be said to be the products of the chemical and mechanical processes and interactions in the four geologic spheres—the lithosphere, the hydrosphere, the atmosphere, and the biosphere. Within the realm of petroleum geochemical exploration, a wide assortment of methods has been used with varying degrees of success. Notable among these are: (1) gasometrics, (2) organic geochemistry, (3) hydrogeochemistry, (4) microbiological, (5) radiometrics, (6) oxidation/reduction potential, (7) luminescence (fluoroanalysis), (8) trace metals, (9) carbonate (CO_2), soil salts, and (10) geobotanical.

All these methods employ measurements of near-surface sediment or soil for effects of continuous or nearly continuous migration of gases and low-molecular-weight hydrocarbons from underlying petroleum accumulations. These gases and their condensates migrate up through the strata by at least three seepage mechanisms: (1) effusion as a free gas migrating along fissures, faults, fractures, and zones of disturbance to produce macroseeps; (2) diffusion of gases usually dissolved in ground waters to produce microseeps; (3) vertical movement of low-molecular-weight hydrocarbons dissolved in water through capping shales as a result of hydrodynamic or chemical potential drive. Of these three, diffusion is probably the most important process involving the migration of gases through water-filled pores, along surfaces of mineral particles, and through the crystal lattice structures of minerals.

PROBLEMS OF EARLIER GEOCHEMICAL METHODS

Most geochemical prospecting has been based on the premise that vertical leakage of hydrocarbon gases from an underlying pool is detectable in one form or another at the surface. The most direct way of detecting such a condition would be to analyze soil samples for the gases themselves. This procedure, however, is difficult and expensive, and the quantities in which the substances of interest occur are measured in parts per billion. Another procedure would be to analyze for substances entrained and carried to the surface by the upward leaking gases. Such substances could include soluble salts such as sulphates and chlorides.

The main problem in most geochemical methods employed to date is that the condition being measured results from a balance between factors tending to increase its magnitude and factors tending to diminish it. For example, the quantity of volatile hydrocarbons present in the soil at any given time is a function of the rate of upward leakage into the soil where the sample is taken, compared to the rate of leakage into the atmosphere and the rates of any other physical and chemical factors tending to eliminate the volatile hydrocarbons. In the geological time frame considered, the amount of volatile hydrocarbons present at any given time will be very small. So also with soluble salts; the quantity present in the soil will be determined by the rate at which the salts are carried up to the sample horizon, and the rate at which the salts are taken away by geological processes.

Clearly, then, if a soil characteristic can be determined that is produced by leaking hydrocarbon gases and is stable and cumulative, an advantage would be gained by analyzing substances that could be processed in the order of parts per hundred or parts per thousand or simply in millivolts.

MEASUREMENT OF SOIL ALTERATION

Some thirty-seven years ago, a property of surface soils was found that is related to the presence of oil and gas accumulations at depth. This property was observed as an abnormally high rate of evolution of carbon dioxide when specially prepared samples were heated to predetermined temperatures. The CO_2 is evolved as a result of thermal decomposition of carbonates or closely related compounds. Soils are composed mainly of silicates and aluminosilicates; calcium carbonate is the only carbonate generally found in soils in an appreciable quantity, because it is the least soluble carbonate in nature. In addition, $CaCO_3$ can occur as cement and can be embedded in the lattice structure of aluminosilicates, where it is protected from leaching and other weathering processes. Other carbonates may occur in greater quantity, but only as a result of special conditions in certain geologic environments. Since measurements of high CO_2 evolution correlated with the presence of oil and gas regardless of the total carbonates present, it was concluded that the soil carbonates had been altered by hydrocarbon leakage from the underlying petroleum accumulation. Because it preserves evidence of leakage after the gases themselves are gone, the soil carbonate alteration is called a fossil anomaly.

Horvitz observed, when measuring ethane and heavier hydrocarbons, that halos produced by migrating hydrocarbon gases disappear even before depletion of a producing field. This is due principally to the apparent continuous flow of hydrocarbons from the underlying accumulations to the surface with appreciable quantities being absorbed along its pathway. As the field continues to be produced toward depletion, the migration rate is lowered and near-surface adsorption ceases. Horvitz reports:

> To note the weakening and ultimate disappearance of an anomaly as the accumulation is withdrawn from its reservoir is of great interest. Equally important, however, is the observation that an anomaly remains in the soil, and its presence can be confirmed even though a period of many years has elapsed since it was first recognized, so long as the accumulation producing the anomaly remains undisturbed [1969].

As long as there have been vertical migrating hydrocarbon gases, and these gases have reached the near-surface sediments, there will be oxidation and a resultant soil alteration. The soil alteration is the fossil carbonate anomaly which will remain in place after the depletion reported by Horvitz has taken place.

HALO EFFECT

Gases and low-molecular-weight hydrocarbons which rise with limited lateral dispersion from underlying petroleum accumulations cause a measurable soil alteration called a "halo." Much debate has centered on its origin and ultimate meaning in geochemical exploration, and the phenomenon of a "halo" pattern seems to have no simple explanation. The common type of halo patterns encountered are the ring and the solid anomalies. There have been many theories advanced in an attempt to satisfy its critics, and many, if not all, may be applicable. But the very foundation of geochemical prospecting is to locate these areas that exhibit a halo pattern.

The following ideas provide a general explanation. A federally funded study showed theoretically and experimentally that cap rocks of shale, limestone, and dolomite will allow upward migration of hydrocarbons (Pandely et al., 1974). Most investigators attribute halos both to slow vertical leakage of hydrocarbons through cap rocks and to the release of hydrocarbons from migrating ground fluids. These fluid mixtures come together in a trap that acts as a focal mechanism to gather, retain, and process water carrying gases, low-molecular-weight hydrocarbons, and other organics. In a structural or stratigraphic trap these mixtures become distinct oil and gas fluids under their own temperature and pressure regimes. The trap also discharges water upward, laterally, or downward, or in a combination of these directions. Overton (1978) points out that the movement of water by mechanical forces is essentially horizontal, whereas the net chemical movement is vertical. This seems to agree with the views presented here. Enclosing membranes of the trapping rock act as wicks; when the membranes (wicks) are water-soaked, they leak water, which travels up to the surface around the capping mechanism of the reservoir. This is referred to as vertical by-pass leakage. Bubbles or globules of hydrocarbons are entrained in the water, and when these hydrocarbons reach the near-surface sediments, they become oxidized. The resulting carbon dioxide reacts with water and carbon dioxide to produce bicarbonate ions which are soluble and hence mobile. At the surface, degasing may occur with loss of CO_2 and precipitation of calcium carbonate as distinctive pore-filling cements. The precipitation and dissolution of calcium carbonate are determined by the chemical exchange equilibria:

CO_2 (gas) $\rightleftarrows$ CO_2 (dissolved)

CO_2 (dissolved) + $H_2O \rightleftarrows H^+ + HCO_3^-$

$CaCO_3$ (solid) + $H^+ \rightleftarrows Ca^{++} + HCO_3$

CO_2 (gas) + H_2O + $CaCO_3 \rightleftarrows Ca^{++} + 2HCO_3^-$

These pore-filling cements retain an isotopic peculiarity which reflects hydrocarbon leakage and subsequent oxidation in the near-surface sediment and soils. This distinctive form of cementation in the near-surface rocks produces a ring halo pattern around the edge of the petroleum accumulation rather than over its center (Figure 1). The solid halo pattern usually results over steeply folded structures where there has been extreme microfissuring through the cap rocks to the surface. This relatively unrestrained leakage has enabled rapid evaporation of the migrating hydrocarbons through the near-surface sediment and soils with much less attendant oxidation resulting in a solid halo.

Theories about halos have been modified or refuted by many, but the weight of empirical evidence confirms the existence of the halo effect. Some other interesting ideas of halo formation are presented by Soli (1954, 1957), MacElwain (1963), and Johnson (1970).

OVERCOMING OBJECTIONS TO USE OF GEOCHEMISTRY

Most geologists and petroleum engineers are skeptical of any concept involving vertical movement of gases. This may be due principally to the belief that lateral migration of water would disturb the gaseous flow and cause erroneous surface geochemical anomalies. In reality, however, such lateral migration is probably extremely slow relative to the velocity of vertical migration of gases during the geological time frame. It must be remembered that we are considering a gravity field where vertical force vectors operate on every individual hydrocarbon molecule (Stegna, 1961). If there are routes open upward, the hydrocarbons will rise until they are stopped by some impenetrable zone or altered by chemical processes. In nature, most joints, fractures, and faults are not inclined to a critical degree and, in most instances, are effectively sealed by some form of secondary mineralization. Two references that are enlightening on this subject are Levorsen (1954) and Roof and Rutherford (1958).

The exact mechanism of gas migration to the surface continues to be an enigma. The effects of these migrations are observed by geochemists universally, however, and the phenomenon has been more than validated by empirical data. A geologist or petroleum engineer tends to think in terms of zero permeability, as if rocks were totally impervious to the passage of gas. Among physical chemists and ceramic engineers, however, it has long been recognized that even at room temperatures some gases can pass through any number of materials, including glass, ceramics, plastics, and metals.

It has been proven that there is relatively rapid ascension of hydrocarbon gases through a sedimentary column. Our studies indicate that colloidal-size gases are displaced upward through the sedimentary pile by surrounding water at rates up to several millimeters per second, regardless of sedimentary particles in the way.

Additional study and research are necessary to enable us to understand the overall effects of lateral and vertical ground-water movements, but experience has demonstrated that the movements of subsurface water do not cancel the validity of this geochemical effect.

One finds a wealth of good technical literature on the application, success, and failure of all types of geochemical methods in hydrocarbon prospecting. For accounts of early developments of these methods, one should consult Rosaire (1938, 1939, 1940), Horvitz (1939, 1945, 1954, 1959, 1969), and McDermott (1940). For more recent data, the reader is referred to Fedyvsky and others (1975) and Horvitz (1978).

Geochemical prospecting is a direct approach to hydrocarbon exploration, whereas geophysical methods, no matter how refined they become, will always remain an indirect approach. Geochemical methods used in exploring for hydrocarbons locate only pe-

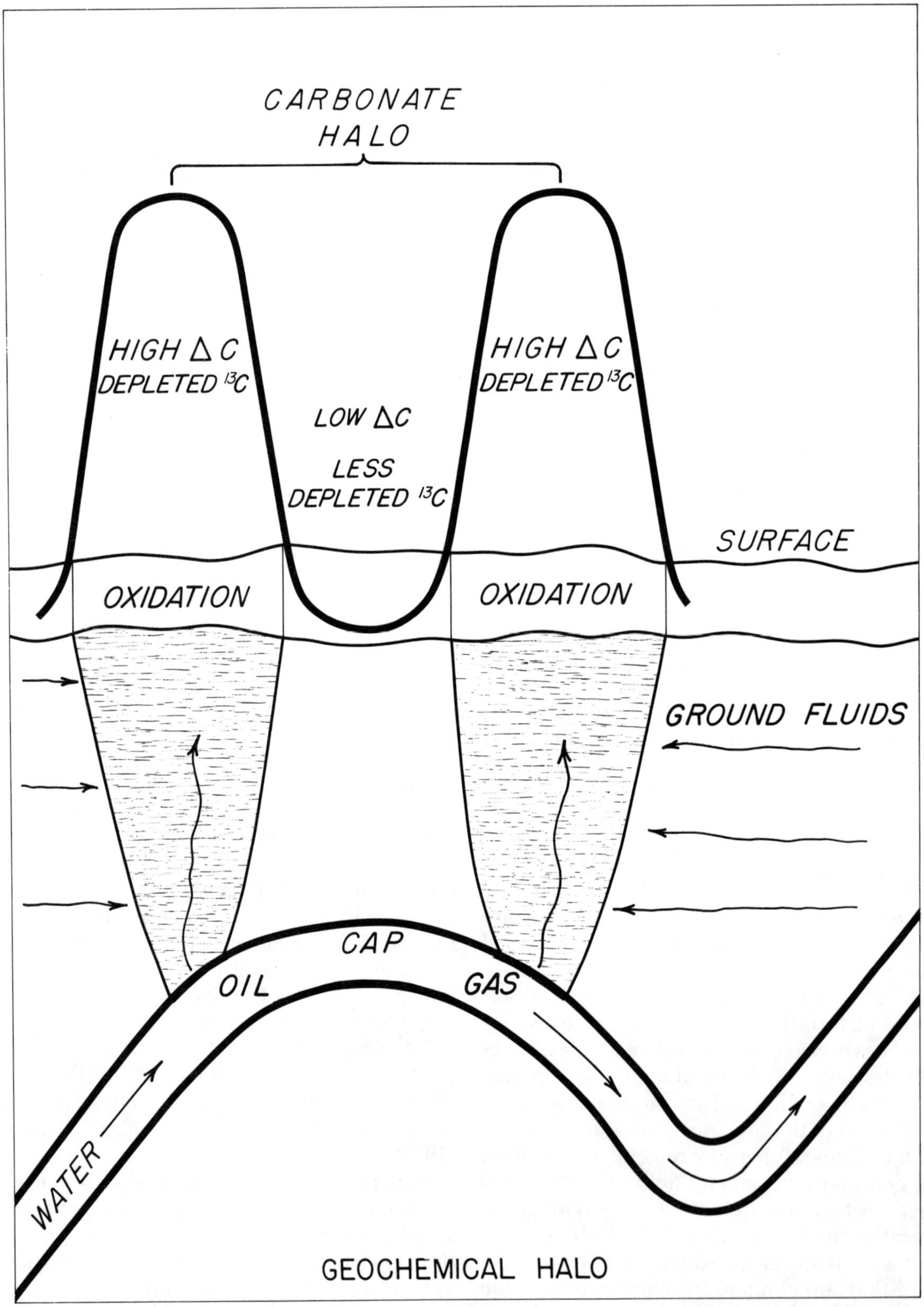
CARBONATE
HALO
HIGH Δ C
DEPLETED 13C
HIGH Δ C
DEPLETED 13C
LOW ΔC
LESS
DEPLETED 13C
SURFACE
OXIDATION
OXIDATION
GROUND FLUIDS
CAP
OIL
GAS
WATER
GEOCHEMICAL HALO

Fig. 1.

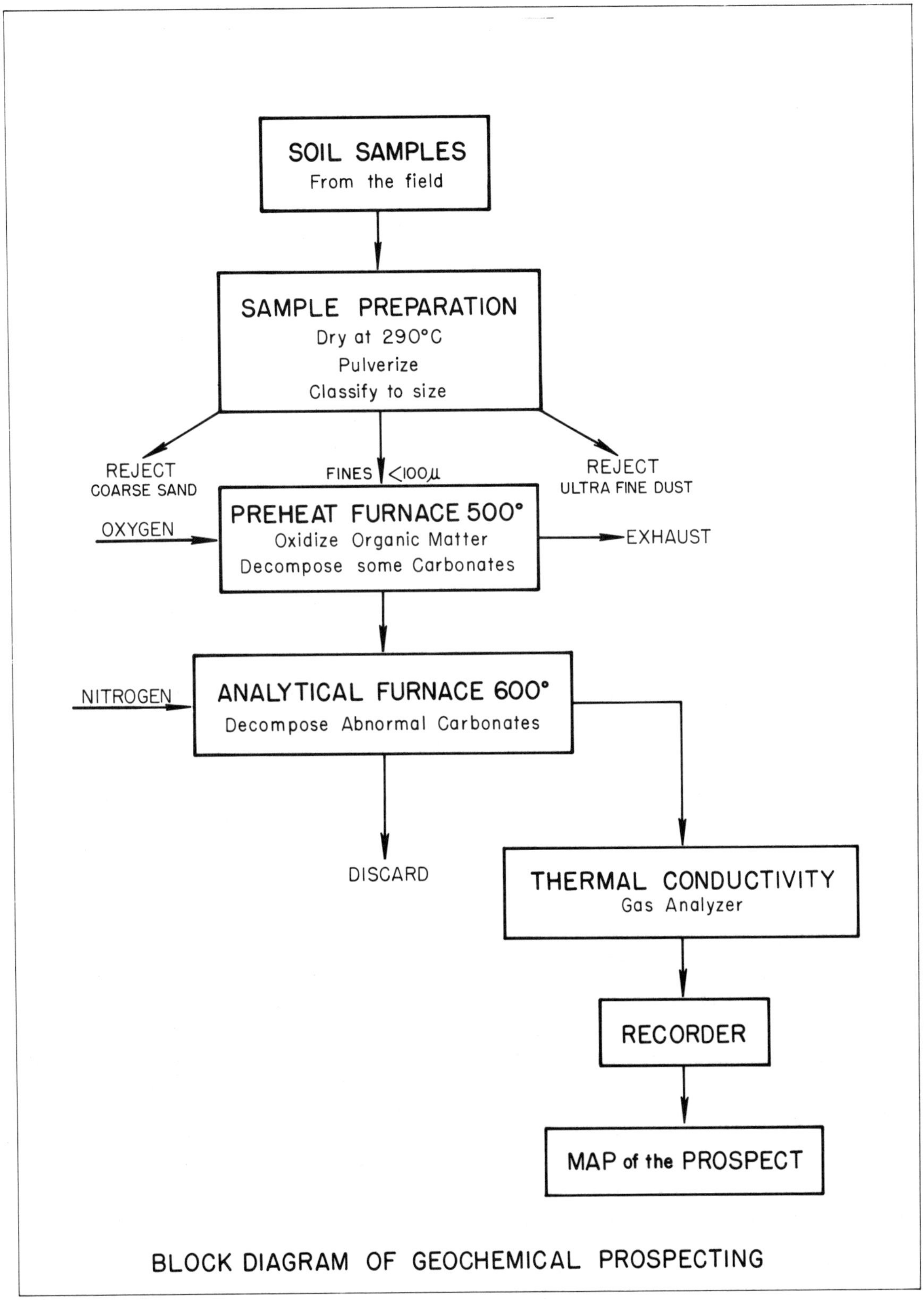
SOIL SAMPLES
From the field
SAMPLE PREPARATION
Dry at 290°C
Pulverize
Classify to size
REJECT
COARSE SAND
FINES <100μ
REJECT
ULTRA FINE DUST
OXYGEN
PREHEAT FURNACE 500°
Oxidize Organic Matter
Decompose some Carbonates
EXHAUST
NITROGEN
ANALYTICAL FURNACE 600°
Decompose Abnormal Carbonates
DISCARD
THERMAL CONDUCTIVITY
Gas Analyzer
RECORDER
MAP of the PROSPECT
BLOCK DIAGRAM OF GEOCHEMICAL PROSPECTING

Fig. 2.

troleum accumulations that give a geochemical anomaly, just as the magnetometer and gravimeter locate only magnetic and gravity anomalies. Likewise, the seismograph locates structural traps but misses most stratigraphic traps.

Since 1942 Geochemical Surveys has surveyed approximately 70,000,000 acres using its unique geochemical techniques. This has resulted in 875,000 individual soil sample analyses. From these surveys, the drilling of 174 geochemical anomalies resulted in 38 field discoveries. The significant factor of these discoveries is that 22 fields were stratigraphic traps.

CARBONATE ΔC ANALYSIS

Geochemical Surveys, in its long history, published only descriptive data that was not technical in nature (Ransone, 1947, 1958, 1968). With the advent of the energy shortage, and following the publication of some interesting and advanced geochemical exploration techniques (Donovan, 1974; Donovan and Dalyiel, 1977; Donovan et al., 1975; Hudson, 1977; Murata et al., 1969), Geochemical Surveys decided to publish, for the first time, the methods used in its exploration.

Geochemical Surveys collects soil samples from 5 ft to 7 ft deep across a grid of eight samples per section on half-mile lines. An average crew will collect approximately 1,000 samples per month for a survey coverage of 125 sq mi. In the laboratory, these soil samples are dried at 290°C, pulverized, and classified by size. We learned early in our experience that sand in surface soil has little attraction for hydrocarbons, while clay has great capacity for adsorption of gases. Sand acts as a diluent and only reduces the magnitude of an anomaly. For these reasons the soil samples are classified to less than 100 microns by removal of sand and ultrafine soil dust, and only the clay fraction is used in analysis. The samples are then preheated in an oxygen atmosphere at 500°C to decompose any low-temperature carbonates and carbonaceous material. Organic matter in an oxygen atmosphere produces CO_2; the samples are preheated to eliminate this possibility. After a predetermined time cycle, the samples are moved to another oven with a nitrogen atmosphere at a temperature of 600°C. At this temperature the clays holding hydrocarbons in adsorption or absorption dehydrate, and the evolved gas passes through a thermal conducting cell where it is analyzed (Figure 2). The evolved gas is carbon dioxide, the quantity of which varies as the carbon content of the soil sample. The CO_2 so evolved is referred to in our analysis as ΔC. Figure 3 is a plot of a number of carbonates illustrating their thermal dissociation at various temperatives. The carbonates shown will either dissociate at a lower temperature than 600°C or will not dissociate appreciably until a temperature higher than 600°C is reached. Our research through the years indicates that there is no known massive carbonate mineral that occurs in soil or sediments that will dissociate to any considerable extent at 600°C. The CO_2 produced at this temperature has its origin in the oxidation of methane gases and can be a significant component of underlying petroleum accumulations.

The only carbonate that is frequently associated with petroleum-bearing sediments not shown in Figure 3 is siderite ($FeCO_3$). Our research and that reported by Rowland and Jonas (1949) shows that the effect of heating on siderite, the most labile carbonate, results in decomposition characteristics not generally encountered in thermal analysis of other carbonate minerals. When siderite is heated in an air atmosphere, thermal decomposition takes place between 500°C and 570°C with the liberation of carbon dioxide. In the absence of dilution by inert substances, siderite dissociation begins at about 530°C. At approximately 570°C, the evolution of carbon dioxide decreases to a point where the resulting FeO is susceptible to oxidation. The rate of oxidation is dependent on the rate at which carbon dioxide can escape. The FeO, however, will become totally oxidized immediately after the loss of all carbon dioxide, based on the following reaction:

$$FeCO_3 \rightarrow FeO + CO_2$$
$$2FeO + \tfrac{1}{2}O_2 \rightarrow Fe_2O_3$$

In the ΔC analysis at the critical temperature of 600°C, all the carbon dioxide has been evolved

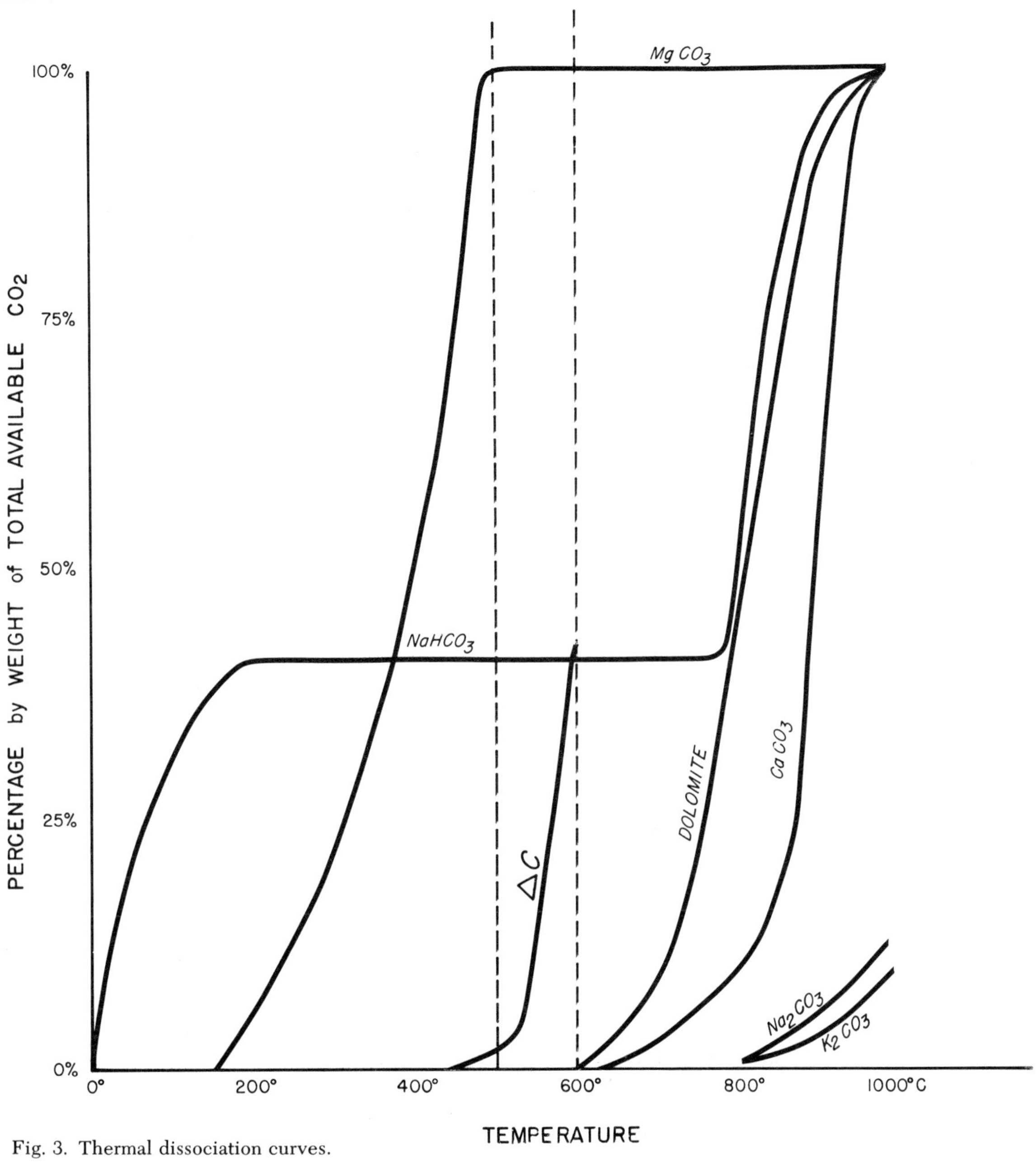

Fig. 3. Thermal dissociation curves.

and siderite will present no appreciable problem to our measurements. Other important research on this subject can be found in Laurence et al. (1951), Rowland and Lewis (1951), and Tissot and Welte (1978).

At present, the exact nature of the carbonate and its mechanism of emplacement in the soil remains unknown. On Figure 3, it has been given the notation ΔC.

It should be remembered that the most important practical requirement of a prospecting tool is that it be successful in locating petroleum accumulations. Any technique that will do this consistently, as Geochemical Survey's does, will be accepted whether or not the underlying theory is completely understood. The ΔC method is justified pragmatically by close correlation with other methods and by the location of known producing gas and oil fields.

Figure 4 illustrates the amount of carbon

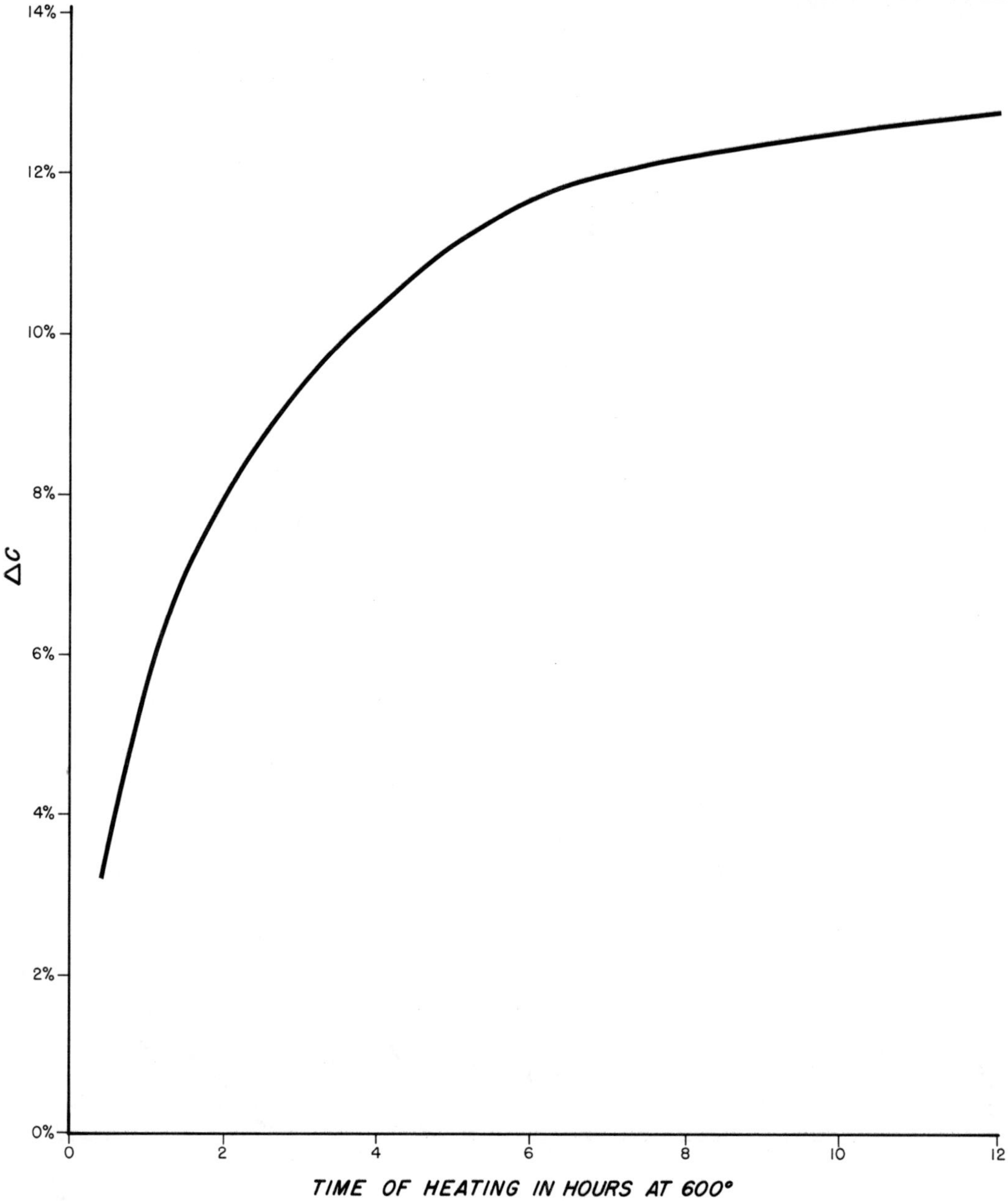

Fig. 4.

dioxide produced by heating a representative prepared sample at 600°C. The amount of CO_2 produced on prolonged heating approaches 13.0% by weight of the sample. It has been determined by experiment that prolonged heating does not alter the significance of the result, but only increases its magnitude. In the sample represented in Figure 4, nearly half the final quantity of CO_2 (6.5%) is evolved after one hour's heating.

The effect of normal carbonates in producing a ΔC value was further tested by adding varying percentages of $CaCO_3$, MgCO3, FeCO3, and CaMg $(CO_3)_2$ to soil samples. The results indicated that the carbonate percentages in themselves were not a significant factor in the ΔC determination. In actual measurement of soil, the type or the amount of carbonate reflects the type of soil and is not indicative of hydrocarbon accumulation at depth. The ΔC analysis was discovered by sampling over known oil and gas fields by measuring different ranges of temperature for the dissociation effect of soil carbonates that were hydrocarbon derived. A temperature range was discovered that yielded anomalous thermal dissociation above the edges or around the perimeters or as a solid halo over underlying petroleum accumulations producing the characteristic halo pattern. This effect will be discussed further in the portion of this paper dealing with carbon-isotope ratios.

It was also determined, after surveying more than four hundred known oil and gas fields, that not all fields give this carbonate dissociation effect. However, the number of petroleum accumulations not giving a ΔC anomaly was found to be small.

One negative aspect of the ΔC method is that it offers no depth perception. Ideally, every geochemical anomaly drilled on a ΔC prospect should be drilled to the Precambrian basement in order to test all the geochemical potential of the prospect. There are a number of instances where geochemical anomalies previously condemned by shallow drilling are now producing fields as a result of drilling deeper.

DATA INTERPRETATION

When ΔC values are obtained, they are recorded at the sample locations on base maps that are usually at the scale of 1 inch = 1 mile. Prior to contouring, the ΔC values are evaluated statistically to separate true anomalies from the relatively large fluctuations of normal background values. Geochemical prospecting is concerned only with an abnormal distribution of values which may be considered as normal background variations with anomalous values superimposed.

Normal frequency plots and cumulative frequency plots are the methods of statistical calculation used to determine normal background and arrive at the most probable value for the contour interval. A single contour is drawn at a value usually equal to the mode + 1.5 sigma. It is sometimes necessary when surveying large areas to change the contour value where soil types change. Considering these areas individually enables better overall interpretation of the data.

For more detailed information on the theory and application of statistics, Elkins (1940), Goulden (1952), and Sinclair (1976) may be considered.

USE OF CARBON-ISOTOPE RATIOS IN HYDROCARBON EXPLORATION

Carbon has two stable isotopes, ^{12}C and ^{13}C. The abundance in nature of ^{12}C and ^{13}C is 98.89% and 1.11%, respectively. Since the work of Craig (1953, 1957) and Silverman (1964), it is possible to make accurate measurements of carbon-isotope ratios to determine the source of different carbonaceous materials. With these advances, investigators can determine carbon-isotope ratios of hydrocarbons in the ranges of concentration found in near-surface sediments. This new information has led to the discernment of unusual isotopic compositions that result from interactions between petroleum, natural gas (mainly methane), carbon dioxide, and ground water.

One of these unusual compositions is an isotopically distinctive pore-filling carbonate cement that occurs in near-surface sediment and soils. This cement is formed principally by hydrocarbons that have leaked from underlying petroleum accumulations and have been oxidized in accordance with the equation:

$$CH_4 + 2O_2 \rightarrow 2H_2O + CO_2$$

The gas used in all $^{13}C/^{12}C$ ratio measurements is CO_2. The procedures for preparation of the CO_2 follow after McCrea (1950). Obtaining the CO_2 involves removal of any CO_2 initially present in a sample and conversion of the resulting carbonates to CO_2 by treatment with phosphoric acid. Carbon-isotopic ratios of the resultant CO_2 are determined by mass spectrometry and are reported in δ terminology as parts per thousand (0/00 = per mil) relative to a standard. The $^{13}C/^{12}C$ ratio is expressed by the equation:

$$\delta^{13}C = 1000 \left[\frac{C^{13}/C^{12} \text{ Sample}}{C^{13}/C^{12} \text{ Standard}} - 1 \right]$$

The most widely used standard is belemnite

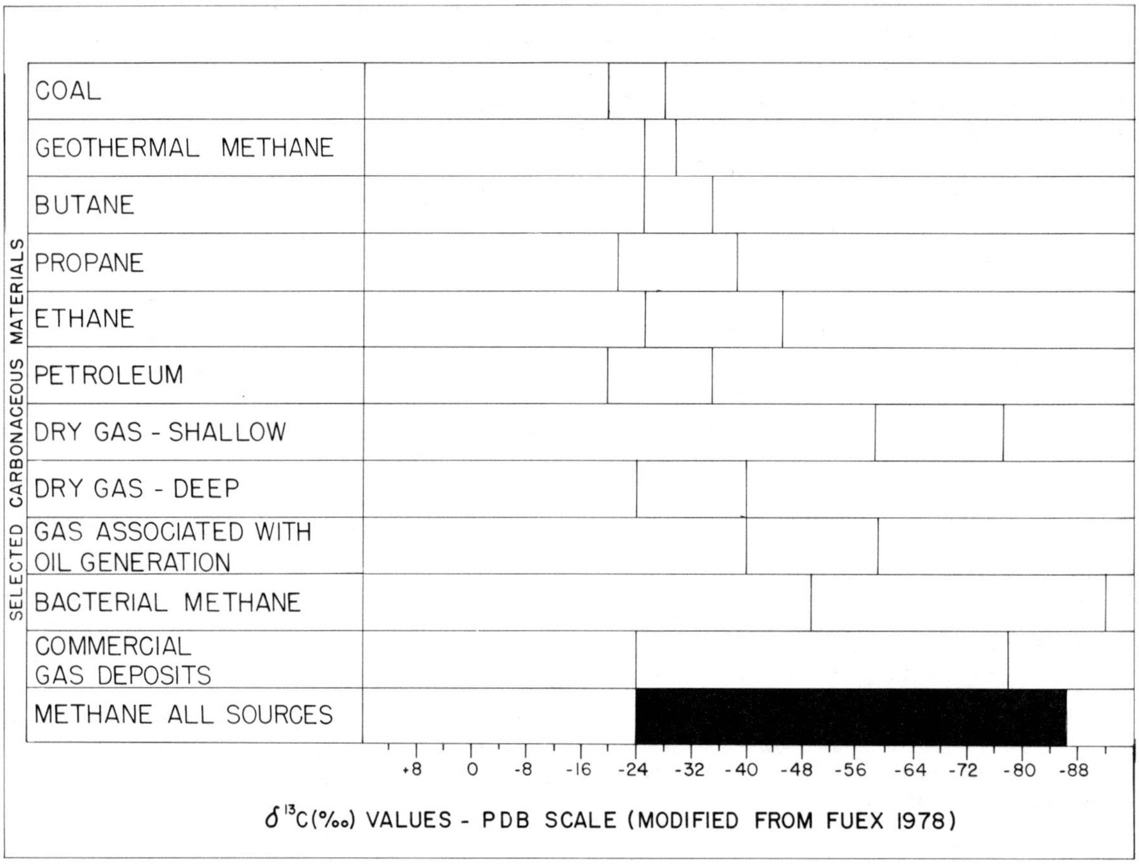

Fig. 5.

(PDB), a carbonate from the Peedee Formation (Upper Cretaceous) of South Carolina. Its isotopic composition, by definition, is $\delta^{13}C = O$.

When $\delta^{13}C$ values are negative, they indicate enrichment in ^{12}C relative to the PDB standard, and positive values indicate depletion in ^{12}C or enrichment in ^{13}C. Figure 5 presents a PDB scale modified from Fuex (1977), illustrating values and their range for selected carbonaceous materials. Additional background data for values may be obtained from Degens (1969), Hoefs (1973), and Milliman (1974).

FIELD OBSERVATIONS

Work by Donovan (Donovan, 1974; Donovan and Dalyiel, 1977; Donovan et al., 1974, 1975; Henry and Donovan, 1978) has furnished such excellent theory and application to actual producing fields that not much is left to conjecture.

Donovan noted the $\delta^{13}C$ values in carbonate cement become increasingly negative toward the crests of buried structural petroleum accumulations. He suggested that the distribution of $\delta^{13}C$ values in the carbonate cements of the surface rocks resulted from the chemical or biochemical oxidation of light hydrocarbons or their condensates. He also studied the oxygen-isotope ratios to determine the fractionation mechanism of the associated ground water that was controlled by the upward seeping hydrocarbons over buried petroleum accumulations. This paper is concerned only with the carbon-isotope ratios and their relationship to the ΔC analysis.

Donovan's work indicates that areally limited, interrelated cementation processes cause a halo around producing fields:

> The range, distribution pattern, and lightness of the carbon isotopic composition of calcite cements in the surface rocks is consistent with oxidation (biochemical) of hydrocarbon

and subsequent conversion of the resulting CO_2 to carbonate (after possible admixture with isotopically heavier primary carbonates). Surface halos are believed to develop above and peripheral to buried hydrocarbon deposits as a result of vertical bypass leakage at the edge-water/oil contact. The halo may be a reflection of chemical activity at the edge-water/oil contact of a buried hydrocarbon deposit. Alternatively, the surface anomaly may not be directly related to fresh-water circulation, but may instead reflect vertical bypass leakage and diffusion of low-molecular-weight hydrocarbons around the reservoir seal with subsequent surface oxidation. (Donovan et al., 1975)

Thus, seeping hydrocarbons are oxidized in the near-surface sediment and soil, and the combining of the resulting carbon dioxide with dissolved salts in the ground water will result in the precipitation of carbonate cement that gives characteristic carbon-isotope ratios. From these distinctive ratios, halos that reflect the subsurface distribution of petroleum can be interpreted around oil and gas fields.

The above publications by Donovan and others offer a more complete grasp of the implications of diagenetic carbon-isotope indicators that overlie oil and gas fields.

Field Example: Fox-Graham Field, Carter County, Oklahoma

With Donovan's approval, I have chosen one of his field studies, the Fox-Graham Field, Carter County, Oklahoma (Henry and Donovan, 1978) as my example to combine the carbon-isotope and the ΔC analysis methods in a geochemical exploration model. This field lies a few kilometers southeast of the larger Velma Field and northwest of the Wheeler Field (Figure 6)*. The Fox-Graham field was discovered as two separate fields; Fox in 1915 and Graham in 1917. Since the mid-1920s the two fields have been considered as one. The Fox structure is an irregular domal anticline of low relief separated from the elongate northwest-southeast trending Graham anticline by a deep-seated fault. The Fox-Graham field produces gas and oil from 742 producing wells from multiple zones in the Permian, Pennsylvanian, and pre-Pennsylvanian sections. API gravities of the oils increase with depth ranging from 19 API to more than 40 API (George and Bunn, 1924; Tomilson, 1952).

Figure 7 illustrates the isopleth map of the carbon-isotope values for carbonate cements from surface rocks overlying the Fox-Graham field and part of the Wheeler Field. Note the high negative or "light" $\delta^{13}C$ values overlying the apex of the structure. The values form a definite carbon-isotope anomaly over the field, and if this map had been used to locate the first wildcat well, it would have discovered the field.

The highly negative $\delta^{13}C$ values found over the petroleum accumulations at Fox-Graham and Wheeler are readily explainable in terms of the distribution of $\delta^{13}C$ values of petroleum. Petroleum is strongly enriched in ^{12}C with an average $\delta^{13}C$ value of - 28 0/00 relative to the PDB standard. Carbon in methane is, however, even more strongly enriched in ^{12}C than the petroleum with which it is usually associated. Although they may vary widely, $\delta^{13}C$ values as low as -40 to -70 0/00 have been reported for methane of some oil fields (Figure 5). Moreover, it has been demonstrated that the heavier hydrocarbon fractions concentrate ^{13}C relative to the lighter petroleum fraction, such that the $\delta^{13}C$ values increase (become less negative) as one proceeds upward in the series from C_1 through C_{15}. Therefore, the lightest and hence most volatile hydrocarbon fractions are the more strongly enriched in ^{12}C and will exhibit the highest δ ^{13}C anomaly. As the light hydrocarbons migrate into the soil, they may undergo chemical and physical reactions that incorporate some of the carbon into carbonate (e.g., calcite, aragonite, or ankerite); or they may be adsorbed onto the surface or absorbed into the lattice structure of clay minerals. Therefore, soil samples collected over a petroleum accumulation in which the associated methane is capable of migrating into the overlying soil horizons will through enrichment exhibit highly negative δ ^{13}C values.

At Fox-Graham, Geochemical Surveys collected soil samples using our regular grid to obtain ΔC values. The calculated value for the contour interval above normal background was 10 millivolts. Figure 8, contoured on the ΔC value, shows a geochemical halo around the

*See atlas for Figs. 6, 7, 8, 9.

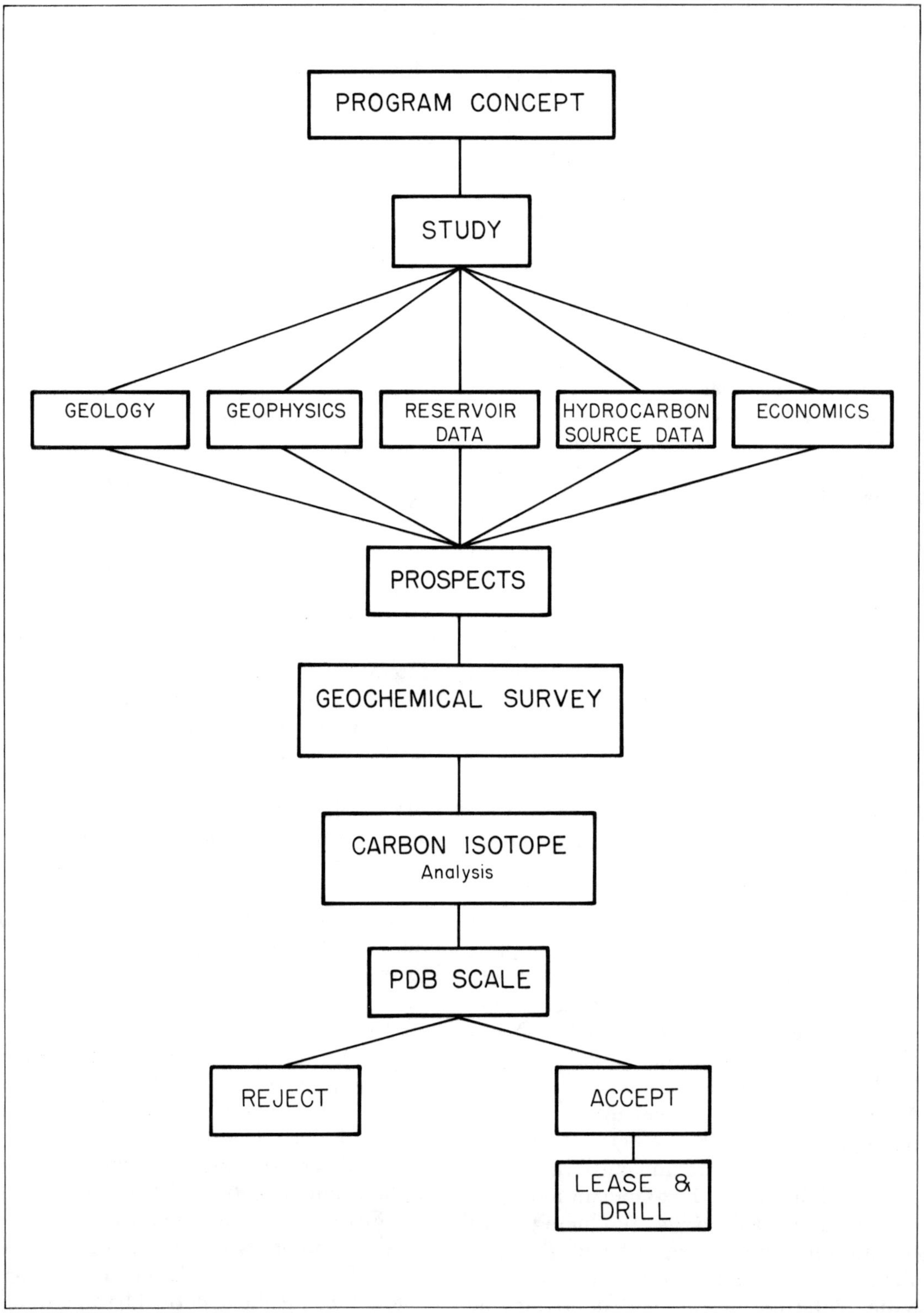

Fig. 10. The model.

Fox-Graham Field and indicates a halo developing around the Wheeler Field to the southeast.

Figure 7 shows ^{13}C increasing systematically relative to ^{12}C away from the center of both fields. In other words, the pore-filling carbonate cement over the anticline's crest is isotopically lighter with respect to carbon toward the anticlinal flanks. This phenomenon is interpreted to be the result of evaporation of formation waters carrying abundant hydrocarbons which are readily available for oxidation in the near-surface sediments and soils. This mechanism was first proposed by Mills and Wells (1919) and later expanded by Nisle (1941). Nisle used a thermodynamic approach and showed that the amount of aqueous vapor in an expanded volume of gas increases substantially at depths less than 300 m. He arrived at the conclusion that the gas-induced evaporation of ground water would result in the concentration of minerals at shallow depths. Donovan (1974) elaborates further and in greater detail on this premise. For our purposes, we conclude that $^{13}C/^{12}C$ values reflect the source of carbon, natural gas and/or its condensates, and the efficiency of its migration into the near-surface ground water system with subsequent oxidation.

Likewise, Figure 8 shows the ΔC values increasing where $\delta\,^{13}C$ is becoming heavier producing a well-defined ΔC halo around the main area of production. Away from the producing area, past the edges of the field, the low ΔC values are in the normal geochemical background.

Figure 9 verifies the coincidence of the two methods and illustrates that the two analyses agree exactly with the original premise.

A GEOCHEMICAL SURVEY MODEL

The following is presented to illustrate the exploration process used by Geochemical Surveys in its prospecting for hydrocarbons.

After an area or region of interest is selected for a soil-analysis program, data are gathered for study of the geologic, geophysical, organic geochemistry, potential reservoir information and economics. The study will point up interesting prospects and trends where a soil-sampling party will be employed to collect samples for analysis.

As soil samples are collected in the field and sent to the laboratory for analysis, ΔC values are obtained across the prospect area. Frequency curves are plotted and a contour value is chosen. As more ΔC data are obtained from approximately 1,000 samples per month, prospects begin to develop. As interesting geochemical anomalies appear and are substantiated by the geologic and geophysical information, selected samples are chosen for carbon-isotope ratio measurements. In choosing these samples, several dip and strike lines are selected across the ΔC halo. If the ΔC values are caused by leaking hydrocarbons from underlying petroleum accumulations, the carbon-isotope values, measured from the near-surface samples, could be very close to those of the values from the actual reservoir. These can be checked by referring to a PDB scale for $\delta^{13}C$ (Figure 5). This is a most important step in the procedure because the high negative $\delta^{13}C$ values in the near-surface soils over a geochemical anomaly are generally similar to that of the hydrocarbon phase of the subsurface petroleum accumulation which the anomaly reflects.

At this juncture of our survey, all available information is assessed and the individual prospects are either selected for leasing and drilling or rejected. The block diagram (Figure 10) summarizes this model.

SUMMARY

While geochemical prospecting for petroleum has been practiced for many years, it is underutilized in U.S. exploration.

Of the numerous methods, some have severe limitations. For instance, outright measurement of gases or entrained hydrocarbons in soil is not always effective because: (1) samples are difficult to handle without contamination or escape of the substance of interest; (2) their presence in the soil diminishes as reservoir pressure decreases; (3) these substances are measured in parts per billion, an order of magnitude difficult and expensive to handle; and (4) their source is often difficult to identify, as they may be generated by soil microorganisms.

Geochemical Surveys, since 1942, has developed and refined a method that overcomes these limitations because it measures a residual, stable, and cumulative effect of hydrocarbon migration, and this effect can be measured in parts per thousand, parts per hundred, or simply millivolts.

It is our opinion that the high carbon dioxide (ΔC) values in the geochemical halo result directly from the carbon-isotope ratios that exhibit themselves as low or less negative $\delta^{13}C$ values.

A prospect can be assessed prior to leasing and drilling by referring to a standard PDB scale (Figure 5) for $\delta^{13}C$. If these values fall in the range for hydrocarbons or are significantly lighter than normal soil carbonates, a drilling venture will almost certainly have a better chance of success.

The two methods, ΔC carbonate and carbon-isotope ratio, most certainly serve as a basis for geochemical prospecting for oil and gas accumulations. These two methods are an important addition to the geochemists' hydrocarbon exploration tools. In this regard, geochemical prospecting for hydrocarbons should offer to geophysicists, geochemists, and geologists common ground to interpret data properly and without prejudice in order to achieve optimum results.

ACKNOWLEDGMENTS

Personally, the writer wishes to express his indebtedness to Chal Pittman, for if this paper has any value, the credit belongs to him for having kept the spirit of geochemical exploration alive in me through his constant encouragement and his pursuit of truth.

Additional recognition must be given to Dr. Ian Kaplan and Louis Pandolfi of Global Geochemistry Corp., Los Angeles, for their support in its writing. Also, many helpful ideas were interchanged with Dr. Douglas Pride and his associates at Ohio State University.

Thanks are also due to Vic Ferguson, who drafted the figures. Last, but not least, much gratitude is expressed for Betty Heuer for her patience during the many hours of typing.

REFERENCES

Craig, H., 1953, The geochemistry of the stable carbon isotopes: Geochimica et Cosmochimica Acta, v. 3, p. 53-92.

———, 1957, Isotopic standards for carbon and oxygen and correction factors for mass spectrometric analyses of carbon dioxide: Geochimica et Cosmochimica Acta, v. 12, p. 133-149.

Degens, E. T., 1969, Biochemistry of stable carbon isotopes, *in* Eglinton, G., and Murphy, M. T. J., eds., Organic Geochemistry: New York, Springer-Verlag, p. 304-329.

Donovan, T. J., 1974, Petroleum microseepage at Cement, Oklahoma: Evidence and mechanism: American Association of Petroleum Geologists Bulletin, v. 58, no. 3, p. 429-446.

———, and Dalziel, M. C., 1977, Late diagenetic indicators of buried oil and gas: U.S. Geological Survey Open-File Report 77-817, 38 p.

———, Friedman, J., and Gleason, J. D., 1974, Recognition of petroleum-bearing traps by unusual isotopic compositions of carbonate-cemented surface rocks: Geology, v. 2, p. 351-354.

———, Noble, R. L., Friedman, J., and Gleason, J. D., 1975, A possible petroleum related geochemical anomaly in surface rocks, Boulder and Weld Counties, Colorado: U.S. Geological Survey Open-File Report 75-47, 10 p.

Elkins, T. A., 1940, The reliability of geophysical anomalies on the basis of probability considerations: Geophysics, v. 5, no. 4, p. 321-336.

Fedynsky, V. V., Karus, E. V., and Polshko, M. K., 1975, Geophysical and geochemical surface surveys for detection of oil and gas pools: Essex, England, Applied Science Publishers, Ltd., World Petroleum Congress, 9th, Proceedings, v. 3, p. 279-288.

Fuex, A. N., 1977, The use of stable carbon isotopes in hydrocarbon exploration: Journal of Geochemical Exploration, v. 7, p. 155-188.

George, H. C., and Bunn, J. R., 1924, Petroleum engineering in the Fox and Graham oil and gas fields, Carter County, Oklahoma: U.S. Bureau of Mines, 81 p.

Goulden, C. H., 1952, Methods of statistical analysis: New York, John Wiley.

Henry, M. E., and Donovan, T. J., 1978, Isotopic and chemical data from carbonate cements in surface rocks over and near four Oklahoma oil fields: U.S. Geological Survey Open-File Report 78-927, 17 p.

Hoefs, J., 1973, Stable isotope geochemistry, New York, Springer-Verlag.

Horvitz, L., 1939, On geochemical prospecting: Geophysics, v. 4, p. 210-228.

———, 1945, Recent developments in geochemical prospecting for petroleum: Geophysics, v. 10, p. 487-493.

———, 1954, Near-surface hydrocarbons and petroleum accumulation at depth: Mining Engineering, v. 6, p. 1205-1209.

———, 1959, Geochemical prospecting for petroleum, *in* Symposium on geochemical exploration: International Geological Congress, 20th, Mexico City, v. 2, p. 303-319.

———, 1969, Hydrocarbon geochemical prospecting after

thirty years, *in* Unconventional methods in exploration for petroleum and natural gas: Dallas, Southern Methodist University, p. 205-218.

———, 1978, Near-surface evidence of hydrocarbon movement from depth: American Association of Petroleum Geologists Bulletin, v. 62, no. 3, p. 524.

Hudson, J. D., 1977, Stable isotopes and limestone lithification: Journal of the Geological Society of London, v. 133, p. 637-660.

Johnson, A. C., 1970, Inorganic geochemical prospecting for oil and gas accumulation, *in* Kentucky Oil and Gas Association Meeting, 34th and 35th, Proceedings of the technical sessions, Special Publication 21, p. 62-66.

Laubmeyer, G., 1933, A new geophysical prospecting method, especially for deposits of hydrocarbons: Petroleum, v. 29, no. 18, p. 1-4.

Laurence, J., Kent, P., and Kerr, P. F., 1951, Thermal study of the Ca-Mg-Fe carbonate minerals: American Mineralogist, v. 36, nos. 9, 10, p. 643-670.

Levorsen, A. I., 1954, Geology of petroleum: San Francisco, W. H. Freeman & Co., 724 p.

Mac Elwain, R. C., 1963, What do near-surface signs really mean in oil finding? Oil and Gas Journal, Feb. 1963, p. 132-146.

McCrea, J. M., 1950, On the isotopic chemistry of carbonates and a paleotemperature scale: Journal of Chemical Physics, v. 18, p. 849.

McDermott, E., 1940, Geochemical exploration (soil analysis): American Association of Petroleum Geologists Bulletin, v. 24, p. 859-81.

Milliman, J. D., 1974, Marine carbonates: New York, Springer-Verlag, 375 p.

Murata, K. J., Friedman, I., and Madsen, B. M., 1969, Isotopic composition of diagenetic carbonates in marine miocene formations of California and Oregon: U.S. Geological Survey Professional Paper, 614-B, 24 p.

Nisle, R. G., 1941, Considerations on the vertical migration of gases: Geophysics, v. 6, no. 4, p. 449-454.

Overton, H. L., 1978, Appraisal of water movement from subsurface data: Log Analyst, v. 19, no. 5, p. 3-19.

Pandey, G. N., Tek, M. R., and Katy, D. L., 1974, Diffusion of fluids through porous media with implications in petroleum geology: American Association of Petroleum Geologists Bulletin, Feb. 1974, p. 291-303.

Ransone, W. R., 1947, Geochemical history of the Hardy Oil Field, Jones County, Texas: Geophysics, v. 12, no. 3, p. 384-392.

———, 1958, Case history of the Sojourner Oil Field, Haskell County, Texas, a geochemical confirmation: Geophysics, v. 23, no. 3, p. 574-578.

———, 1968, Case history of the Sixto Oil Field, Starr County, Texas, a geochemical discovery, *in* Unconventional methods in exploration for petroleum and natural gas: Dallas, Southern Methodist University, p. 187-196.

Roof, J. G., and Rutherford, W. M., 1958, Rate of migration of petroleum by proposed mechanisms: American Association of Petroleum Geologists Bulletin, v. 42, no. 5, p. 963-980.

Rosaire, E. E., 1938, Shallow stratigraphic variations over Gulf Coast structures: Geophysics, v. 3, p. 96-115.

———, 1939, Discussions and communications: Geophysics, v. 4, p. 300-305.

———, 1940, Geochemical prospecting for petroleum: American Association of Petroleum Geologists Bulletin, v. 24, no. 8, p. 1400-1433.

Rowland, R. A., and Jonas, E. C., 1949, Variations in differential thermal analysis curves of siderite: American Mineralogist, v. 34, July-Aug. 1949, p. 550.

Rowland, R. A., and Lewis, D. R., 1951, Furnace atmosphere control in differential thermal analysis: American Mineralogist, v. 35, p. 80.

Silverman, S. R., 1964, Investigations of petroleum origin and evolution mechanisms by carbon isotope studies, *in* Isotopic and cosmic chemistry: Amsterdam, North-Holland Publishing Co., p. 92-102.

Sinclair, A. J., 1976, Applications of probability graphs in mineral exploration: Association of Exploration Geochemists, Special v. no. 4, 95 p.

Sokolov, V. A., 1933[a], The gas survey as a method of prospecting for oil and gas formations: Informatsionnyi sbomik Neftyanogo Gedogo-Razvedochnogo Instituta (NGRI) Union of Scientific Publishing Houses (ONTI).

———, 1933[b], New prospecting methods for petroleum and gas: Technika, Neftyanogo Geologo-Razvedochnogo Instituta no. 1.

Soli, G. G., 1954, Geomicrobiological prospecting: American Association of Petroleum Geologists Bulletin, v. 38, p. 2555-2558.

———, 1957, Microorganisms and geochemical methods of oil prospecting: American Association of Petroleum Geologists Bulletin, v. 41, no. 1, p. 134-145.

Stegna, L., 1961, On the principles of geochemical oil prospecting: Geophysics, v. 26, no. 4, p. 447-451.

Tissot, B. D., and Welte, D. H., 1978, Petroleum formation and occurrences: New York, Springer-Verlag, 538 p.

Tomilson, C. W., 1952, Odd geological structures of southern Oklahoma: American Association of Petroleum Geologists Bulletin, v. 36, no. 9, p. 1820-1840.

SUGGESTED READING

Allan, J. R., and Matthews, R. K., 1977, Carbon and oxygen isotopes as diagenetic and stratigraphic tools—surface and subsurface data, West Indies: Geology, v. 5, no. 1, p. 16-21.

Burst, J. F., 1969, Diagenesis of Gulf Coast clayey sediments and its possible relation to petroleum migration: American Association of Petroleum Geologists Bulletin, v. 53, no. 1, p. 73-93.

Cheney, E. S., 1964, Stable isotopic geology of the Gas Hills Uranium District, Wyoming: Yale University, Ph.D. dissertation, p. 342.

Coplen, T. B., and Hanshaw, B. B., 1973, Ultrafiltration by a compacted clay membrane; oxygen and hydrogen isotopic fractionation: Geochimica et Cosmochimica Acta, v. 37, p. 2295-2310.

Graf, D. L., 1960, Geochemistry of carbonate sediments and sedimentary carbonate rocks; pt. 4-A, isotopic composition—chemical analysis: Illinois Geological Survey Division Circular 308, p. 42.

Hathaway, J. C., and Degen, E. T., 1968, Methane-derived marine carbonates of Pleistocene age: Science, v. 165, p. 690-692.

Hedberg, H. D., 1936, Gravitational compaction of clays and shales: American Journal of Science, v. 231, p. 241-287.

Hsu, K. J., and Siegenthaler, C., 1971, Preliminary experiments on hydrology of supratidal dolomitization and cementation, *in* O. P. Bricker, ed., Carbonate cements: Johns Hopkins University Studies in Geology, no. 19, p. 315-318.

Kartsev, A. A., Tabosaramskii, Z. A., et al., 1959, *in* Geochemical methods of prospecting and exploration for petroleum and natural gas (English trans.), P. A. Witherspoon and W. D. Romey, eds.: Berkeley, University of California Press, p. 347.

McAuliffe, C., 1963, Solubility in water of C_1-C_9 hydrocarbons: Nature, v. 200, p. 1092-1093.

McCulloh, T. H., 1967, Mass properties of sedimentary rocks and gravimetric effects of petroleum and natural gas reservoirs: U.S. Geological Survey Professional Paper 528-A, p. 50.

Pirson, S. J., Pirson, J. E., and Alsinbili, M. B., 1969, Exploration by surface electrical measurements: Oil and Gas Journal, v. 67, p. 108-111.

Sackett, W. M., 1964, The depositional history and isotopic carbon composition of marine sediments: Marine Geology, v. 2, p. 173-185.

Silverman, S. R., and Epstein, S., 1958, Carbon isotopic composition of petroleum and other sedimentary organic materials: American Association of Petroleum Geologists Bulletin, v. 42, p. 998-1012.

Smith, J. E., Erdman, J. G., and Morris, D. A., 1971, Migration, accumulation, and retention of petroleum in the earth: World Petroleum Congress, 8th, Proceedings, no. 8, v. 2, p. 13-26.

Trask, P. D., 1932, Origin and environment of source sediments of petroleum: Houston, American Petroleum Institute, Gulf Publishing Co., p. 323.

Verger, J., and Hoefs, J., 1976, The nature of O^{18}/O^{16} and C^{13}/C^{12} secular trends in sedimentary carbonate rocks: Geochimica et Cosmochimica Acta, v. 40, p. 1387-1395.

Zartman, R. E., Wasserburg, G. J., and Reynolds, J. H., 1961, Helium, argon, and carbon in some natural gases: Journal of Geophysical Research, v. 66, no. 1, p. 277-306.

AIRBORNE GEOCHEMICAL RECONNAISSANCE METHODS FOR OIL AND GAS EXPLORATION

A. R. Barringer

BARRINGER RESEARCH, INC.

INTRODUCTION

The development of airborne exploration methods has had a major impact in the field of mining exploration and has played a key role in the discovery of dozens of mineral deposits having a combined value of many billions of dollars (Paterson, 1973). The success of airborne methods in mining and their low cost-per-unit area inevitably leads to the question as to whether any comparable techniques could be developed for oil exploration. Effective airborne reconnaissance techniques that could overfly ground where access is difficult would obviously be very attractive. Their usefulness would be even greater if they could be employed offshore. The methods outlined in this paper are an attempt to meet these requirements and are based on the use of geochemical phenomena relating both to surface microlayer effects and to compaction and microseepage in sedimentary basins.

It is important to stress at the outset that the methods described in this paper are certainly not stand-alone techniques, and it is necessary to use them in conjunction with surface follow-up procedures using appropriate complementary geological, geochemical, and geophysical methods.

TERRESTRIAL SURFACE MICROLAYER GEOCHEMISTRY

The two airborne geochemical methods that will be described rely upon the geochemical characteristics of the surface microlayer that covers the entire surface of the earth, including soils, vegetation, and the oceans. Over land areas, this surface microlayer is the air/earth interface at which evaporative processes tend to deposit saline residues from upward migrating moisture. A portion of this evaporation relates to the normal cycle of rainfall and evaporation that carries salts to the surface of soils by capillary rise and to the surface of vegetation by evapo-transpiration. An important additional phenomenon, however, that affects the surface microlayer in sedimentary basins is that of the thermal dewatering of sediments caused by deep burial at elevated temperatures. A high proportion of the pore water trapped within marine sediments is gravitationally expelled during the first kilometer of burial and compaction (Athy, 1930; Hedberg, 1926, 1936). Thereafter, the process of dewatering is partly related to the expulsion of water contained within hydrated minerals and the interlayer structure of expanded clays, such as montmorillonite.

This process is a very gradual one associated with increasing descent within the geothermal gradient of the earth's crust. Hydrated minerals, such as gypsum associated with evaporites, lose their water of hydration at temperatures in the vicinity of 40°C-60°C corresponding to burial depths between 1 and 2 km (Heard and Ruby, 1966), whereas the loss of interlayer water from the expanded clays only begins at depths in the vicinity of 2 km where temperatures reach 50°C. Thereafter it continues on to the point where the sediments are buried 5 km or more (Burst, 1959, 1969, 1976; Weaver, 1959; Dunoyer de Segonzac, 1969, 1970).

An additional important factor affecting the expulsion of water from sediments undergoing progressive burial within the geothermal gradient is that of expansion caused by heating of trapped water. Migration related to this cause will always be present even in the absence of expanded clay minerals (Barker, 1972; Magara, 1974).

The net result of gravitational and thermal dewatering is that very substantial quantities of water are released, amounting, for example, to a total of about 1,000 to 2,000 tons of water per sq m of surface area in a sedimentary basin of 10

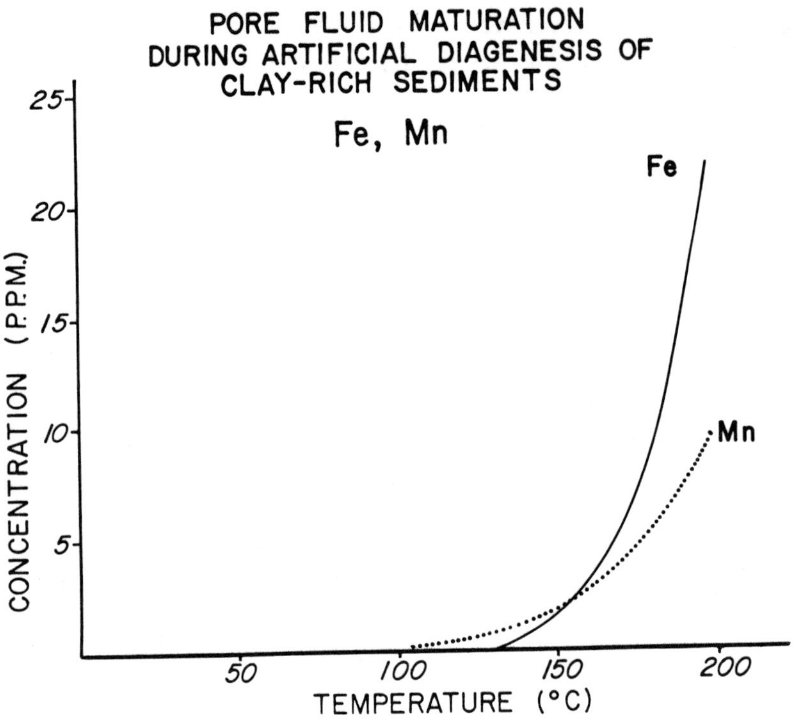
PORE FLUID MATURATION
DURING ARTIFICIAL DIAGENESIS OF
CLAY-RICH SEDIMENTS
Fe, Mn
Fe
Mn
CONCENTRATION (P.P.M.)
25
20
15
10
5
50
100
150
200
TEMPERATURE (°C)

Fig. 1A.

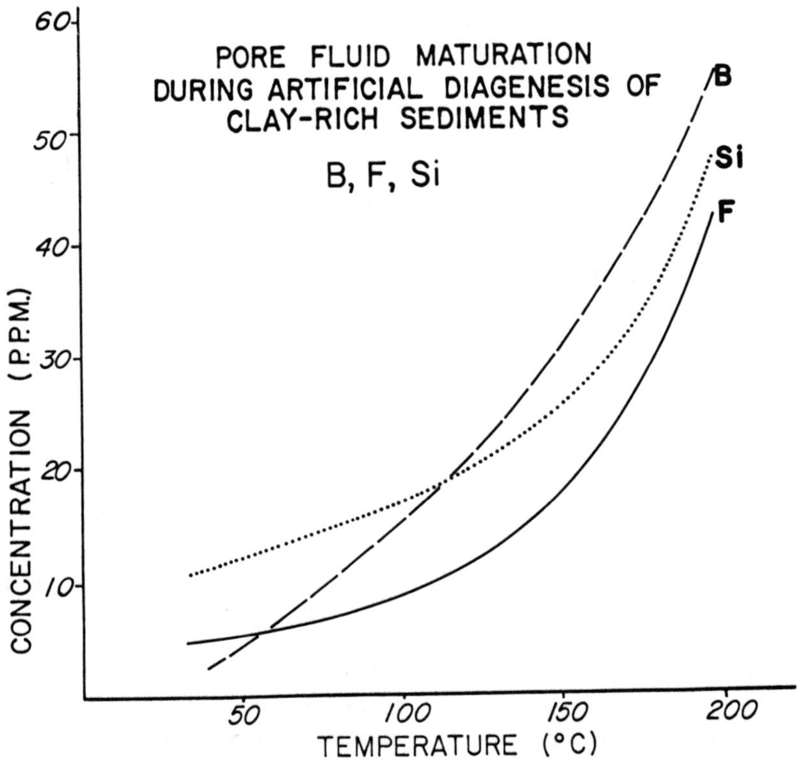
PORE FLUID MATURATION
DURING ARTIFICIAL DIAGENESIS OF
CLAY-RICH SEDIMENTS
B, F, Si
B
Si
F
CONCENTRATION (P.P.M.)
60
50
40
30
20
10
50
100
150
200
TEMPERATURE (°C)

Fig. 1B.

km depth. This large scale extrusion of water from sedimentary basins is distributed in time throughout the sedimentation and compaction history of the basin.

The process of thermal dehydration of clay minerals is accompanied by the release of large amounts of cations and anions which contribute to the composition of the formation waters and provide them with saline concentrations which normally greatly exceed those of the connate seawater originally trapped within the sediments. Experimental work carried out in high temperature, high pressure vessels has graphically demonstrated this progressive release of cations and anions from natural marine sediments as the temperatures are elevated through a range corresponding to normal burial within the geothermal gradient (Figures 1a and 1b). (Personal communication, work in progress—Geology Department, Imperial College of Science and Technology, London, 1977-1979).

The dewatering of sedimentary basins is both gradual and continuous in a general sense, but it is probable that superimposed on this trend there will be minor pulsations related to extensive hydrofracturing associated with the damming up of upward migrating fluids beneath impermeable horizons (Fyfe, Price, and Thompson, 1978).

This phenomenon apparently results in periodic failure and fracturing of such impermeable horizons due to the differential pressures that build up across these barriers. Supporting field evidence, such as vertical and near-vertical jointing and fracturing carrying carbonate in-filling, is very abundant in sedimentary basins where good rock exposures exist.

In addition to these phenomena of damming and hydrofracturing there is also the development of preferred migration pathways associated with nonhydraulic faulting, fracturing, folding, etc. Evidence of this localized upward migration of formation waters is provided in the extensive subsurface temperature studies that have been published (Artemenko and Malovitskiy, 1978; Melik-Pashayev, Gattenberger, and Lutkov, 1973). Thus migration of fluids toward the surface is accompanied by an upward transfer of heat and the development of thermal anomalies that are seen as isothermal highs overlying the zones of accelerated upward migration. This subject is fully discussed elsewhere in this volume (Roberts, W. H. III).

The end result, geochemically speaking, of the upward migration of saline formation brines is the development of complex multielement distribution patterns at the surface. These are not geologically consistent with the outcrop of formations and facies but are much more suggestive of the preferred upward extrusion of brines along faults, fractures, jointed zones, and other leakage pathways. This conclusion tends to be supported by a study of three-band color composite imagery from the LANDSAT series of earth resource satellites which frequently show strong tonal and albedo patterns occurring in the soils and vegetation of sedimentary basins. Many of these features are considered to be related to the influence of upward migrating formation waters that modify the albedo and color of soils, change the species distribution of plants, and alter their spectral reflection and emissivity characteristics.

In semiarid sedimentary basins, the surface soils represent a convenient zone for studying multi-element distribution patterns, and they provide information that is a clue to the migration pathways within the basin. Similarly, in vegetated regions the biogeochemical composition of plants is related to the available elements which the plant root systems can extract from near-surface soils. In general, the soil constituents derived from upward migrating brines are readily available to plants, and it appears that the biogeochemistry of vegetation in sedimentary basins is therefore quite responsive to these migration phenomena.

The surface microlayer of both soils and vegetation is a zone that is particularly sensitive geochemically because of the previously mentioned evaporative and saline concentration processes that occur at these surfaces. The sampling of these surface microlayers can be carried out over both vegetation and soils, with the use of suitable equipment, and it is possible to carry out multielement analyses of materials collected, even when these are in milligram or microgram amounts.

The significance of multielement maps

which can be interpreted in terms of identification of migration pathways is that many of these pathways are related to the routes along which oil and gas have also migrated. It is not necessary to assume a migration theory for hydrocarbons that calls upon an aqueous carrier, since even when relating to an oil phase hydrocarbon migration model (Dickey, 1975), the migration pathways for hydrocarbons that pass through fractures, faulting, and porosity will inevitably show a great deal of commonality with any earlier or later aqueous migration. In other words, the channeling of aqueous formation fluids will tend to be along the same or similar pathways to those followed by hydrocarbons, and these routes will lead to the same stratigraphic and structural traps which are likely to be the sites of present-day oil and gas fields. Similarly, the ongoing upward leakage of brines above these traps will tend to be concentrated in a zone overlying the traps, since these are the focal point of accumulation for both migrating brines and hydrocarbons. This is evidenced by the frequently observed relationship between hydrocarbons at depth and the surface distribution of multi-element geochemical anomalies. There remain, however, problems in interpreting these anomalies, such as their frequent displacement away from apical positions above the hydrocarbon accumulations that is possibly related to groundwater movement, and the development of multielement anomalies which have no relationship to underlying oil or gas. It seems that the upward expulsion of formation waters will occur whether or not any hydrocarbon source rocks are present, so that the interpretation of inorganic geochemical patterns on their own and without the backup of complementary data has very distinct limitations. The methods of multielement onshore reconnaissance geochemistry described in this paper are therefore not considered as stand-alone techniques and must be integrated with other methods before they can become fully effective.

MARINE SURFACE MICROLAYER GEOCHEMISTRY

The above discussion has been concerned with the principles of geochemical migration in sedimentary basins and the formation of surface microlayer concentrations on terrestrial soils and vegetation. In the marine environment, upward migration from deep within the sediments will proceed much as on land; however, in the sediments at shallow depths beneath the sea floor there will be a marked difference from the land situation because of the absence of horizontal movements of ground water. It can be expected that lateral displacements of geochemical anomalies will be reduced in the sea floor as compared with near-surface onshore geochemistry (Hitchon, 1974).

Hydrocarbons released from microseepages on the sea floor will be partially dissolved but will also modify the geochemical characteristics of the surface microlayer of the ocean overlying seep areas. (The mechanisms contributing to this upward migration to the surface are discussed later in the section on Offshore Applications of the AIRTRACE System.) The surface microlayer is a discrete zone which may be only a molecular monolayer in thickness that is normally formed from the biological degradation products of plankton. It is composed of organic materials such as fatty acids, alcohols, and lipids, it is often enriched in trace elements, and it occurs in wide areas of the ocean where marine life flourishes (Garrett, 1967; MacIntyre, 1974; Duce et al., 1972). The quite complex chemistry of the marine surface microlayer is locally modified by the presence of hydrocarbon microseep materials that have migrated to the surface, and also, in a secondary way, by the presence of dissolved hydrocarbons in the water.

A further discussion on the effects of the surface microlayer on ambient aerosols over the ocean is provided in the AIRTRACE section of this paper.

AIRTRACE ® AN EXPLORATION SYSTEM FOR SAMPLING THE SURFACE MICROLAYER VIA THE COLLECTION OF AIRBORNE ATMOSPHERIC PARTICULATES

The earth's lower atmosphere is characterized by a universal dispersion of suspended particulate material. Man-made pollution is a contributing source of material around cities and industrial areas, but in nonurban onshore environments the giant-sized particulates (plus 10 micrometers) in the atmosphere are predominantly derived from the surface layer of soils and vegetation (Hilst and Nickola, 1959). In general the coarser and heavier the particles, the more local they are in origin. Under good atmospheric mixing conditions, there is a continual flux of particles being lifted off the ground as well as a steady sedimentation and deposition of particles back onto the ground. Particles tend to be lifted by thermal plumes which constantly rise off the surface in the presence of solar radiant heating and achieve vertical velocities of several meters per second (Taylor, 1958; Kaimal and Bussinger, 1970; Bussinger, 1972). These plumes are one of the principal sources of the familiar atmospheric turbulence that affects light aircraft flying at low altitude.

The AIRTRACE equipment (Barringer, 1973a, 1973b, 1975, 1976a, 1976b, 1976c, 1976d, 1977a) collects atmospheric particulates using either cyclones or louvered plate inertial devices. Thermal and particle flow rate sensors identify rising plumes, and a fast switching technique separates the sample stream derived from these thermal plumes from particulates suspended in stagnant air (Barringer, 1976c, 1977). This provides an effective technique for selectively sampling the underlying terrain and minimizing the dilution of anomaly contrast by particles that have drifted from nearby or distant regions. An early proponent of the collection and analysis of atmospheric particles for prospecting was Weiss, who variously proposed the collection of particles on filter paper, on nylon fishing line wound on frames dragged behind the aircraft, and on webs of window-screen-like material (Weiss, 1967, 1969, 1971). None of these methods, however, is selective of particles in rising air plumes.

The AIRTRACE system has been mounted in both fixed wing aircraft and helicopters, and is operated at survey altitudes that are normally less than 100 m above the ground or vegetation surface. This low terrain clearance is used in order to optimize the collection of local particles in thermal plumes and minimize the effects of wind displacement.

The system has been adapted for both mineral and oil exploration, but hydrocarbon targets are more easily detected by the AIRTRACE system because of the much larger aerial dimensions of their anomalies as compared with the geochemical anomalies that occur over most mineral deposits. Since the lifting of particles into the atmosphere is closely related to the development of thermal plumes, as discussed above, the upward flux of particles is not a continuous movement but rather is a discontinuous and somewhat erratic process. Typical plume intervals in good mixing conditions are every 150 to 300 m, and the statistical chances of encountering a satisfactory number of rising thermal plumes across an oil field anomaly which is likely to be at least 1 km wide are significantly better than those of finding a mineral deposit anomaly that is typically only 100 to 300 m wide.

The particle collectors and separators used in the AIRTRACE system can be mounted on each side of a helicopter, as shown in Plate 1 and Plate 2, or on a fixed-wing aircraft, as shown in Plate 2. The particulate sample stream is split, and one portion of the sample stream is analyzed in flight for volatile hydrocarbon content with a continuous output provided to a recorder. The remaining portion of the sample stream is stored by impacting the particles on adhesive tape, using a specially designed tape deck that advances the tape every few seconds. Particle counters monitor the amount of material coming in, and the tape is advanced automatically when a preset threshold particle count has been reached. The thermal plume sensor technique described above is used to switch the incoming particles from rising plumes onto "hot spots" on adhesive tape, while the remaining particles that are extracted from stagnant background air are stored on "cold spots" on the tape. As mentioned, this concept is vital in achieving localization of collection and relating the particles to the underlying terrain. A typical tape deck assembly is shown in Plate 3, and the appearance of the samples collected on a transparent adhesive tape from both rising and stagnant air is shown in Plate 4.

When operating overland the system func-

Plate 1. AIRTRACE® and Magnetometer equipment on survey in a Bell Jet Ranger.

Plate 2. Helicopter and fixed wing AIRTRACE® installations.

Plate 3. Adhesive tape deck assembly for collection of particulate samples.

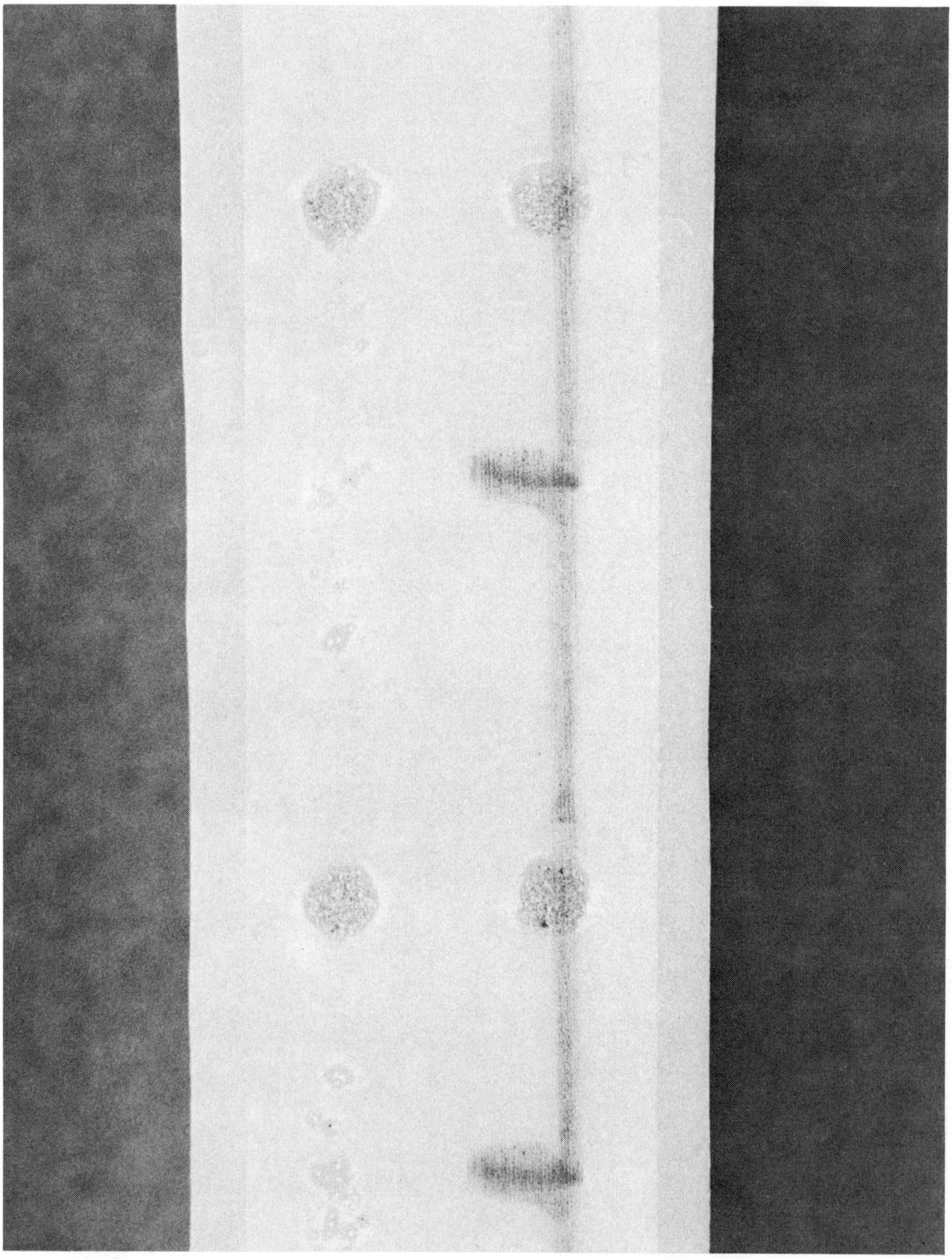

Plate 4. AIRTRACE® adhesive tape. Upper spots are impacted particulates from rising plumes (hot spots). Lower spots are particulates from stagnant air (cold spots).

tions best under sunny conditions where there is a slight wind. Radiant heating of the surface generates strong mixing activity, and such mixing is further accentuated by light winds. Strong winds can cause excessive smearing of the anomalies and create difficult flying conditions. Helicopter pilots who are skilled in low flying (e.g., crop dusting pilots) find no difficulty, however, in carrying out many hours of survey during the course of a day at heights that vary between 10 and 100 m above the ground surface or vegetation canopy, the clearance depending upon topography. Surveys have been carried out in extremely rough mountainous terrain, where the mixing is often very good, allowing somewhat increased terrain clearance. Survey production rates vary between 100 and 500 line km per day, depending on weather conditions and the ferry distances involved. In the case of helicopters, fueling is normally carried out within the survey area using fuel trucks or fuel caches.

An analytical system of unique design has been developed especially for handling the adhesive tape samples collected during flight. This system, known as LASERTRACE(TM), is able to analyze one thousand samples per day for twenty-five elements and is described under a separate heading below. An airborne organic analytical system has also been developed that is capable of providing high sensitivity analyses during flights of volatile hydrocarbons absorbed in the atmospheric particulates.

Survey experience with the AIRTRACE system for hydrocarbon exploration applications has so far been limited, but it appears to have a good future. Tests carried out over the Mereenie and Palm Valley fields in the Amadeus Basin of Central Australia yielded very strong hydrocarbon anomalies over the fields and other associated structures. The results from a portion of the survey over the Mereenie field are shown in Figure 2*. Analytical data are superimposed over a LANDSAT interpretation of the structures and lineaments. All thermal particulate plumes encountered along the flight lines are recorded, and the concentration of volatile hydrocarbons in the particulates within the plumes is represented by circles whose areas are proportional to concentration.

*See atlas for Figs. 2, 3, 4.

It will be noted that highly anomalous concentrations occur over the Mereenie anticline, which is reported to contain 300 million barrels of oil and 500 million cu ft of gas (Benbow, 1968). Lesser anomalies occur over portions of the Johnny Creek Anticline that are some miles in distance from locations that have yielded dry holes.

In general, the correlation between the AIRTRACE anomalies, LANDSAT features, and known hydrocarbons appears to be very good.

The Mereenie field had not been placed in production at the time of the survey because of its distance from markets, and contamination possibilities were therefore minimal since all the wells had been capped.

Multielement AIRTRACE traverses across the Cement Field in Oklahoma (Barringer, 1977) also yielded a pattern which correlated well with the position of the field and with the results published by Donovan on anomalous manganese and C_{12}/C_{13} ratios in surface carbonate cement materials over the field (Donovan, 1974, 1978). More recently an AIRTRACE installation made in a USGS Skymaster aircraft has yielded positive correlative data in manganese and iron over the Recluse Field in the Powder River Basin of Wyoming, and this survey is described in a paper by Donovan and Dalziel given at this symposium and published in this volume. They have shown that the inorganic AIRTRACE anomalies correlate more closely with the geochemistry of the vegetation than they do with the soils, and that the atmospheric particulate anomaly is therefore probably mainly attributable to the vegetation.

Very extensive engineering and development has been carried out on the AIRTRACE system, with major support from the Anglo-American Corporation on the mineral version of the system. Experience has shown that at the present stage of development, the system can work reliably in mining exploration only when the targets are large and there is strong mixing in the lower atmosphere. The uncertainty involved in detecting small mineral anomalies with the present equipment is a constraint that does not apply, however, to hydrocarbons, particularly in view of recent advances, and it now seems that the full and immediate potential of the system will best be realized in its application to the search for oil and gas. Furthermore, not only do the onshore possibilities seem good, but there is also a definite potential for carrying out large-scale off-shore reconnaissance programs, as discussed below.

OFFSHORE APPLICATIONS OF THE AIRTRACE SYSTEM

The basis for the use of AIRTRACE in the exploration for offshore hydrocarbons is the fact that microseepages of oil on the sea floor are generally accompanied by the evolution of gas, and the gas bubbles act as a very effective carrier for the rapid transport of traces of oil to the surface (Strickland, 1979; Warnstedt, 1972). Furthermore, there is a natural migration of the immiscible oil toward the surface as a result of its low density and hence buoyancy. Thus the ocean surface over an oil seep tends to develop a surface microlayer rich in hydrocarbons which is displaced laterally by an amount that is dependent upon the depth of water, the velocity of ascent, and the current flow. Even a gas seep over a gas producing region can carry enough heavier hydrocarbons in solution to create a surface microlayer of the above type.

Aerosols are generated above the ocean surface that are enriched in the materials contained in the surface microlayer by the well-known mechanism of bursting gas bubbles (Blanchard and Woodcock, 1957; Blanchard, 1964; Baier, 1972). Columns of gas bubbles generated over a hydrocarbon seep will be very active in generating aerosols at the air/sea interface if the flow of gas reaches the surface. This is a function of water depth, temperature, and gas saturation (Guinasso and Schink, 1975). The fact that such bubble streams do frequently reach the surface is confirmed by the results of bottom profiling and side scan sonar surveys that can image gas seeps (Geyer and Sweet, 1973). It has been shown that rising gas bubbles are particularly effective in scavenging emulsions, particulate materials such as oil droplets, dispersions of bacteria, and surface microlayers, and subsequently injecting them into the atmosphere (Quinn et al., 1975; Strickland, 1979; Warnstedt, 1972; Blanchard and Syzdek, 1972; Hoffman and Duce, 1976, 1977; Simoneit, Chester, and Eglington, 1977).

Thus anomalous accumulations of hydrocarbon rich aerosols can be expected to occur over active gas seeps. It should be noted, however, that even in regions where there are not active gas seeps there appears to be a continual migration of air bubbles toward the surface, at least some of which is associated with air impounded by breaking waves. Thus, the entire ocean surface is a significant generator of sea salt aerosols, and in regions of biological activity these aerosols are generally highly enriched in the natural organic materials produced at the ocean surface by the degradation of plankton.

In locations where there are oil and gas seeps but where the gas seepage is insufficient for bubbles to reach the surface, the upward migration of buoyant microseepage oils will nevertheless create high backgrounds of volatile hydrocarbons both in the sea and its surface microlayer and in the aerosols generated in the overlying atmosphere. We can expect, therefore, to see a general rise in background hydrocarbon levels in aerosols overlying oil seep regions with the superimposition of pockets of aerosols highly enriched in hydrocarbons in the close vicinity of active seeps from which hydrocarbon gas bubbles reach the surface. This pattern has in fact been observed in test surveys.

Survey operations with the AIRTRACE system offshore utilize fixed-wing aircraft operating at low survey altitudes such as 30 m above the water surface. The correct meteorological conditions in terms of relative sea and air temperatures, wind conditions, etc. are a prerequisite in order to insure that suitable mixing activity exists in the lower atmosphere. Adequate forecasting is therefore necessary for the planning and control of surveys.

The results of an offshore test program over the Kingfish, Halibut, and Marlin fields of Gippsland Basin in the Bass Straits of Australia are shown in Figure 3 and Figure 4. All thermal particulate plumes encountered along the flight lanes are recorded, and the measurable

Plate 5. Helicopter mounted SURTRACE® equipment showing flexible sampling probe.

concentrations of volatile hydrocarbons carried by the particulates are indicated by circles whose areas are proportional to the relative concentrations of volatiles in the particulates. It will be noted that anomalous particulate values in volatile hydrocarbons occur in the vicinity of all of the fields, but that activity is much greater over the Marlin field, which is the only one producing gas and condensate. It would appear that strong gas leakage over this field is enhancing the production of hydrocarbon-enriched aerosols.

THE SURTRACE(TM) SYSTEM

The problems encountered with spatial resolution in the mining applications of the AIRTRACE system led to the development of ideas on the direct sampling of the surface microlayer from helicopters using tubular and flexible suction probes (Barringer, 1977b; Bradshaw and Barringer, 1979). A series of prototype systems were developed which quickly established the feasibility of carrying out such sampling, but these experiments also indicated that extensive engineering was required to perfect the technique. A fully practical helicopter system has now been evolved which employs a basic probe with replaceable tips that can be selected according to the type of terrain being surveyed (Plate 5). The upper section of the SURTRACE probe consists of a tough, tapered fiberglass tube that is hinge-mounted beneath a high-performance helicopter on a spring-tensioned assembly. The tube is terminated with a long, replaceable sampling probe constructed of high-impact strength plastic. The type of probe used over bare soils in semiarid regions is simply a flexible tube with a hole at the bottom end. When operating over scrub vegetation, conifers, or deciduous trees, a probe is used that consists of a tube that is perforated with holes over the lower meter of its length. Both types of assembly are approximately 12 m long, and material is vacuumed off the surface of the ground or vegetation canopy using a powerful suction pump. At periodic intervals, a high pressure back-flushing of hot compressed air is used to clear the tubing of any entrained particles.

The sample stream is passed through a particle counter, and the pilot is automatically signaled when a sufficient quantity of particles has passed through the monitor. He then flies on over a predetermined distance to the next sampling point and maintains forward motion while dragging the probe over the surface. The surface microlayer particles are collected on a special tape that is designed to allow for up to 40 mg of material to be collected within sample cells that are spaced at short intervals along a specially manufactured tape. The tape is covered with an adhesive tape similar to that used in the AIRTRACE system, and in post-flight processing the adhesive tape is stripped off along with its adhering particles for analysis on the LASERTRACE equipment.

The tape deck used for the SURTRACE system is substantially different from that used in the AIRTRACE equipment, since the tape is much bulkier (Plate 6). Sizing equipment is also built into the tape deck in order to provide optimum size selection from the incoming sample stream. The comparatively large sample stored within the sample cells (as opposed to the portion of the sample adhering to the adhesive cover tape) is analyzed by equipment that has been designed to measure the concentration of adsorbed volatile components within the samples. Other special types of analyses that are not available on the LASERTRACE equipment can also be carried out on these samples.

The SURTRACE system can be flown in conjunction with a magnetometer mounted on a boom, as is the case for the AIRTRACE system, and the survey altitude of 10 m provides very high resolution magnetic data. A gamma ray spectrometer is also sometimes incorporated into the system. Normally the helicopter is dipped down to touch the surface of the ground, typically about five times per kilometer, and the pilot maintains contact with the ground until the particle count monitor indicates that sufficient sample has been collected. The sample interval, however, can be adjusted according to the expected size of the target. When surveying over soils, contact with the ground is indicated by a microphone sensor on the probe, which gives an audio signal on a loudspeaker

---

(TM—Registered trademark of Barringer Research, Inc.)

Plate 6. SURTRACE® tape deck for collecting both adhesive tape samples and samples of up to 40 mg on special perforated tape.

mounted close to the pilot. He can quickly learn to sense the motion of the probe, and he can also observe the probe on a mirror mounted on the helicopter skids.

When sampling trees, the pilot descends periodically to drag the probe through the foliage with the tip submerged well into the canopy. Good sampling results have been obtained using this technique in mixed forests of conifers and deciduous trees.

A 35 mm strip camera with a very wide angle lens is employed to photograph the ground continuously during survey, and a navigator guides the pilot down survey lines which have been prelocated on aerial photographs. Navigation at this low altitude requires considerable skill, but nevertheless it has been proven feasible.

The helicopter SURTRACE system has already seen extensive usage for mining applications, but at the time of writing has only been

Plate 7. Backpack SURTRACE® sampling equipment.

applied for hydrocarbons using a ground-based backpack version of the system (Plate 7). This ground system essentially duplicates the action of airborne technique and has been used to establish the geochemical viability of the concept. An example of surface microlayer analyses of samples collected in a traverse across the Medio field in the San Juan Basin, New Mexico, is shown in Figure 5.

Survey speeds with the helicopter SURTRACE system are in the range of 30 to 60 km per hour, depending upon the terrain; and, as such, are about half those obtained with the AIRTRACE equipment. The system has the advantage of collecting a comparatively large sample from precisely known locations and is operable under most dry weather conditions. Compared with AIRTRACE, it has the disad-

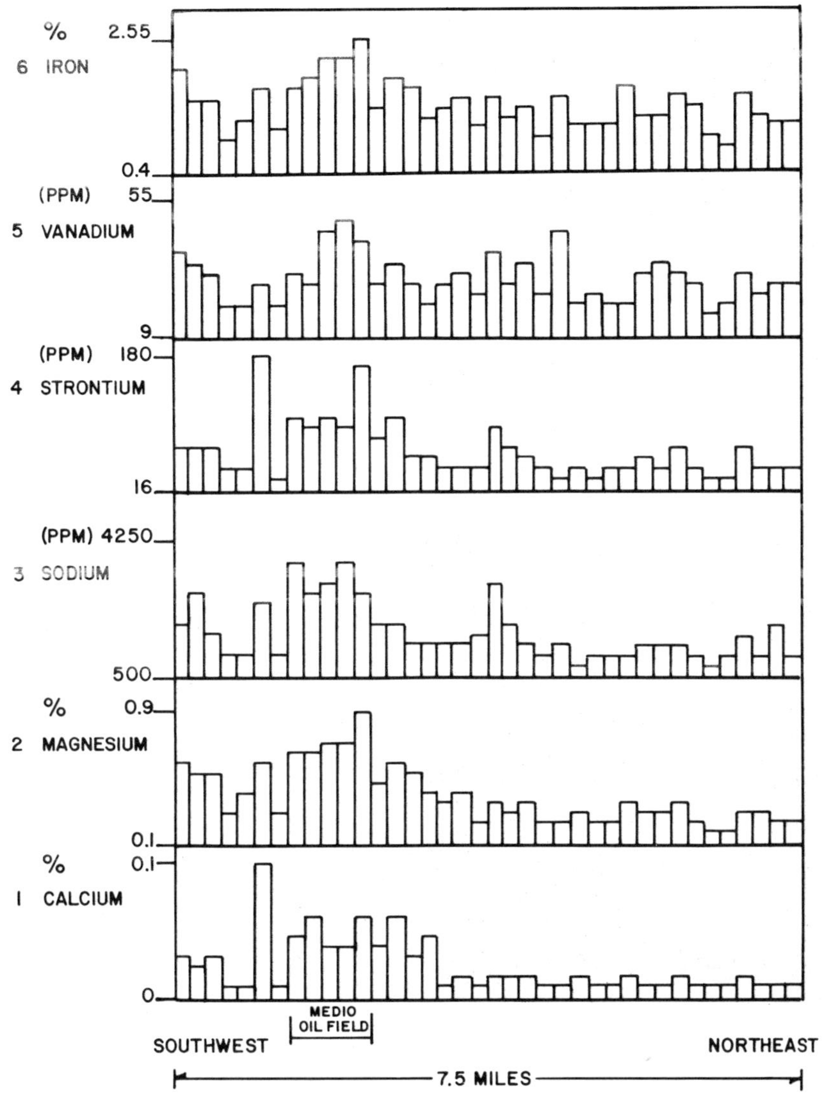

Fig. 5. Analytical results of a ground SURTRACE® surface microlayer sampling traverse across the Medio Field in the San Juan Basin, New Mexico.

vantage of being a ground contact method which is not generally adaptable to fixed-wing operations.

THE LASERTRACE(TM) ANALYTICAL SYSTEM

The LASERTRACE analytical equipment (Barringer, 1977a, 1980); Abercrombie et al., 1978) has been specifically designed to handle the adhesive tapes upon which samples are collected in both the AIRTRACE and SURTRACE systems. It will be noted from earlier descriptions that samples from both of these systems are impacted onto the adhesive so that there are small spots—typically of 3 mm diameter—that are spaced periodically at intervals such as 2 cm all along the tape. The adhesive is covered with a nonadhesive teflon tape which in effect hermetically seals the samples to keep them clean. This cover tape is stripped off in the LASERTRACE tape deck described below.

(TM—Registered trademark of Barringer Research, Inc.)

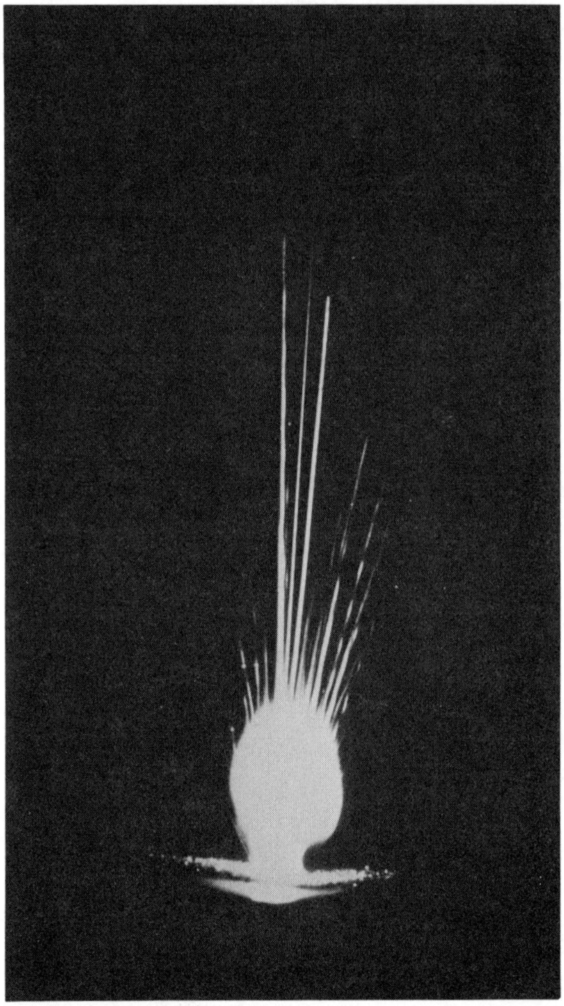

Plate 8. Laser plume generated by laser vaporization of tape samples.

The unique feature of the LASERTRACE system is that the collection tape employed is substantially transparent (both substrate and adhesive) to the infrared laser beam that is used to vaporize the sample off the tape. Thus a focused and very intense pulse of laser light can be projected onto the tape, and if no sample is present, there is no heating or vaporization of either the substrate or the adhesive. On the other hand, if only a single speck of particulate material is present in the laser beam, it is vaporized in a microsecond (Plate 8). This technique enables powdered samples to be vaporized with essentially zero contamination from the substrate, while at the same time providing a very convenient sample carrier which lends itself well to automation.

The LASERTRACE equipment uses a specially fabricated tape deck that provides for the appropriate portion of the tape to be completely immersed in an argon atmosphere. The infrared laser beam is focused onto precisely the correct spot on the tape, and the tape is advanced automatically after the laser has been fired. Each successive sample on the tape is brought into its correct position with optical sensors that detect locating marks which are impressed in flight on the tape. The material vaporized off the tape immediately condenses into a plume of very finely divided particulate material and is carried by an argon stream into an inductively coupled radio frequency argon plasma (Plate 9) (Fassel and Kniseley, 1974). The plasma, which operates at temperatures in excess of 10,000°C, vaporizes the finely divided particle stream with high efficiency, and the resultant strong, spectral emission is analyzed with a dispersive grating spectrometer having 25 slits and photo-multiplier tubes. The output of the photo-multiplier tubes is digitally recorded on magnetic tape for subsequent conversion into absolute elemental values, after appropriate corrections for matrix and scattering effects have been applied in a digital computer.

Standards are inserted onto the tape at periodic intervals, and the entire system operates automatically with a production rate of more than 1,000 samples per working shift. As a result of this exceptionally high productivity, the analytical system is capable of servicing several AIRTRACE or SURTRACE aircraft.

Absolute sensitivities of the LASERTRACE equipment vary between 10^{-10} and 10^{-13} gm according to the element, with typical sensitivities lying in the region of 10^{-11} gm. Analysis for the following elements is provided: Al, Ca, Fe, Mg, Si, A, Ti, K, Na, Mn, Ba, Sr, Pb, Cu, Cr, Be, Zn, Ni, V, Co, Ag, Mo, Cd, Th, C, Zr, and B. It will be noted that analyses are made for all of the major rock-forming elements with the exception of oxygen, and it is therefore possible to compute the mass of sample vaporized by summing up all of the elements and making a computed allowance for

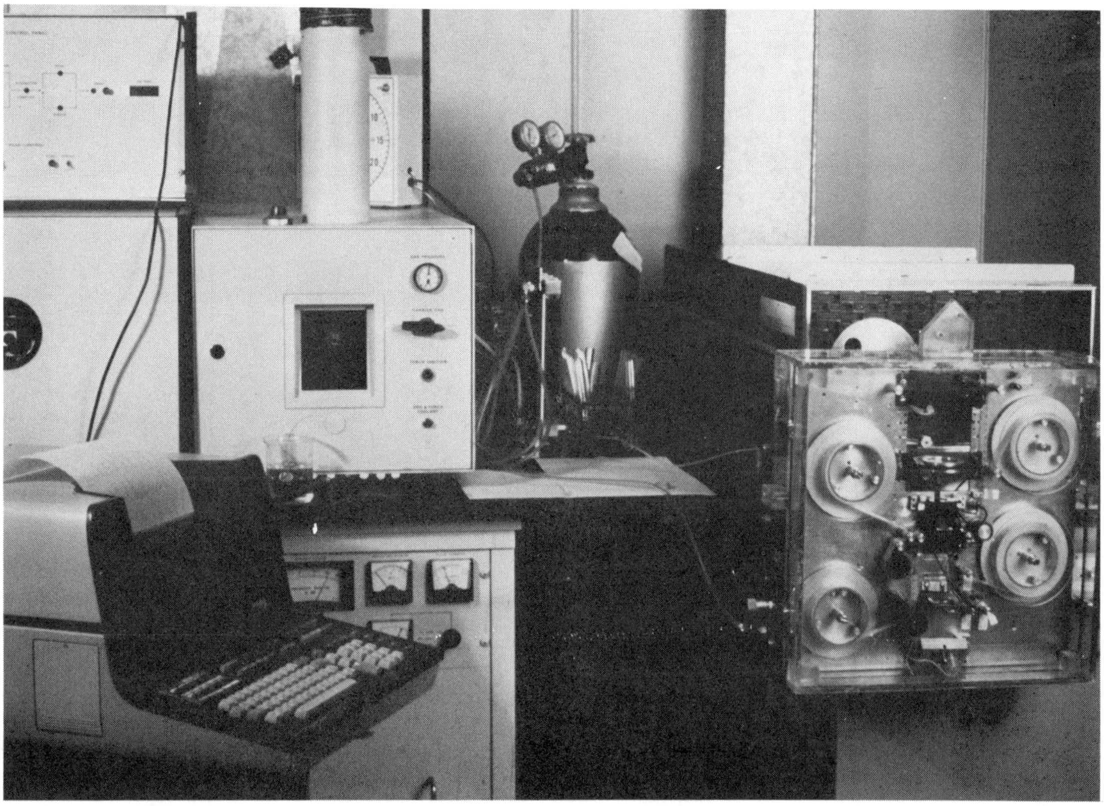

Plate 9. Part of LASERTRACE™ equipment showing laser, tape playback deck, and inductively coupled plasma.

described airborne equipment known as VAPORTRACE(TM) has been developed for lab use with SURTRACE tapes. This equipment is designed for the analysis of 1 mg samples or less for a range of adsorbed gases, vapors, and liquids. These include volatile hydrocarbons, sulfur compounds, mercury, etc.

CONCLUSIONS

The two airborne systems that have been outlined, AIRTRACE and SURTRACE, can be used for reconnaissance geochemical coverage of areas large or small where for reasons of access or economics it is better to explore initially with an airborne system rather than to use surface coverage. The AIRTRACE system is suitable for both onshore and offshore detection of hydrocarbon microseepages where meteorological conditions provide for good mixing in the lower atmosphere.

Since the AIRTRACE system can be flown along with high quality magnetic and other oxygen. This allows the analytical results to be determined in parts per million concentrations for each element analyzed, and although the ppm values are not precisely equal to those obtained by conventional wet analysis, they are nevertheless very reproducible. Repeat analyses of a given soil that has been smeared on the tape will give a reproducibility in ppm that is generally within ±5%.

The linearity of the system of Mn and Si is demonstrated in Figure 6, which shows varying amounts of USGS standard soil samples that have been dispersed upon an adhesive tape and analyzed with the LASERTRACE system. The graphical representation shows the degree of linearity which can be achieved over a fiftyfold variation in tape loading.

AUXILIARY ANALYTICAL SYSTEMS

An addition to the LASERTRACE equipment which carries out analyses for the elements listed above, a variant of the previously

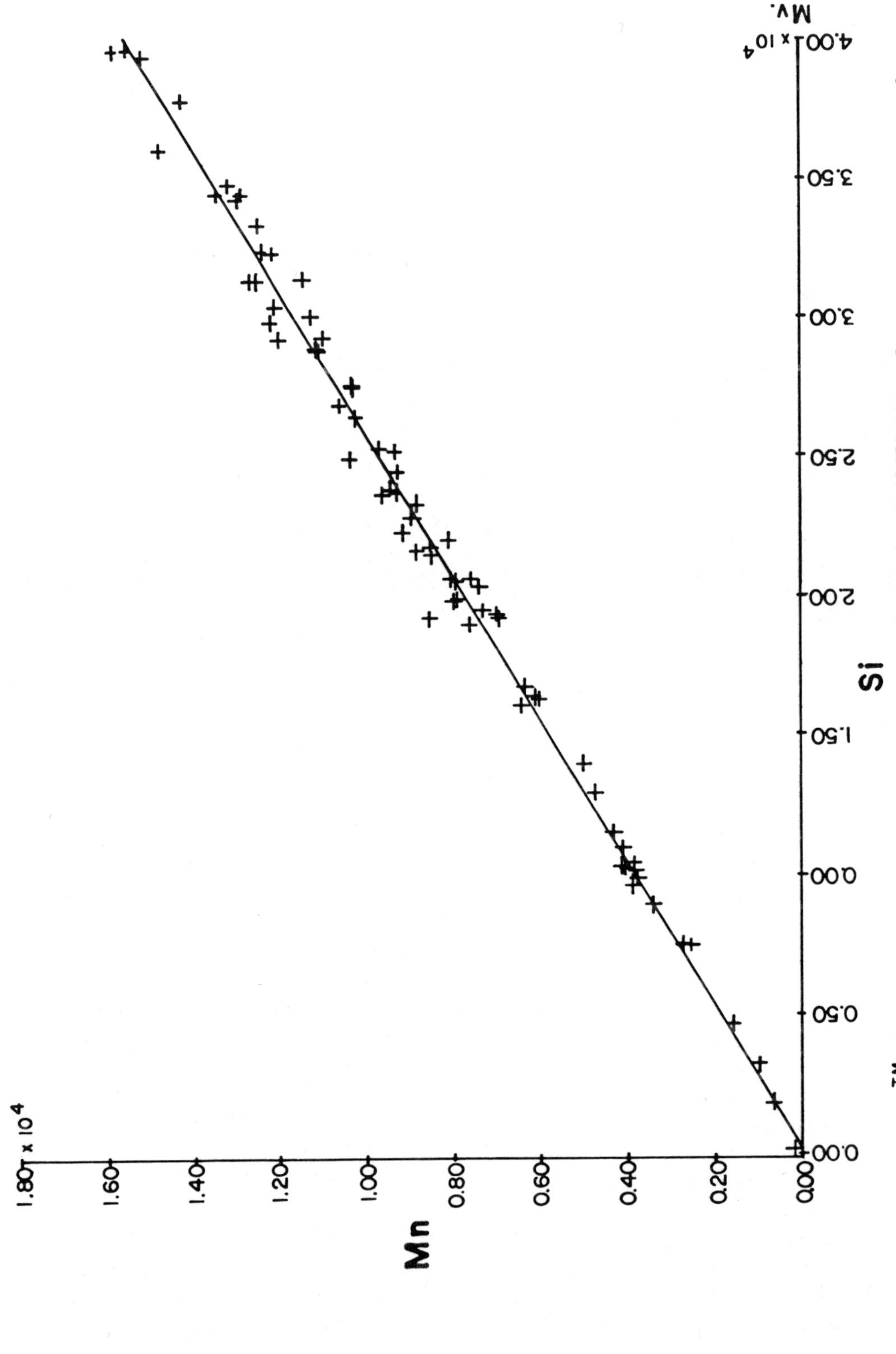

LASERTRACE™ – PLOT INDICATING LINEARITY OF SYSTEM. Shown are outputs for Mn and Si in U.S.G.S. standards for 100fold variation in sample loading.

Fig. 6.

geophysical surveys, it is also possible to derive a great deal of basement and intersedimental geological/geophysical information at the same time geochemical surveys are carried out. The AIRTRACE system is thus a powerful generator of hydrocarbon exploration data, since it can provide simultaneous multi-parameter information on inorganic and organic geochemistry as well as geophysical survey data.

The SURTRACE system, like the AIRTRACE equipment, also operates in conjunction with airborne geophysical systems, and it can be used in place of the AIRTRACE system in regions where the weather is considered to be unsuitable for AIRTRACE, or where there are reasons for requiring high spatial resolution in the data. Alternatively, it can be operated as a follow-up tool for the AIRTRACE system for providing confirmation and detailed mapping of anomalies located with the AIRTRACE equipment. When used in this fashion, it is possible to switch SURTRACE equipment into an AIRTRACE aircraft fairly rapidly so that a single helicopter can be mobilized for sequential AIRTRACE and SURTRACE surveys.

REFERENCES

INTRODUCTION

Paterson, Norman R., 1973, Twenty years of airborne EM in perspective: Canadian Mining Journal, v. 94, no. 4, p. 101-105, illus.

TERRESTRIAL SURFACE MICROLAYER GEOCHEMISTRY

Artemenko, V. I., Malovitskiy, Ya. P., 1978, Results of a bottom geothermal survey in the Bakhar oil-gas field: Petroleum Geology (USA) (PEGEAX), v. 15, no. 4, p. 150-152, illus.

———, 1978, Results of a bottom geotemperature survey in the Bakhar petroleum deposit: International Geology Review (USA) (IGREAP), v. 20, no. 3, p. 367-370, 1978, 11 ref., illus., geological sketch map.

Athy, L. F. 1930, Compaction and oil migration: American Association of Petroleum Geologists Bulletin, v. 14, pp. 25-35.

Barker, C., 1972, Aquathermal pressuring—role of temperature in development of abnormal-pressure zones: American Association of Petroleum Geologists Bulletin, v. 56, no. 10, p. 2068-2071.

Burst, J. F., 1959, Postdiagenetic clay mineral environmental relationships in the Gulf Coast Eocene, *in* Swineford, A., ed., Clays and clay minerals: National Conference of Clays and Clay Minerals, 6th, Proceedings, London-New York, Pergamon Press, p. 411.

———, 1969, Diagenesis of Gulf Coast clayey sediments and its possible relation to petroleum migration: American Association of Petroleum Geologists Bulletin, v. 53, p. 73-93.

———, 1976, Argillaceous sediment dewatering: Earth and Planetary Science, annual review, v. 4, p. 293-318.

Dickey, P. A., 1975, Possible primary migration of oil from source rock in oil phase: American Association of Petroleum Geologists Bulletin, v. 59, p. 337-345.

Dunoyer de Segonzac, G., 1969, Les mineraux argileux dans la diagenese passage au metamorphisme: Mem. Serv. Carte Geol. Als. Lorr., v. 29, p. 320.

———, 1970, The transformation of clay minerals during diagenesis and low grade metamorphism, a review: Sedimentology v. 15, p. 281-286.

Fyfe, W. S., Price, N. J., and Thompson, A. B., 1978, Fluid in the earth's crust: Amsterdam-Oxford-New York, Elsevier Science Pub. Co., p. 383.

Heard, H. C., and Rubey, W. W., 1966, Tectonic implications of gypsum dehydration: Geological Society of America Bulletin, v. 77, p. 741-760.

Hedberg, H. D., 1926, The effect of gravitational compaction on the structure of sedimentary rocks: American Association of Petroleum Geologists Bulletin, v. 10, p. 1035-1072.

———, 1936, Gravitational compaction of clays and shales: American Journal of Science, v. 31, p. 241-287.

Magara, K., 1974, Aquathermal fluid migration: American Association of Petroleum Geologists Bulletin, v. 58, no. 12, p. 2513-2526.

———, 1975, Re-evaluation of montmorillonite dehydration as cause of abnormal pressure and hydrocarbon migration: American Association of Petroleum Geologists Bulletin, v. 59, p. 292-302.

Melik-Pashayev, V. S., Gattenberger, Yu. P., and Lutkov, V. A., 1973, Geothermal conditions of Uzen' oil field: International Geology Review (IGREAP), v. 15, no. 12, p. 1365-1372, illus. (incl. sketch map).

Roberts, W. H. III, 1981, Some uses of temperature data in petroleum exploration (this volume).

Weaver, C. E., 1959, The clay petrology of sediments, *in* Clays and clay minerals: National Conference on Clays and Clay Minerals, 6th, Proceedings, p. 154-187.

MARINE SURFACE MICROLAYER GEOCHEMISTRY

Duce, Robert A., Quinn, James G., Olney, Charles E., Piotrowicz, Stephen R., Ray, Barbara J., and Wade, Terry L., 1972, Enrichment of heavy metals and organic compounds in the surface microlayer of Narragansett Bay, Rhode Island: Science (AAAS), v. 176, no. 4031, p. 161-163.

Fischer, Peter J., and Stevenson, Andrew J., 1973, Natural hydrocarbon seeps along the northern shelf of the Santa Barbara Basin, California: Offshore Tech. Conference, Preprints, v. 1, no. 5, p. 159-164, illus. (incl. sketch maps).

Garrett, W. D., 1967, Organic chemical composition of ocean surface: Deep-Sea Research, v. 14, p. 221.

Hansen, H. P., 1975, Photochemical degradation of petroleum hydrocarbon surface films on seawater: Marine Chemistry (MRCHBD), v. 3, no. 3, p. 183-195, illus., tables.

Hitchon, B., 1974, Application of geochemistry to the search for crude oil and natural gas, *in* Levinson, A. A., Introduction to exploration geochemistry: Applied Publishing Ltd., Calgary (BP509-505).

MacIntyre, Ferren, 1974, Chemical fractionation and sea-surface microlayer processes, *in* The sea, v. 5, Marine chemistry, ed. Edward D. Goldberg, air-sea interactions: New York, John Wiley, p. 245-299, illus.

Morris, R. J., and Culkin, F., 1974, Lipid chemistry of eastern Mediterranean surface layers: Nature, v. 250, no. 5468, p. 640-642, sketch map.

AIRTRACE: A SYSTEM FOR SAMPLING THE SURFACE MICROLAYER VIA THE COLLECTION OF AIRBORNE ATMOSPHERIC PARTICULATES

Barringer, A. R., U.S. Patents:

1973a, Methods and apparatus for geochemical surveying, patent #3,759,617.

1973b, Method and apparatus for sensing substances by analysis of adsorbed matter associated with atmospheric particulates, Patent #3,768,302.

1975, High resolution geochemical prospecting method and apparatus, Patent #3,868,222.

1976a, Method and apparatus for transferring particles from one fluid stream into another, Patent #3,998,734.

1976b, Remote sensing of marine hydrocarbon seeps, Patent #3,961,187.

1976c, Method and apparatus for collecting atmospheric samples in atmospheric geochemical prospecting, Patent #3,970,428.

1976d, Exploration method and apparatus utilizing atmospheric micro-organic particulates, patent #3,985,619.

———, 1977, AIRTRACE: an airborne geochemical exploration technique: Annual William T. Pecora memorial symposium, 1st, Sioux Falls, S. D., 1975, U.S. Geological Survey Professional Paper (USA) (XGPPA9), no. 1015, p. 231-251, 1977a, 66 ref., illus., table, sketch maps.

Benbow, D. D., 1968, Case history, Mereenie field: Australia Petroleum Exploration Association Journal, v. 8, pt 2, p. 114-119, illus. (incl. sketch maps).

Bussinger, J. A., 1972, Remote sensing of the troposphere: U.S. Department of Commerce National Oceanic and Atmospheric Administration, University of Colorado.

Donovan, T., 1974, Petroleum microseepage at Cement, Oklahoma, evidence and mechanism: American Association of Petroleum Geologists Bulletin, v. 58, no. 3.

———, and Dalziel, M., 1977, Late diagenetic indicators of buried oil and gas: U.S. Department of the Interior, Geological Survey, Open-File Report 77-817, 44 p.

Hilst, G. R., and Nickola, P. W., 1959, On the wind erosion of small particles: American Meteorological Society Bulletin, v. 40, p. 73-77.

Kaimal, J. C., and Bussinger, J. A., 1970, Case studies of a convective plume and a dust devil: Journal of Applied Meteorology, v. 9, p. 612-620.

Taylor, R. J., 1958, Thermal structures in the lowest layers of the atmosphere: Australian Journal of Physics, v. 11, p. 168-176.

Weiss, O., 1967, Method of aerial prospecting which includes a step of analyzing each sample for element content, number and size of particles: U.S. Patent No. 3,309,518.

———, 1969, Method of aerial prospecting: U.S. Patent No. 3,462,995.

———, 1971, Airborne geochemical prospecting: Geochemical Exploration, CIM special v. 11, p. 502-514.

OFFSHORE APPLICATIONS OF THE AIRTRACE SYSTEM

Baier, R. E., 1972, Organic films on natural waters. Their retrieval, identification, and modes of elimination: Journal of Geophysical Research (JGREA2), v. 77, no. 27, p. 5062-5075.

Blanchard, D. C., and Woodcock, A. H., 1957, Bubble formation and modification in the sea and its meteorological significance: Tellus, v. 9, no. 2, p. 145-158.

Blanchard, D. C., 1964, Sea-to-air transport of surface active material: Science, v. 146, p. 396-397.

———, and Syzdek, Lawrence D., 1972, Concentration of bacteria in jet drops from bursting bubbles: Journal of Geophysical Research, v. 77, no. 27, p. 5087-5099.

Bloch, M. R., and Luecke, W., 1972, Geochemistry of ocean water bubble spray: Journal of Geophysical Research (JGREA2), v. 77, no. 27, p. 5100-5105.

Geyer, R. A., and Sweet, W. M., 1973, Natural hydrocarbon seepage in the Gulf of Mexico: Gulf Coast Association of Geological Societies, Transactions, 23, p. 158-169.

Guinasso, N. L., and Schink, D. R., 1975, Chemical-acoustic model of methane bubbles rising from a marine gas seep (abstract): Geological Society of America Annual Meeting, 1975, p. 1093-1094.

Hansen, H. P., 1975, Photochemical degradation of petroleum hydrocarbon surface films on sea water: Marine Chemistry (MRCHBD), v. 3, no. 3, p. 183-195, illus., tables.

Hoffman, E. J., and Duce, Robert A., 1976, Factors influencing the organic carbon content of marine aerosols—a laboratory study: Journal of Geophysical Research, v. 81, no. 21, p. 3667-3670 (JGREA2).

———, 1977, Oceanic carbon in marine atmospheric particulate matter; concentration and particle size distribution: Geophysical Research Letters (USA) (GPRLAJ), v. 4, no. 10, p. 449-452, 17 ref., illus., table.

Quinn, John A., Steinbrook, Richard A., and Anderson, John L., 1975, Breaking bubbles and the water-to-air transport of particulate matter: Chemical Engineering Science (CESCAC), v. 30, no. 9, p. 1177-1183.

Simoneit, B. R. T., Chester, R., and Eglington, G., 1977, Biosenic lipids in particulates from the lower atmosphere over the eastern Atlantic: Nature (GBR) (NATUAS), v. 267, no. 5613, p. 682-685, 34 ref., tables, sketch, map.

Strickland, W. T. Jr., 1979, Laboratory results of cleaning produced water by gas flotation: AIME, Society of Petroleum Engineers (SPE), Prod. Oper. Symposium, 1st, Proceedings, Oklahoma City, Okla., Publ. by AIME, New York, p. 57-58.

Warnstedt, H. C., 1972, Investigation of crude-petroleum removal from brines by treatment with small gaseous bubbles (in German): Erdoel-Erdsas-Z, v. 88, no. 4, p. 115-129.

Wozniak, D., Kiwak, M., Cahoon M. C., Edgerton, Robert H., 1976, Water to air bacterial transfer by bursting bubbles: American Society of Civil Engineers ASCE Environmental Engineering Division Journal, v. 102, p. 567-570 (JEEGAV).

THE SURTRACE SYSTEM

Barringer, A. R., 1977b, Detection of concealed metalliferous deposits, hydrocarbons and explosives: Patent #4,056,969.

Bradshaw, P. M. D., and Barringer, A. R., 1979: SURTRACE™ — An airborne and ground exploration technique based on surface microlayer geochemistry (abstract), International Geochemical Exploration Symposium, 7th, Golden, CO, 1978, Proceedings, p. 497.

LASERTRACE ANALYTICAL SYSTEM

Abercrombie, F. N., Barringer, A. R., Bradshaw, P. M. D., Cruz, R. D., and Murray, A. D., 1978, Multi-element geochemical analysis using the radio frequency argon plasma with conventional acid digestion and laser vaporization: Symposium on the Analysis of Geological Material, Geological Survey of Canada and the Canada Center for Mineral and Energy Technology (CANMET), Ottawa, Canada, May 15-17, 1978, invited paper.

Barringer, A. R., 1980, Laser spectrochemical analysis (LASERTRACE™): U. S. patent application allowed.

Fassel, V. A., 1976, Current and potential applications of inductively coupled plasma atomic emission spectroscopy in the exploration of mining and processing of materials: International Symposium on Analytical Chemistry in the Exploration, Mining and Processing of Materials, Johnannesburg, Republic of So. Africa, Aug. 24, 1976, plenary session, NTIS ISM-85.

———, and Kniseley, R., 1974, Inductively coupled plasma–optical emission spectroscopy: Analytical Chemistry, v. 46, no. 13, p. 1110.

Merryfield, R. N., and Lloyd, R. C., 1979, Simultaneous determination of metals in oil by inductively coupled plasma emission spectrometry: Analytical Chemistry, v. 51, no. 12, Oct., 1979, p. 1965-1968.

THE AIRBORNE GAS SENSOR AND ITS ROLE IN EXPLORATION RECONNAISSANCE

C. Keith Thompson

GEOPHYSICAL SYSTEMS CORPORATION AND GEOCHEMICAL SYSTEMS CORPORATION

ABSTRACT

An introduction to hydrocarbon microseepage reconnaissance and its relationship to petroleum reservoirs must include a brief history of the significance of oil and gas seeps in exploration worldwide. Studies presented on this subject have generally confirmed that most of the oil-producing regions of the world were discovered from oil and gas seeps. Such evidence emphasizes the importance of hydrocarbon gas reconnaissance as an idicator of oil/gas reserves and exemplifies how it can be effectively utilized in exploration programs.

New evidence from research (Department of Energy, U.S. Geological Survey, and Horvitz, 1978) substantiates and further supports the concept of microseepage and, thus, the importance of remote gas sensing as a valuable reconnaissance exploration tool to be employed in the early stages of exploration.

The Systems Approach to exploration includes: (1) reconnaissance studies utilizing the cost-savings benefit of LANDSAT and Airborne Gas Sensor surveys, and (2) geochemical and geophysical validation of anomalies to develop and delineate prospects. The most promising anomalous areas can then be subjected to rigorous geochemical as well as geophysical and geological study as a result of the concentration of efforts on limited areas. Inherent savings are available through focusing of efforts only on anomalies discovered utilizing remote reconnaissance in the early stages of the exploration program. Evidence that the Systems Approach to exploration may offer improved success in exploration includes: (1) better-than-average success by companies using reconnaissance and near-surface geochemical techniques, and (2) recent studies providing a better understanding of the dynamics of hydrocarbon microseepage from petroleum reservoirs.

Future conventional methods of exploration must be economically sound and must offer an improved success ratio. The airborne gas sensor employed as a part of the Systems Approach to petroleum exploration and the subsequent focusing of geochemical, geophysical, and geological efforts on a limited number of prospective areas can yield these benefits.

INTRODUCTION

Science and technology have made vast advances during the past three decades, while petroleum exploration techniques actually employed in the field have not substantially changed from those methods used thirty years ago. That hydrocarbon gases seep vertically through the sedimentary rocks above oil and gas reservoirs has been known for many years. Their vertical migration can no longer be considered an issue, as research continues to explain the mechanism of hydrocarbon microseepage and migration.

In the 1970 symposium, "Geology of Giant Petroleum Fields" (AAPG Memoir no. 14), the history of oil and gas exploration was briefly outlined, with the progress of exploration methods charted from oil seeps, surface geology, and emerging geophysics to current sophisticated concepts in subsurface geology and reflection seismology. One conclusion drawn about current exploration methods and presented by Michel T. Halbouty in the papers of that symposium was accurately summarized by Halbouty: "Since 1952, exploratory results, unaided by the introduction of new methods of exploration, have declined drastically" (Halbouty, 1970b). Although improvements in subsurface mapping technology and equipment have been devised in recent years, little has been done to assist in exploration or in the identification of stratigraphic accumulations of oil and gas. As a result, the overall success ratio of wildcat wells drilled to those completed as new field discoveries has not changed substantially from year to year.

While predominantly employed conventional methods of exploration continue to offer less than promising results in the United States, Soviet research institutes have spent years attempting to develop and refine geochemical techniques for locating oil and gas reservoirs. In fact, more research effort in this area has probably been expended in the Soviet Union than in any other nation. It has been reported (Sweet, 1966) that in the 1950s the Soviets maintained as many as three hundred geochemical crews in the field at one time. The reason for this intensive Soviet investigation is clear—success in developing direct and indirect detection methods could effect huge economic impact in exploration (after Davidson, 1967). Although recent information on the accuracy of Soviet prospecting success is not available, in 1959 V. A. Sokolov stated before the fifth World Petroleum Congress in New York: "Under favorable geologic conditions, the proportion of the correct predictions is rather high—about 70%. For instance, in the North Caucasus (Kuban), predictions made by gas surveys were confirmed in thirteen cases out of seventeen (Sokolov et al., 1959). More recent comments have been made on Russian work in the field of geochemistry by Frederick R. Siegel (1974) in his book *Applied Geochemistry*. Similarly, two Soviet publications, "Researches in the Geology and Geochemistry of Oil and Gas" (Vyshemirskii, 1977) and "Major Achievements of Siberian Petroleum Geochemists" (Vassoevich, 1977), offer current references to Soviet success with geochemical methods.

Historically, oil seeps have proven to be significant indicators of petroleum reservoirs. Walter K. Link, past chief geologist of Standard Oil Company, New Jersey, in his paper entitled "Significance of Oil and Gas Seeps in World Oil Exploration," presented his views on the importance of seeps; and in his opinion, "most of the important oil-producing regions of the world were detected or discovered by surface oil and gas seeps" (Link, 1952). Where direct (or surface) seeps once readily identified the location of oil and gas reservoirs, those macroseeps have, for the most part, been found. We must now focus our attention on the more subtle manifestations of those same types of seeps in the form of microseeps. Robert C. MacElvain, who presented a paper at Symposium I, did much to enhance geochemical exploration by defining the mechanics of gaseous ascension through a sedimentary column. Horvitz (1978) has presented data clearly validating the fact that lighter hydrocarbons move upward from petroleum accumulations. Evidence of this phenomenon may be found in carbon isotopic comparison of near-surface hydrocarbons to reservoir hydrocarbons and in the analyses of well cuttings collected from productive areas. New field wildcats drilled on prospects where geochemistry played a vital role have been reported to yield significant improvements over the national discovery average (Davidson, 1967). The challenge confronting the explorationist thus seems to be one of detecting the subtle indications of hydrocarbon microseepage and, therefore, the vertical or near-vertical migration of hydrocarbon gases that mark the location of prospective reservoirs, and to do so economically.

Professor Martin J. Davidson in his 1967 paper "Petroleum Geochemical Exploration from Space: A Radical New Technology" was the first to observe that geochemical soil sampling could be preceded or even replaced by remote sensing from aircraft or earth orbit using spectrographic recognition. Donald F. Saunders (1969) of Texas Instruments, Inc., also with great foresight, presented a paper at Symposium I entitled "Airborne Sensing as an Oil Reconnaissance Tool." In this paper he explored the possibilities and economic impact of airborne reconnaissance and the importance of the development of a practical system for remotely identifying geochemical indications of the presence of petroleum. Now, seven years after the premier orbit of LANDSAT I, it seems obvious that the ultimate success of any efficient oil and gas exploration program must be based on the premise that subsurface accumulations of hydrocarbons subtly manifest themselves at the surface of the earth. By remotely detecting evidence of the presence of these hydrocarbons, the airborne gas sensor system provides vital information when employed as a reconnaissance tool early in the exploration program.

The airborne gas sensor and its role in the overall Systems Approach to exploration will be discussed. The Systems Approach is considered to be a comprehensive exploration program including LANDSAT studies and/or airborne gas sensor reconnaissance surveys that are validated by detailed geochemical and geophysical data. This approach will be developed and examples of typical exploration programs presented. Finally, the future of reconnaissance geochemical exploration will be considered.

THE HYDROCARBON MICROSEEPAGE MECHANISM

In a recent U.S. Geological Survey study entitled "Late Diagenetic Indicators of Buried Oil and Gas," Terrence J. Donovan and Mary C. Dalziel (1977) investigated the mechanism of hydrocarbon microseepage, whereby hydrocarbons may escape entrapment and migrate to the surface level. The authors concluded that at least three hydrocarbon microseepage mechanisms operate over oil and gas fields; these are: (1) effusion of hydrocarbons through inadequate caprocks and along faults and fractures; (2) low-molecular-weight hydrocarbons dissolved in water moving vertically through capping shales as a result of hydrodynamic or chemical potential drive; and (3) diffusion of gases dissolved in water. Combinations of these mechanisms may also occur. This research study also pointed up the overall significance of using these subtle indicators as reconnaissance targets by stating: "Changes in the chemical and physical properties of surface rocks correlate with the subsurface distribution of petroleum and potentially can be detected from both airborne and spaceborne platforms" (Donovan and Dalziel, 1977).

In Project "Halo," a research effort extending over five years and conducted under the auspices of the Energy Research and Development Administration (Department of Energy), exploration techniques were investigated and findings were published in a January 1979 paper entitled "A Critical Laboratory and Field Evaluation of Selected Surface Prospecting Techniques for Locating Oil and Natural Gas" (Heemstra et al., 1979). Of the various exploration methods investigated, the resulting data provide evidence that the lighter hydrocarbons, that is, methane through pentane, when measured at near-surface levels, correlate with subsurface petroleum reservoirs. The report's recommendations include additional field work in the form of airborne remote reconnaissance and near-surface gas sample collection to determine the level of free hydrocarbon gases that exists above proven oil and gas reservoirs and prospective areas. This research also recommends that color-intensified or otherwise enhanced LANDSAT imagery be combined with airborne gas sensing and hydrocarbon geochemistry to create a valuable exploration tool.

Additional evidence substantiating microseepage was recently presented by Victor T. Jones and R. J. Drozd of Gulf Research and Development Co., in a paper entitled "Predictions of Oil or Gas Potential by Near-Surface Geochemistry" (1979). This report presented comparative ratios of the lighter hydrocarbons, such as methane through pentane, that have been used successfully to predict the type of reservoir present—i.e., oil, gas, or gas condensate—by employing near-surface gas survey techniques prior to drilling. Likewise, Geophysical Systems Corporation's research indicates that a definite relationship between remote gas sensor anomalies and hydrocarbon microseepage exists over both proven oil and gas fields and prospective areas. These conclusions are drawn from numerous remote gas sensor surveys performed over known production and are supported by additional evidence derived from detailed geochemical sampling, resulting in the validation of anomalous microseepage patterns over these existing reservoirs.

THE AIRBORNE GAS SENSOR SYSTEM

The remote gas sensor is an electronic installation on either airborne or ground vehicle platform (refer to Figure 1)* that detects the presence of hydrocarbon gases escaping at the surface of the earth. Conclusive proof does not exist that the gases being detected by the sensor are low molecular hydrocarbons and nothing else. The weight of existing evidence is strong enough, however, to permit consideration of the gases as hydrocarbons for operational purposes. Highly directional pulses of specific frequency are transmitted from a rotating beacon antenna on the resonant absorption spectral lines of the gas molecules to be detected. A portion of these pulse transmissions is absorbed by the hydrocarbon gas molecules, which then move to a higher, or excited, energy state. In the time between electronic pulses, the excited gas molecules collapse and return

*See atlas for Figs. 1, 3ii, 4, 5, 6, 7, 8, 9.

Fig. 2. Typical airborne installation of gas sensor.

to their original energy state, concurrently emitting the energy absorbed as radiation at a different frequency than the incident exciting frequency. A receiver is tuned to detect this characteristic signature, or echo, from the gases. Thus, signals detected by the remote gas sensor indicate where the gases of interest are present at the surface and their relative intensities as a function of the amount of gas present. The angular position of the antenna indicates the direction, while the distance to the gas seep area is indicated by the time difference between transmitted and returned signals. Considered together, these parameters display an aerial picture of the gas concentration on a cathode ray tube video display screen. The displayed indications of the presence of hydrocarbon gases are then recorded on maps of the surveyed area.

In a typical reconnaissance gas survey, the airborne gas sensor is mounted in a helicopter and flown over the specified area at low altitude (Figure 2). The video output display is monitored to determine where gases are escaping from the subsurface reservoirs. The anomalous indications are recorded on maps of the area for further geologic consideration. It should be understood that the combination of gas sensor survey results with subsurface geologic information, geochemical survey validation, geophysics, or additional information pertinent to the specific area under consideration is necessary for an adequate evaluation of any given prospect.

The most significant advantage of the airborne gas sensor when compared with traditional surveying methods is cost-savings. The airborne gas sensor can survey as much as thirty thousand acres a day when employed over clear or partly clear terrain, with flat or gently rolling surface conditions. The effective range of the gas sensor under ideal reconnaissance surveying conditions is approximately one-half mile radius. Once anomalous areas are identified, a more detailed evaluation is required at close range. By hovering the helicopter at various locations, the microseepage patterns of the anomalous areas can be readily and accurately mapped. Then, the balance of the reconnaissance survey can be resumed at a more rapid rate. While crops, vegetation, trees, and plowed fields present a barrier to the efficient transmission and identification of return signature signals, the net result is merely a slowdown of the overall survey. Under these circumstances, gas microseepage anomalies of sufficient area can usually be identified in open fields located adjacent to and among the forested or cultivated areas. These partially defined anomalies may be mapped in more detail through the assistance of near-surface geochemical gas sampling and analysis.

Following a typical gas sensor survey and prior to other detailed geological and geophysical investigations, soil gas geochemical samples are collected in profiles or grid patterns over the anomalous areas, employing specially developed equipment and techniques. Free gas samples are extracted, carefully controlled, and submitted to laboratory analysis by hydrogen flame gas chromatography. Analysis is of lighter hydrocarbons, including methane, ethane, propane, isobutane, normal butane, isopentane, and normal pentane (C_1-C_5). Accuracy of the typical analysis is ± 2% in the 1-10 ppm range (by volume) (Figure 3). The results of the sample analyses are evaluated to determine the likely source of microseepage. Of course, excessive amounts of methane (CH_4) are known to exist in areas where glacial till, shale gas, or other decomposing material may be buried. If the analysis yields confirming anomalous amounts of ethane through pentane (C_2-C_5) and heavier hydrocarbons, and the concentrations are sufficiently high when compared with background samples collected from the same area, then geochemical validation of the prospect is considered to be complete. Here it is important to note once again the results of recent investigations reported by Gulf Oil Corp. researchers Jones and Drozd (1979), which indicate a relationship between the ratios of light hydrocarbons present in the near-surface and the prediction of the type of reservoir prior to drilling.

The Systems Approach to exploration, including LANDSAT geomorphology, airborne gas sensor reconnaissance, geochemical sampling, geologic studies, and geophysics, may be greatly enhanced by the use of the airborne gas sensor. Large scale reconnaissance surveys may be accomplished rapidly and inexpensively through regional geological studies aided by LANDSAT geomorphic maps and gas sensor reconnaissance. Anomalous leads followed up by soil gas sampling can be thoroughly analyzed to determine the most promising prospective areas. These areas should exhibit both significant hydrocarbon microseepage as well as a defined geologic and/or geophysical trap. Geophysical exploration budgets may thus be focused upon the most promising areas.

EXAMPLES OF REMOTE GAS SENSOR AND SOIL GAS SURVEYS

One example of this type of exploration program is a survey conducted in the Matador Arch area of West Texas. As an initial step, enhanced LANDSAT maps of the area were acquired and interpreted to relate the significance of known surface geologic leads over a 250,000-acre survey area. The airborne gas sensor system was next employed to evaluate these leads; it confirmed the presence of a significant amount of microseepage in several cases. Follow-up soil gas sampling validated the gas sensor's anomalies, thereby forcusing efforts on specific prospects. In order to establish anomalous as well as background levels of hydrocarbon microseepage in the area, existing production was surveyed at the Ross Ranch field in King County, Texas. The airborne gas sensor mapped the apparent microseepage limits of the field and a follow-up free gas sample profile was employed to establish the background and validate the reconnaissance mapping (Figure 4). The resultant information was then used as a

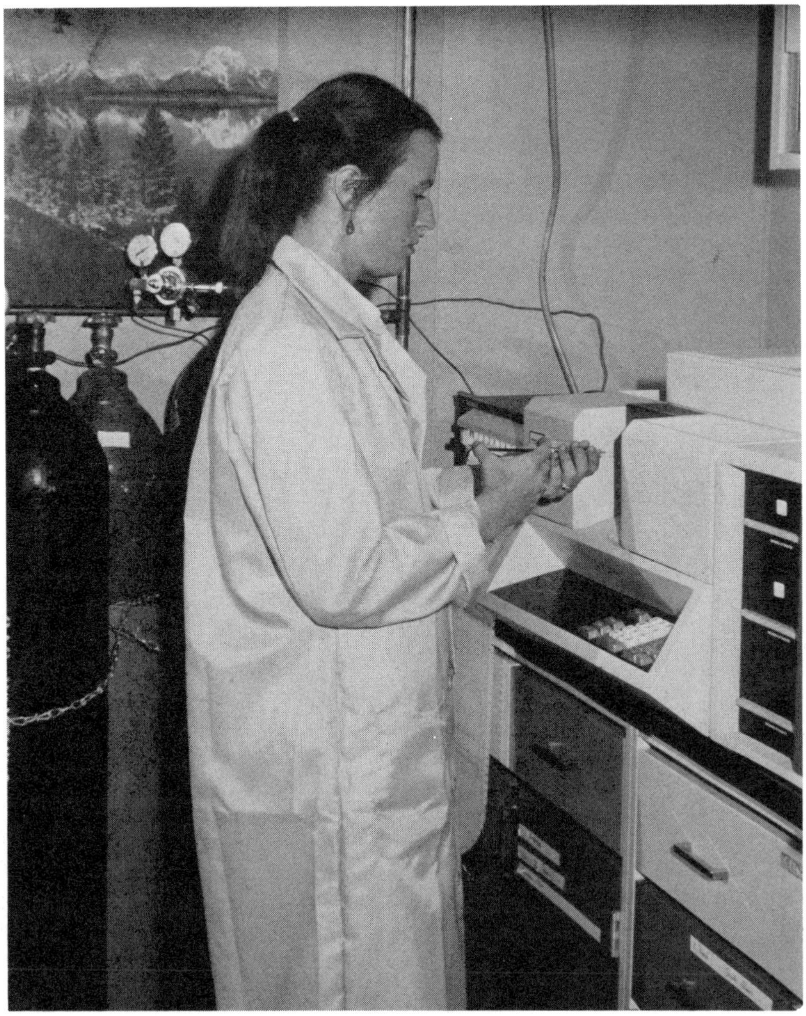

Fig. 3. Interstitial hydrocarbon gas sample injection into hydrogen flame gas chromatograph.

basis for comparison with subsequent airborne and free gas surveys in the area.

A second useful role of the gas sensor is that of trend definition. Prior to drilling in new areas, and therefore creating interest in surrounding leases, a gas sensor may be utilized to help confirm assumptions and to determine the most promising adjacent areas. These same procedures have been effectively employed in evaluating extensions to existing production that have not been completely defined by drilling. For example, in another typical survey conducted by Geophysical Systems Corp., the remote gas sensor system was employed for reconnaissance in an area of over 100,000 acres. While surveying the existing production and adjacent areas to determine if they were fully developed, this survey turned up possible extensions to two producing areas as well as an additional prospective area. Detailed geochemical sampling of the gas sensor anomalies validated the existence of extension possibilities and substantiated the new prospect. Free gas sample profiles were again used initially to validate the gas sensor survey results (Figure 5). Subsequent drilling has proved a high degree of correlation between the microseepage patterns mapped and the actual limits of the stratigraphic reservoir in the subsurface.

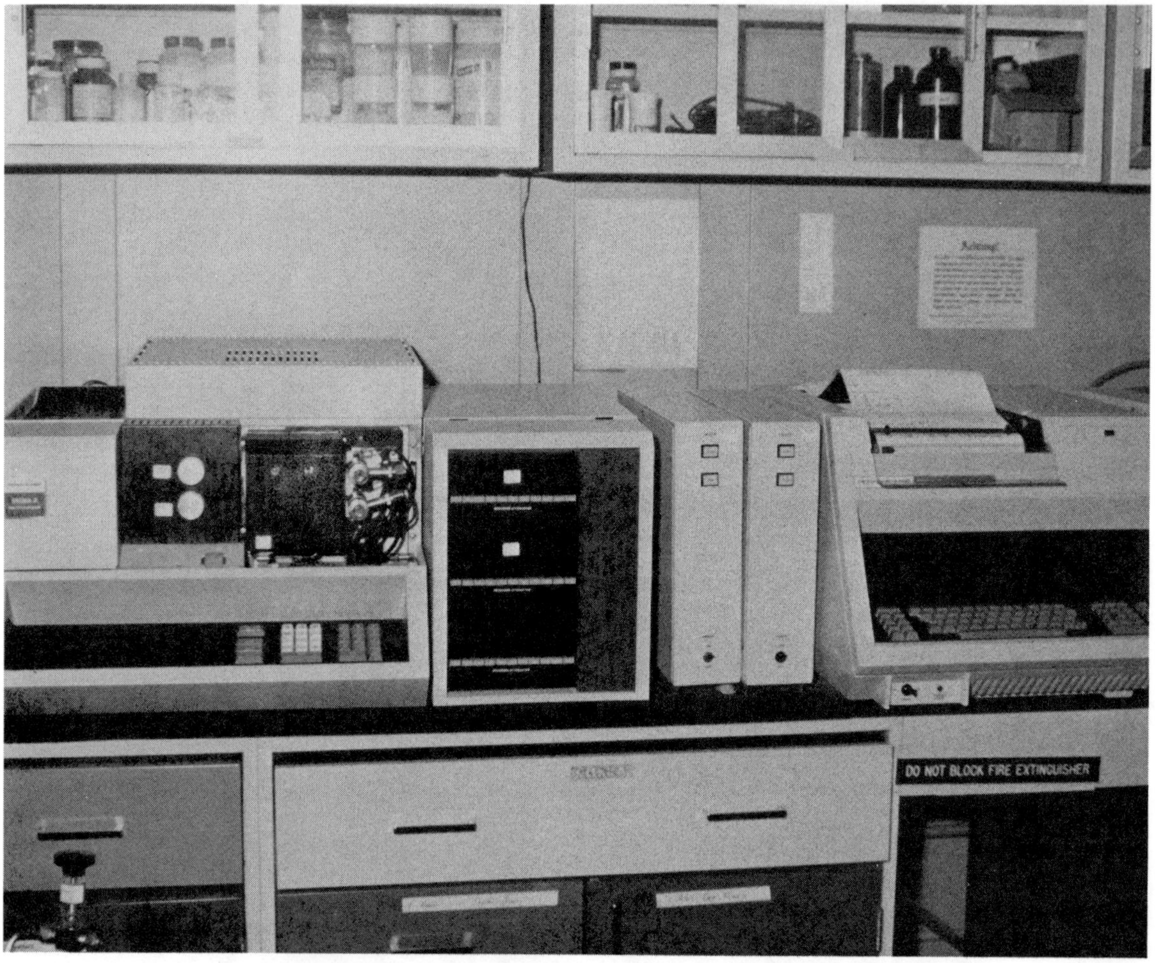

Fig. 3i. Laboratory equipment. Hydrogen flame gas chromatograph with integrating attachments.

Similarly, in surveys conducted in North Central Texas, as well as in other areas, the remote gas sensor identified stratigraphic channel sand reservoirs that had been previously overlooked by drilling. Detailed free (interstitial) gas sampling pinpointed the reservoir microseepage limits, and subsequent leasing and drilling proved to be productive in areas that had been previously abandoned (Figures 5, 6, and 7). It is important to comment on our experience with the airborne gas sensor as a negative indicator of petroleum. No exploratory well drilled at a location condemned by the sensor has been completed as a commercial producer. Although a more definitive explanation, including specific locations and detailed results, would serve to illustrate these concepts even more clearly, many of the examples of surveys performed by Geophysical Systems Corp. must be held in confidence for proprietary reasons.

It is of major importance to recognize that stratigraphic reservoirs may be exploited with these methods as readily as structurally controlled reservoirs. This point alone could be vital. Halbouty et al., in Part II of a 1970 paper entitled "World's Giant Oil and Gas Fields: Geologic Factors Affecting Their Formation, and Basin Classification," states "that the new, large, economic petroleum reserves of the future will be found in stratigraphic traps—in features which are not easily found by present day exploration methods" (Halbouty et al., 1970). We believe the systematic use of LANDSAT information and airborne gas sensor reconnaissance, along with detailed geochemistry and geophysics, can achieve a high degree of success in identifying stratigraphic traps as well as structurally controlled reservoirs.

The preponderance of evidence from research and development over the years seems conclusive. There is significant, although sometimes subtle, hydrocarbon microseepage at the surface above oil and gas reservoirs (Horvitz, 1978; Davidson, 1967; and others). This fact points up, more than ever before, the importance of geochemical exploration and the cost-saving benefits of the reconnaissance geochemical methods described in this report.

In the early studies of Earth Resources Technology Satellite (ERTS) data, satellite imagery was evaluated for its exploration potential in an investigation conducted by the Eason Oil Co. These investigators compiled one of the most comprehensive and analytical reports in the geology phase of the ERTS (satellite) Program. In the evaluation, the cost effectiveness of the ERTS approach to petroleum exploration is discussed—major savings in the initial reconnaissance stages of exploration are indicated, and further savings are suggested in the later stages. For example, significant cost-savings may be realized by reducing the number of seismic lines that may otherwise have been planned and which could subsequently be eliminated over areas rated as unfavorable for accumulation (Short, 1977). In his National Aeronautics and Space Administration (NASA) study, Martin J. Davidson (1967) reported that the effect of a remote detection direct-oil-finding tool upon petroleum exploration could be as great as that of the seismograph.

Another phenomenon reported by ERTS-1 (LANDSAT) investigators (Collins et al., 1974) is the so-called "hazy" anomaly, described as a blurring or smudging of the photo-image (Figure 8). A high degree of correlation between these hazy anomalies and producing fields was observed (of thirty-five "hazy" anomalies, thirty-three correlate with producing fields or drilled structures). Donald Saunders (1969) discussed at Symposium I the importance of the correlation of surface "tonal" anomalies observed in high altitude photographs and their apparent relation to oil and gas reservoirs. One explanation of these observations, favored by this writer, postulates that hydrocarbon gases leaking (or microseeping) from subsurface traps introduce chemical changes that are observed in the surface layers and as free gases at the near-surface, thus manifesting themselves as "hazy" or "tonal" anomalies (Figure 9).

Terrence J. Donovan of the U.S. Geological Survey in 1974 reported a plausible geochemical model based on a study of the Cement Field in Oklahoma dependent on the reducing action of methane (CH_4) and other hydrocarbon gases. The model of microseeping hydrocarbon gases recently suggested by Donovan and Dalziel (1978) may explain the relationship of "hazy" anomalies observed in LANDSAT imagery and, in many cases, subsequently mapped by the airborne gas sensor over oil and gas reservoirs.

The exploration approach suggested, therefore, is first one of searching rapidly for surface expressions of hydrocarbons over large areas of the earth and at a modest cost. Then, detailed confirmation work may be concentrated on areas exhibiting evidence of hydrocarbon microseepage and focused on locations most likely to be petroliferous and productive. Future conventional methods of exploration must be capable of providing economic impact through exploration cost-savings and an improved success ratio. The airborne gas sensor employed as a part of the Systems Approach to petroleum exploration, and the subsequent focusing of geochemical, geological, and geophysical efforts on a limited number of prospective areas can yield these benefits. We must pursue continued research and development in the laboratory as well as in the field in order to understand fully the mechanism by which hydrocarbons migrate from subsurface petroleum deposits to the earth's surface.

REFERENCES

Collins, Robert J., et al., 1974, An evaluation of the suitability of ERTS data for the purposes of petroleum exploration: Report-Type III Final, Contract NAS5-21735, National Aeronautics and Space Administration.

Davidson, Martin J., 1967, Petroleum geochemical exploration from space: a radical new technology: National Aeronautics and Space Administration Manned Spacecraft Center, Houston, 78 p.

Donovan, Terrence J., 1974, Petroleum microseepage at Cement, Oklahoma: evidence and mechanism: American Association of Petroleum Geologists Bulletin, v. 58, no. 3, p. 429-446.

———, and Dalziel, Mary C., 1977, Late diagenetic indicators of buried oil and gas: United States Geological Survey, Open-File Report 77-817.

Halbouty, Michel T., 1970a, Geology of giant petroleum fields (Introduction): American Association of Petroleum Geologists Memoir no. 14, p. 1-7.

———, 1970b, Giant oil and gas fields in United States: American Association of Petroleum Geologists Memoir no. 14, p. 126, 92.

———, et al., 1970, World's giant oil and gas fields, geologic factors affecting their formation, and basin classification, Part II: American Association of Petroleum Geologists Memoir no. 14, p. 450.

Heemstra, R. J., Ray, R. M., Wesson, T. C., Abrams, J. R., and Moore, G. A., 1979, A critical laboratory and field evaluation of selected surface prospecting techniques for locating oil and natural gas: U.S. Department of Energy, Report BETC/R1-78/18, p. 51, 72, and 80.

Horvitz, Leo, 1978, Near-surface evidence of hydrocarbon movement from depth: Annual Meeting of the American Association of Petroleum Geologists, 63rd, Oklahoma City, 42 p.

Jones, Victor T., and Drozd, R. J., 1979, Predictions of oil or gas potential by near-surface geochemistry: Annual Meeting of the American Association of Petroleum Geologists, 64th, Houston.

Link, W. K., 1952, Significance of oil and gas seeps in world oil exploration: American Association of Petroleum Geologists Bulletin, v. 36, no. 8, p. 1506.

MacElvain, Robert C., 1963, What do near-surface signs really mean in oil finding, Parts I and II: Oil and Gas Journal, Feb. 18, 1963, p. 132; Feb. 25, 1963, p. 139.

Saunders, Donald F., 1969, Airborne sensing as an oil reconnaissance tool, *in* W. B. Heroy, ed., Unconventional methods in exploration for petroleum and natural gas, Dallas, Southern Methodist University, p. 105-124.

Short, Nicholas M., 1977, Remote sensing energy-related studies: New York, John Wiley & Sons, p. 219-233.

Siegel, Frederick R., 1974, Applied Geochemistry: New York, John Wiley & Sons, p. 353.

Sokolov, V. A., et al., 1959, Investigations into direct oil detection methods, World Petroleum Congress, 5th, New York.

Sweet, George Elliot, 1966, The history of geophysical prospecting: Los Angeles, p. 249-250.

Vassoevich, N. B., 1977, Major achievements of Siberian petroleum geochemists: Geologiya i Geofizika, v. 18, no. 5, p. 150-154.

Vyshemirskii, V. S., 1977, Researches in the geology and geochemistry of oil and gas: Geologiya i Geofizika, v. 18, no. 12, p. 81-87.

FUTURE DIRECTION OF PETROLEUM EXPLORATION

Lewis R. Tucker

LAMONT-DOHERTY GEOLOGICAL OBSERVATORY OF COLUMBIA UNIVERSITY

For the past half-century the mainstay of petroleum exploration prior to drilling has been the seismograph, either refraction or reflection, and it is a high statistical probability that drilling locations have been determined only after the results of the interpretation of a seismic survey have been examined.

Conventional and even advanced exploratory methods for the discovery of oil and gas have failed to find enough of these fuels to meet our growing and seemingly insatiable demands. In every year since 1965, with the exception of 1968 when the Prudhoe oil field was discovered, we have found less oil than we have produced. Part of our failure to keep even with the demand has been the sobering fact that recently discovered fields have been *smaller* than those found in the past. Giant fields that can yield anywhere from 50 to 100 million barrels of oil (or gas in oil equivalents), or even more depending on the definition of such a field, were once fairly *easy* to find. They have produced and still contain about one-half of our known oil and gas reserves.

In any discussion on exploration and exploration success (or lack of success) it must be understood that when we quote statistics we are talking about at least two different types of success. The first is the success achieved through the thousands of wildcat wells which are productive but which yield only a restricted quantity of crude or natural gas. These successes are a real tribute to the skill, perseverance, or maybe luck of those instrumental in the drilling of the well. This measure of statistical success seems to be acceptable to the industry, and that this success ratio has stayed more or less the same over a fairly long period of time, when considered in the context of an obviously decreasing quantity of conventional prospects, shows that someone must be doing something right. Unfortunately, for our long-term needs the total quantity found is too small to make this method sufficient in itself. In the early days it was the really big "fish" that gave us the quantity of oil that was needed to fulfill our demands, but these large fields are becoming harder and harder to find and are found more and more rarely. In recent decades, most of these fields have been in new exploratory areas, while discoveries of large fields in mature basins have been few. We must direct our exploration efforts to finding *large* fields in order to boost our reserve picture. That is what we hope can be done through the application of different exploration techniques. It is not that we would want to throw back the little fish. We need them. But we also need some whales. Further, it is logical to assume that, structural mapping being as precise as it is now, if any big ones were overlooked they will *have* to be found by methods other than structural ones.

In this discussion of future discovery methods, we will confine ourselves to some of the more or less unconventional gas and oil accumulations—unconventional in the sense that they may have accumuluated in formations other than normal anticlines, salt domes, fault closures, or the more typical stratigraphic traps. We want to see if we can identify, *prior* to drilling, the locations of the possible large accumulations that have *no* known structural manifestations. We will not talk about the tremendous quantities of oil that are known to exist in the Athabasca tar sands and other similar areas. We know the oil is there, and we know how to recover it; but we also realize that the magnitude of investment necessary will be staggering. We know of the gas that seems to be trapped in coal seams, in water in the deep geothermal formations in the coastal basins along the Gulf of Mexico, and in shales. This resource may exist for the taking in almost astronomical quantities, but once again the taking price is almost beyond reach given the present value of energy. Maybe some day we will be able to exploit it commercially, in large quantities and at a profit, but today most of it remains an unreachable goal.

Large quantities of gas may also be entrapped in gas hydrates, or possibly just below them. It is well known that there are large areas which may be underlain by these hydrates and that many of them are in areas that can be exploited. They are actually reported as being drilled and exploited in Russia. These features can be discovered by seismic methods. The areas where they might exist are generally limited to regions usually requiring high pressure and low temperatures, a combination to be found in the arctic regions and possibly in some marine areas, neither of which could be considered an optimum area for an economic recovery system.

Referring again to economics, a small operator—or a large one for that matter—can find and produce a small field and sometimes make money on it. His success can be duplicated by hundreds of other operators. Statistics of discovery rates, i.e., the number of fields found per well drilled, seem to be acceptable and often seem to have caused some writers to paint an optimistic picture of our future. When considered in the framework of the quantity of oil discovered, however, the picture is not so optimistic. Many studies have shown how the field sizes in new discoveries have been decreasing. AAPG data and other statistics show this quite clearly. Is there a better way than our present exploration methods afford to find whatever giants may still exist?

Most of our giant fields have been found by examining their structural features. Discovery methods have varied, but most have been based on structural anomalies. Fields could thus be found through the use of surface geological methods, core drilling, and, in the last fifty years or so, geophysical methods—mainly seismic, but also including gravity, magnetics, and other less well-known systems. The few large fields and the many smaller ones that are interpreted today to be of a stratigraphic or a lithologic nature were not always found through exploration based on conventional techniques. Many of them were stumbled across while drilling was being done on supposed structures that did not in fact exist. Or else they were found by that catchall, serendipity.

Today, in most of the still promising but fairly well-drilled exploratory areas of the world, it can reasonably be stated that most of the existing structures, especially the large and shallow ones, have been found. The ones remaining either are so deep as to be close to the limits of the depth capabilities of present-day geophysical discovery techniques, or if they are discoverable, would entail a discovery cost that might be unacceptable.

Seismic methods, which have been the mainstay in exploration, are reaching a cost level that is almost out of hand. Table 1 shows the continuing increasing cost of seismic exploration over the past years. Since 1966, it has doubled to tripled in cost per mile, the most appropriate statistical yardstick. Part of this increase includes the cost of better and more expensive equipment, and part (not always shown in the statistics) the cost of expensive computer systems with extensive playback facilities. These newer methods are good, but they do increase costs.

In the past it was possible to shoot a large grid with widespaced lines and pick up any substantial structures. Now, the search is for smaller and smaller features, and ground coverage must be much more extensive—e.g., quarter-mile grids. Putting this into dollars and cents: for detailing a ten-mile-square exploratory area, costs for data acquisition and reduction alone can be calculated at about $25,000 per sq mi. This is to be compared with costs of $1,930 per sq mi in 1966.

According to statistics gathered by the Society of Exploration Geophysicists and shown in Table 1, it now costs about $2,500 per mi to obtain multicoverage seismic data on land in the United States. The processing of this data, depending on the number of times it is restudied, can add from a minimum of 10% up to 50% or more to that cost. The conventional concept of structural closure in a fairly mature exploration area will require coverage on a quarter-mile grid. Putting this into estimated costs, field work would run about $20,000 per sq mi and interpretation would add from $2,000

TABLE 1

Comparative Seismic Cost, Representative Years

Year	Acquisition Costs per Mile	Reduction Costs per Mile	Total Cost per Mile	Line Miles Shot in Average 100 sq m Grid
1966	$875	$88	$965	200
1976	$2,127	$200-$1,000	$2,327-$3,027	800
1978	$2,500 (est.)	$250-$1,250	$2,750-$3,750	800
	Total costs per square mile			
1966	$1,930			
1976	$18,600-$28,216			
1978	$22,000-$30,000			

SOURCE: Society of Exploration Geophysicists, yearly statistics report.

to $10,000, providing a total of $22,000 to $30,000 per sq mi.

While it is impossible to state how large an area must be surveyed before one can come up with a prospect, a ten-mile by ten-mile grid with quarter-mile line spacing is not out of line. Costs now are $2,200,000 minimum, $3,000,000 maximum just to cover an area well enough to be sure that no identifiable structure is missed. Drilling costs in the United States for an average 10,000-ft hole is around $50.00 per ft, or $500,000 per well. This amounts to around 1/3 to 1/4 seismic costs without including lease cost, which could cover a wide spectrum. In other words, it might cost more to do a survey than to drill three or four wells. The numbers I have used for seismic costs are average figures. If one were to use more exotic surveys in extremely difficult terrain, the cost could be higher.

What I am trying to demonstrate with numbers is that seismic costs of determining the possible existence of a drillable structure are quite high, and there is no reason to believe they will get any lower. In marine work, such increased capabilities as noise-free cables and other improvements have permitted faster shooting speeds and lower costs per mile. Land work, however, is still "foot slogging."

What this does point out is that costs of data acquisition and reduction by conventional seismic methods that are almost completely oriented toward structure are getting almost prohibitively high. The success ratio is good but not many large fields are discovered, only numerous small ones. Given enough time, effort, and money to do the job, all findable structures could be discovered by the seismic method. In fact, the old saying about learning more and more about less and less till we know everything about nothing might possibly apply.

It has been only in recent years that a determined effort has been made to extract stratigraphic or lithologic information from the seismic record. These newer methods are still only secondary to the primary use of seismic data for the purpose of acquiring structural information. The reason for this is fairly clear cut: the time on a seismic record is a "fact" and as such does not lend itself to manipulation, improvement, enhancement, or modification. The depth determined for this "time" is subject to some change, depending on the conversion systems used and the actual location of a subsurface depth point, which can vary widely not only because of velocity but also because of dips of the reflecting bed.

Presented below is a brief outline of the positive and negative aspects of the added interpretive methods, beginning with the more positive and "surest" method. A more detailed account has been written in many lengthy texts and papers. (See especially the American Association of Petroleum Geologists Memoir 27, 1977.)

First there was the use of velocity to determine lithology. This method was based on the assumption that low velocities meant more recent, unconsolidated sections, while higher velocities meant more consolidated beds, with the very high velocities probably representing basement, i.e., granites or basalts. These very high velocities could also, however, be representing limestones, quartzites, or evaporites. Therefore, unless one had some knowledge of the actual geologic section, obtained from either outcrops or wells, the determination of what might be drilled into in the regions of higher velocities had a high degree of speculation and was often the reason for tremendous disappointment and failures when a wildcat well was drilled.

A method related to that of velocity involved the use of the actual appearance of the reflections on a record, dealing with constant amplitude settings on the amplifiers, or (if there was an amplitude control) on restored records. Here, too, some fairly basic conclusions could be drawn. If a reflection was very strong, with a sharp definition from a very low noise background, it could mean a transition from a low velocity medium into one of high velocity, the reflection coefficient of these two adjacent rock beds being quite high. This method could and did have many variations, but it usually told only one thing: that one was going from a low velocity to a higher one, not what the velocities represented. It could also be reversed so that the transition was from high to low, something that might be told from the direction, on the record, of the first impulse.

The appearance of many reflections on a record could be interpreted as indicating a section with rapidly changing lithology. Conversely and unfortunately, it could also indicate a lot of multiple reflections. The divergence or convergence of reflections could mean the thickening or pinching out of various beds. This method was almost foolproof, except that it was influenced greatly by the filter response of the instruments. Phase changes or alterations of the appearance of the composite reflections could be overshadowed by using a low frequency response; the use of a higher band could result in a lot of unwanted noise.

The beauty of record sections has been improved by the application of increasing amounts of mixing, either on the ground or in the instruments, that finally reached its zenith (if in fact it has been reached) in common depth-point systems where as many as ninety-six traces are combined to get a resultant, simulated, single-trace record. Whether this always improves the actual interpretiveness of a section has been debated. At present, most people think it does; only a few diehards still have doubts. And very importantly, the similarity in appearance of a seismic cross section to a geological cross section made instant interpreters out of a whole generation of oilmen.

"Bright Spots" are a further advance in the use of seismic data in interpreting stratigraphy; this method goes a step or so farther and purportedly gives some idea of the fluid content of the beds adjacent to the reflecting beds. It is not a foolproof method either, although it does add a new tool to the explorationist's kit. The fairly common use of velocity logs derived from seismic data has added immeasurably to the knowledge of stratigraphy prior to drilling, and their routine application has been of considerable assistance in interpretation.

There is probably not a single seismic group today that is not doing considerable research work using many different types of investigation into the use of seismic data for both stratigraphic and fluid-content determinations. Some of these are and will be successful to a degree. In this difficult job of trying to find oil, any improvement in the odds is desirable.

In this discussion of the advantages and limitations of the seismic system, I am not trying to discount or downgrade what many researchers and investigators have done in attempting to derive lithologic or stratigraphic information from seismic data. Most of these methods mentioned are actually not new processes; many of them started as far back as the time of refraction-fan-shooting days in the Gulf Coast, at least fifty years ago. The search for, and the identification of, the hoped-for "lead" or precursor in the recorded seismic refraction trace was then the goal of all interpretation. While this "lead" was actually dependent on the velocity of the medium in which the waves were traveling, it was interpreted as a direct indication of salt and, it was hoped, a salt dome. This, of course, was a lithologic interpretation.

From that day till now there have been constant improvements in the ability of the interpreter to derive something other than structural information from the seismic data. In addition to arrival time of the refracted wave on a refraction record, amplitude studies permitted the fairly positive determination of the types of rocks through which the seismic wave had traveled. For example, in eastern Venezuela it was possible to state with a fair degree of assurance the type of basement one was shooting over just from the appearance of the recorded refracted event—and also, with somewhat less assurance, what the wave went through in arriving there. Not that this aided much in the discovery of oil, but it was an intellectual and geological achievement.

From these early refraction beginnings, the same sort of interpretation spilled over into reflection shooting. For a generation and more of interpreters, the amplitude, persistence, and general appearance of the wide band, relatively unfiltered, single detector reflection record gave stratigraphic and lithologic clues that were of considerable value in determining the probable geological section. This has progressed much farther in the past decade or so, but I feel that it is essential now to put in a word of warning. A plain vanilla, unvarnished, wide band, true amplitude recording, if taken in an area where there is little background noise or unwanted extraneous energy, is fairly easy to interpret, but almost impossible to find today. Over the years, attenuation of unwanted energy has been the objective of all sorts of efforts. Multiple phones, mixing, filtering, focusing, and even common depth point systems are all used to get rid of this unwanted energy or never to record it at all. Many times this has been partly or completely successful. But in doing this manipulating (as one does to get the pure, unique, one-point reflection) one can often degrade the desired data in the process of eliminating the unwanted energy, even to the extent of producing something that was not wanted in the first place. The whole system has been described sometimes as "throwing out the baby with the wash water." We don't want to do that, but neither do we want "manufactured" data. What we need, and should strive for, is the ability to distinguish between the manufactured and the true data. Our present-day techniques do not always do that.

An extreme example, probably one not too familiar to many, is the "Laser" system of interpretation that was being studied some fifteen or so years ago. I will not describe it but only discuss one of the reasons why it was not accepted by the industry. By manipulation of a series of optical slits and filter response, one could achieve an almost infinite series of apparent reflection "lineups." The resolving power of the system was not considered adequate, and the system was dropped from the inventory of interpretive techniques. Yet often what it showed was correct and demonstrated concepts that were alien to many geologists' thinking.

Even today, with our present methods of treating data, we could be manipulating data in a synthetic manner to get what we want to see, not necessarily what is true and what is actually there. We must guard against this, not only in seismic interpretations but in every other method of exploration. I am not convinced that we always do.

So far no one in the seismic profession has come up with a conclusive method or system to find stratigraphic traps. If oil is to be found in the future, it will have to be found in traps other than structural traps, possibly in stratigraphic or lithologic traps. The literature deals quite optimistically with stratigraphic traps, but statistics as to the a priori determination of quantities of petroleum in stratigraphic traps are not so favorable. We must always temper our projections and conclusions by this fact. Also, we must remember that none of our past geophysical methods have been conclusively proven capable of finding stratigraphic traps; the traps that have been found that are productive have most often been more or less fortuitous discoveries. We have had production that we say today was from stratigraphic traps; we may have drilled hundreds or thousands of the same sorts of features without even knowing we were drilling a "strat" test. In other words, we acclaim our successes but we are ignorant of or ignore our failures.

To throw a wrench into the thinking process that goes into our trying to theorize where some of our future oil may come from, let's examine in a pessimistic vein the idea that there are tremendous volumes of oil locked up in stratigraphic traps that have not yet been found. If we are to find these traps, we must devise some successful geophysical or geochemical way to look for them; something other than random drilling.

That there have been large stratigraphic fields found is well known, though in almost all cases there has been some structural deformation that could also have contributed to the presence of oil. Some of these fields have been found by the mapping of pinch outs, others have been discovered as the result of the application of the serendipity concept, but they have been found.

Consider, however, the tremendous number of dry holes that have been drilled in some areas, presumably based on interpreted structural evidence that turned out to be unsuccessful. If a presumed structure was small, it is not hard to believe that perhaps the interpretation was wrong and that no structure existed, so that what was really being drilled was a random stratigraphic test. How many of the thousands upon thousands of dry holes that have been drilled since wildcatting began could have been on this sort of supposed structure is not known, and there is no way of putting this into a statistic. I firmly believe many of them were valid strat tests, and since they have not found any substantial quantities of oil we should not be too sanguine about finding the oil we need by just switching from structural tests to strat tests. We have been using this method for years. Of course, by using new alternate methods we hope to make this search less of a gamble, but we still have no firm concept of the magnitude of the prize that we are looking for, or even whether there is a prize.

At the risk of seeming to go too far back in fundamentals, I would like to review, in a few paragraphs, what we know and what we believe we know about the generation, migration, and accumulation of petroleum and natural gas. I emphasize *what we think we know,* as even now, after a hundred years or more of scientific investigation, using the imagination and minds of countless hundreds of scientists and investigators, geologists and engineers, there is still a lot that is not known.

We have a very imperfect understanding of how oil and gas are generated. We know the apparent general rules: organic sediments, rapid and deep burial, with resulting pressure and heat being applied to the sediments. Given these essentials and allowing enough time, oil or gas or both will emerge, sometimes as a chemically incomplete product and sometimes as the finished material. Except for a better scientific understanding of what actually happened, the origin of oil and gas should not concern us too much. We want to know where it is today and how to find it.

The same lack of facts holds for oil migration. We know the oil moves, sometimes, it seems, as a result of hydraulic pressure, sometimes by gravity, or sometimes even by "squeezing." Again we need only to accept the fact that it does move; we should not be too concerned about how it moved, but we should be greatly concerned about where it moved.

It is in the accumulation phase where we know the most, not from scientfic investigation especially, but just from drilling enough wells on enough structures and in enough stratigraphic areas so that we could arrive at good, valid conclusions. Once we arrived at the conclusion that structural or stratigraphic traps contained the oil that we had already found, we followed a normal course and started looking for other traps of a similar nature. That is what we have done and are continuing to do in much of our geologic field work and in magnetic, gravity, seismic, and other geophysical studies. We are trying to discover some structural or stratigraphic anomalies similar to already known productive structures, with the hope that these newly discovered anomalies might contain oil or gas.

There is an important variation of this method. If there was no anticlinal or other geological trap, then the oil, in its migration, could continue to move, possibly even to the surface where it would be dissipated and evaporated. Alternatively, the residue could form large tar deposits. Maybe along the line there was something that was not a structural barrier which caused the oil to stop moving. It could be an impervious series of beds, or perhaps the driving force ceased to exist. Then the oil could collect in large stratigraphic traps. So we can see that a structural trap calls for some dislocation of the rocks to provide a trap; for stratigraphic traps, this is not needed. I won't deny that there are all graduations from one extreme to the other, but it has to be admitted that often the placement and the retaining of oil in a structural trap and a stratigraphic trap follow different physical rules. So if we are to find stratigraphic accumulations, always assuming that there are undiscovered ones, a different sort of exploration system has to be considered.

There have been some additional methods of structural or stratigraphic studies, other than those discussed in this volume, that have helped in the more direct determination of the presence of oil or gas. One method is based on detailed stratigraphic studies, often derived from the minute examination of well logs and samples that have aided in the possible determination of the direction and magnitude of lithologic changes that could indicate the direction to go in a further drilling attempt that would more probably lead to success. A fair amount of success has been achieved by the application of this method.

Other comparable methods have included the mapping of formation water salinity, since it has been concluded from fairly good evidence that oil will more probably be found near high salinity areas rather than in low salinity areas. Or a possible accumulation may be found on a slanting water table. Or, if one wants to go back far enough in exploration history, there was the attempt to obtain some clues as to the presence of oil by examining the lineups of salt domes in the Gulf Coast area. These are examples of indirect methods.

Some of these methods have been tried in the past and have had different degrees of success. They all have one attribute in their favor: they are, when compared to present-day onshore seismic methods, relatively cheap. Costs can be estimated for only some of these actual and proposed methods. An inexpensive approach, however, is not a valid criterion if the method gets minimal or no results. Many of these methods have had some modicum of success, and the premature abandonment or replacement of some of these unconventional methods by many of the oil and exploration companies may have been a little shortsighted.

It is hoped that explorationists will think more about the whole range of petroleum exploration. A tried and true method, seismic reflection, is approaching a point of no return. There is nothing in the basic system that is susceptible to major changes. There may be many minor and striking improvements, yes; but the targets are becoming more and more scarce, and smaller and smaller, so that they are approaching the vanishing point. Part of the reluctance to change methods is based on two factors. The first is a purely economic and selfish one. It is more profitable for contractors and computer manufacturers to maintain large money-making operations using extensive seismic equipment, and more satisfying and rewarding to exploration managers and their people to have large, active seismic operations than it is to have a small noncostly unglamorous operation. And second, there is always a reluctance, unfortunately, on the part of any field man to accept a new research tool, and invariably it takes him a long time to accept it. But if we are still to find oil, we have to do something different.

This different approach may require a reorientation in some research groups with new directions toward the development of

TABLE 2

Seismic Crew Costs, Selected Years

1964	Total Land Costs	$152,191,000
	Total Months	4,102
	Cost/Month	37,103
1965	Total Land Costs	$163,355,000
	Total Months	4,247
	Cost/Month	38,462
1966	Total Land Costs	$112,420,000
	Total Months	3,211
	Cost/Mile	875
	Cost/Month	35,000
1976	Total Land Costs	$304,030,303
	Total Months	2,843.4
	Line Miles	142,926
	Cost/Mile	2,127
	Cost/Month	106,901

Reference-SEG Publications

some of these more unconventional tools. This has already been done by many research groups, but a further and more widespread reorientation may be necessary if we are to maintain a reasonable oil-finding rate.

The methods with the most apparent potential are those that could directly determine the presence of oil or gas. Because they have had some success, they deserve more extensive experimentation. The majority of future petroleum discoveries will be credited to these or similar methods.

APPENDIX
EXPLORATION GEOLOGY FROM SPACE

Frederick B. Henderson III

THE GEOSAT COMMITTEE

LANDSAT, the present principal civilian earth sensing satellite remote sensing system, was designed primarily to sense vegetation and water for agricultural, hydrological, and land planning needs. Despite much that has been learned using LANDSAT data in the field of exploration geology, its remote sensing capabilities are not optimal for our purposes.

Other planned satellite remote sensing systems, such as SEASAT, SIR A and SIR B radars, MAGSAT, HCMM, and SMIRR, will variously yield some basic geological data, but with the possible exceptions of SMIRR and MAGSAT, the fact remains that no existing or planned satellite was specifically designed for exploration remote sensing from space.

The Geosat Committee, sponsored by one hundred oil, gas, mineral, and engineering companies, has worked diligently with NASA and other pertinent U.S. and foreign governmental agencies in studying what systems could be added to those presently planned to improve satellite remote sensing capabilities for exploration. All potential remote sensing systems which could be utilized in exploration geology, whether as satellite or airborne sensing systems, are currently being evaluated by the joint Geosat-JPL/NASA Test Case Program. These studies are being conducted over Geosat-provided oil and gas, uranium, and porphyry copper sites.

The current recommendations of the Geosat Committee with regard to supplemental satellite earth remote sensing systems during the LANDSAT and Shuttle era of the 1980s are: STEREOSAT, for near-term, high resolution, worldwide stereo coverage in LANDSAT compatible digital data mode; 2.2 μm band added to the Thematic Mapper on LANDSAT D to ratio with the 1.6 μm band for optimum rock discrimination; Large Format Camera, for long-term, worldwide stereo film coverage with high resolution and cartographic capabilities; L, X, and C band radar to provide medium-term geological information in areas with continuous weather problems.

In addition, the Geosat Committee is encouraging continued research and development of multispectral thermal IR scanning systems, the Fraunhoffer Line Discriminator, laser fluoroscopes, and other potential remote sensing systems for satellite or airborne exploration applications.

The Geosat Committee recommendations are being considered by appropriate U.S., French, and Japanese governmental agencies. Several institutional problems regarding the proper roles of government and private industry are being studied. The Geosat Committee is working with NASA and various other executive and congressional bodies of the U.S. government to further the implementation of the Geosat Committee recommendations.

We believe these recommendations should be implemented as soon as possible, as their addition to the planned LANDSAT and other systems can provide data to the world geological community in the 1980s which will help alleviate the world dilemma of declining discovery rates for basic energy and mineral resources. The solution to this dilemma will directly benefit all mankind.